FRACTURE MECHANICS OF DUCTILE AND TOUGH MATERIALS
AND ITS APPLICATIONS TO ENERGY RELATED STRUCTURES

H. W. LIU, T. KUNIO, V. WEISS
AND H. OKAMURA
(EDITORS)

FRACTURE MECHANICS OF DUCTILE AND TOUGH MATERIALS AND ITS APPLICATIONS TO ENERGY RELATED STRUCTURES

**PROCEEDINGS OF THE
USA-JAPAN JOINT SEMINAR
HELD AT HYAMA, JAPAN
NOVEMBER 12-16, 1979**

1981

MARTINUS NIJHOFF PUBLISHERS
THE HAGUE / BOSTON / LONDON

Distributors:

for the United States and Canada
Kluwer Boston, Inc.
190 Old Derby Street
Hingham, MA 02043
USA

for all other countries
Kluwer Academic Publishers Group
Distribution Center
P.O. Box 322,
3300 AH Dordrecht
The Netherlands

ISBN-13:978-94-009-7481-4   e-ISBN-13:978-94-009-7479-1
DOI: 10.1007/978-94-009-7479-1

PREFACE

This volume contains the proceedings of the USA-Japan Joint Seminar on "Fracture Mechanics of Ductile and Tough Materials and Its Applications to Energy Related Structures". The seminar was supported jointly by the National Science Foundation of the United States and the Japan Society for the Promotion of Sciences. The seminar was held from November 12th to 16th, 1979, at Hayama, Japan, a picturesque resort town by the beach of Sagami Bay facing Mt. Fugi.

The safety and integrity of the engineering structures for energy exploration, energy production, and energy transportation are of utmost importance to our welfare. Both the United States and Japan are at the forefront of the research on fracture mechanics and its applications to fracture prevention. During the past few years, major research efforts have been made in the areas of non-linear fracture mechanics and its applications to fracture initiation, slow crack growth, creep and fatigue. This joint seminar offered an unique opportunity for detailed exchange of information on current researches and future efforts.

The seminar consisted of eight sessions, including an open forum, a final session, and sessions on theoretical considerations of J-integral; experimental techniques for J-integral; stress, strain, and COD as fracture criteria; material effects on fracture toughness; and creep and fatigue. In all twenty-nine papers were presented and thirty three delegates participated. In the afternoon on November 15, 1979, an open forum was held on the theme of "Researches for the Safety of Energy Related Structures". The seminar covered the recent research studies as well as the practical field applications.

The Editors

The Delegates

Opening of the Seminar
(Prof. Kunio)

Main Business

Discussion

Banquet

Dinner Address

PARTICIPANTS

Asso. Prof. K. Hatanaka
Dept. of Mechanical Engineering, Yamaguchi University
2557 Tokiwadai, Ube City 755, Japan

Mr. M. Ichiki
Central Research Laboratories Ajinomoto Co., Inc.
1-1, Suzuki-cho, Kawasaki-ku, Kawasaki 210, Japan

Mr. N. Iino
Structure and Strength Dept., Research Institute
Ishikawajima-Harima Heavy Industries Co., Ltd.
3-1-15, Toyosu, Koto-ku, Tokyo 135-91, Japan

Mr. M. Ishihara
Niihama Research Laboratory
Sumitomo Heavy Industries, Ltd.
5-2, Sobiraki-cho, Niihama-city, Ehime 792, Japan

Dr. T. Iwadate
Research Laboratory
The Japan Steel Works, Ltd.
4, Chatsu-muchi, Muroran 051, Japan

Dr. M. L. Kanninen
Battelle Columbus Laboratory, Applied Solid Mechanics Section
505 King Avenue
Columbus, Ohio 43210, USA

Mr. Y. Kasai
Mechanical testing section
R&D Division, Daido Steel Co., Ltd.
2-30 Minami-ku, Nagoya 457, Japan

Dr. M. Kawahara
Technical Research Center
Nippon Kokan K.K.
1-1, Minami-watarida-cho, Kawasaki-ku, Kawasaki 210, Japan

Prof. A. S. Kobayashi
Dept. of Mechanical Engineering, University of Washington
Seattle, Washington 98195, USA

Asso. Prof. H. Kobayashi
Dept. of Physical Engineering, Tokyo Institute of Technology
2-12-1, Ohokayama, Meguro-ku, Tokyo 152, Japan

Prof. T. Kunio
Dept. of Mechanical Engineering, Keio University
3-14-1, Hiyoshi, Kohoku-ku, Yokohama 223, Japan

Prof. H. W. Liu
Dept. of Chemical Engineering and Materials Science, Syracuse University
409 Link Hall
Syracuse, New York 13210, USA

Dr. F. J. Loss
Code 6392, Naval Research Laboratory
Washington, D.C. 20375, USA

Prof. F. A. McClintock
Dept. of Mechanical Engineering, Massachusetts Institute of Technology
Cambridge, Massachusetts 02139

Asso. Prof. T. Miyoshi
Dept. of Precision Machinery Engineering, University of Tokyo
7-3-1, Hongo, Bunkyo-ku, Tokyo 113, Japan

Dr. Y. Nakano
Research Laboratories
Kawasaki Steel Corporation
1 Kawasaki-cho, Chiba 260, Japan

Prof. H. Nishitani
Dept. of Strength of Materials, Kyushu University
6-10-1, Hakozaki, Higashi-ku, Fukuoka 812, Japan

Prof. K. Ohji
Dept. of Mechanical Engineering, Osaka University
Yamada-kami, Suita, Osaka 565, Japan

Asso. Prof. R. Ohtani
Dept. of Engineering Science, Kyoto University
Yoshida, Sakyo-ku, Kyoto 606, Japan

Dr. N. Ohtsuka
Civil & Applied Mechanics Research Department
Chiyoda Chemical Engineering & Construction Co., Ltd.
3-1-4 Ikegamishincho, Kawasaki-ku, Kawasaki 210, Japan

Prof. H. Okamura
Dept. of Mechanical Engineering, University of Tokyo
3-7-1 Hongo, Bunkyo-ku, Tokyo 113, Japan

Prof. M. Sakata
Dept. of Physical Engineering, Tokyo Institute of Technology
2-12-1, Ohokayama, Meguro-ku, Tokyo 152, Japan

Dr. S. Sakata
Mechanical Engineering Research Laboratory
Hitachi, Ltd.
3-1-1, Saiwai-cho, Hitachi-shi, Ibaraki 317, Japan

Dr. C. F. Shih
Corporate Research and Development, General Electric Company
Schenectady, New York 12345, USA

Mr. T. Shindo
Power Generation Design Department
Babcock Hitachi Co., Ltd., Kure Works
6-9, Takaramachi, Kure, Hiroshima 737, Japan

Dr. D. A. Shockey
Shock Physics and Geophysics Department
SRI International
333 Ravenswood Ave.
Menlo Park, California 94025, USA

Asso. Prof. M. Shiratori
Dept. of Mechanical Engineering, Yokohama National University
Tokiwadai, Hodogaya-ku, Yokohama 240, Japan

Mr. T. Suzuki
Products Research & Development Laboratories
Nippon Steel Corporation
5-10-1 Fuchinobe, Sagamihara, Kanagawa 229, Japan

Mr. M. Tomimatsu
Takasago Technical Institute
Mitsubishi Heavy Industries, Ltd.
2-1-1, Shinhama Arai, Takasago, Hyogo 676, Japan

Prof. V. Weiss
Vice President for Research and Graduate Affairs
Syracuse University
Administration Building
Syracuse, New York 13210, USA

Assi. Prof. K. Yamada
Dept. of Mechanical Engineering, Keio University
3-14-1, Hiyoshi, Kohoku-ku, Yokohama 223, Japan

Mr. T. Yanuki
Heavy Apparatus Engineering Laboratory
Toshiba Corporation
2-4, Suehiro-cho, Tsurumi-ku, Yokohama 230, Japan

Seminar Assistants

Mr. K. Kageyama
Graduate Student of University of Tokyo
Dept. of Mechanical Engineering
7-3-1, Hongo, Bunkyo-ku, Tokyo 113, Japan

Mr. H. Nakamura
Graduate Student of Tokyo Institute of Technology
Dept. of Physical Engineering
2-12-1, Ohokayama, Meguro-ku, Tokyo 152, Japan

Mr. S. Tsuzuki
Graduate Student of Keio University
Dept. of Mechanical Engineering
3-14-1, Hiyoshi, Kohoku-ku, Yokohama 223, Japan

U.S.-JAPAN JOINT SEMINAR
ON
FRACTURE MECHANICS OF DUCTILE AND TOUGH MATERIALS
AND
ITS APPLICATION TO ENERGY RELATED STRUCTURES
NOVEMBER 12 - 16, 1979
HAYAMA,JAPAN

PARTICIPANTS

K.HATANAKA

M.ICHIKI

N.IINO

M.ISHIHARA

T.IWADATE

M.F.KANNINEN

Y.KASAI

M.KIKUCHI

M.KAWAHARA

A.S.KOBAYASHI

H.KOBAYASHI

T.KUNIO

H.W.LIU

F.J.LOSS

F.A.McCLINTOCK

T.MIYOSHI

Y.NAKANO

H.NISHITANI

K.OHJI

R.OHTANI

N.OHTSUKA

H.OKAMURA

M.SAKATA

S.SAKATA

C.F.SHIH

T.SHINDO

D.A.SHOCKEY

M.SHIRATORI

T.SUZUKI

M.TOMIMATSU

V.WEISS

K.YAMADA

T.YANUKI

SEMINAR ASSISTANTS

K.KAGEYAMA

H.NAKAMURA

S.TSUZUKI

M.IWAKI

CONTENTS

I. AN OVERVIEW

PROBLEMS AND PROGRESS IN THE DESIGN OF DUCTILE STRUCTURES

F. A. McClintock
Professor of Mechanical Engineering
Massachusetts Institute of Technology, Cambridge, Massachusetts 02139, USA

D. M. Parks
Assistant Professor of Mechanical Engineering
Massachusetts Institute of Technology, Cambridge, Massachusetts 02139, USA

J. L. Bassani
Assistant Professor of Mechanical Engineering and Applied Mechanics
University of Pennsylvania, Philadelphia, Pennsylvania 19104, USA

ABSTRACT

Fracture-stable design may require not only that fully plastic conditions be attained before fracture, but also that the load not fall off too rapidly during crack growth.

A review of crack initiation by hole growth and localization shows that while plane strain specimens appear less ductile than axisymmetric ones, the effect may be due more to triaxiality rather than to differences between plane strain and plane stress. Statistical effects followed by rapid hole growth of the worst few voids are shown to be important in some but not all cases.

For low-cycle fatigue crack initiation in piping subject to out-of-phase bending and pressurization, the stress-strain histories on individual slip planes must be followed, including the Bauschinger effect. Calculations for Mode III crack growth under arbitrary histories and turbogenerator shafts are outlined.

In crack growth under monotonic loading, the loss of triaxiality means that 1-parameter descriptions of stress and strain fields around crack tips may well be inadequate. Likewise near-by opposing cracks may lead to less ductility than expected from common tests.

Stress and strain fields around static and growing creep cracks are reviewed, and numerical procedures are outlined by which they can be calculated numerically, not only for power-law flow relations, but also for more realistic flow relations which must be used numerically.

INTRODUCTION

Linear elastic fracture mechanics is a useful tool for assuring the safety of structures, provided that they are loaded in the elastic range and that inspection procedures are adequate to detect cracks that will not grow to failure before the next inspection. Significant plastic regions are often encountered, however, as in thermal transients or long-time creep in power plants, electrical overloads in central station equipment, or impacts of many kinds.

A central goal is then to design large structures that will withstand plastic deformation even in the presence of reasonable defects. If fracture occurs, it should be stable. This means, as analyzed in detail by Tada, et al. (1), that the drop in "load" across any region containing a crack, associated with the conjugate load point "displacement" required to grow the crack, should be less than the compliance of the surroundings:

$$\left| \frac{d(\text{Load})}{d(\text{Displacement})} \right| < \text{Compliance of surroundings} \qquad (1)$$

An understanding of ductile fracture is helpful in predicting such stability from test results, and also in designing tougher materials.

Recent work and further needs considered here include the initiation of fracture by the growth of holes and by flow localization, low cycle fatigue under arbitrary load histories, stable crack growth, transition from hole growth to cleavage, stress and strain fields around static or growing cracks in creeping material, and underlying needs in plastic flow relations and computation.

CRACK INITIATION BY HOLE GROWTH AND LOCALIZATION

Early models of the limits to ductility due to void growth and coalescence tended to overestimate the deformation required for fracture. One likely reason for this is that these models consider only the growth to complete coalescence of uniformly spaced voids, with no allowance for the possible localization of plastic deformation. Recently, Needleman and Rice (2) have reviewed aspects of localization phenomena in setting limits to ductility.

One approach which they present is based on the work of Rudnicki and Rice (3) and considers a "fracture strain" as closely related to the point at which the macroscopic constitutive laws permit a bifurcation from homogeneous deformation into a local shear band. Conditions for such a bifurcation depend sensitively on details of the constitutive models such as the presence of yield surface vertices, and on the macroscopic state of strain, e.g. plane strain vs. axisymmetric tension. By using deformation theory plasticity as a model of a vertex-forming constitutive law, Needleman and Rice were able to qualitatively match Clausing's (4) experimental observation that the ratio of plane strain to axisymmetric tensile ductility in structural steels is less for stronger steels, and by inference increases with strain hardening.

Conditions for localization in such non-dilating, vertex-forming materials should not be affected by stress triaxiality. Recently Hancock (5), in a continuation of ductile fracture work at Glasgow University (6), varied triaxiality by performing notched tensile tests on HY 80 and HY 130 steel using plane strain and axisymmetric specimens of the same projected notch profile. Finite element analysis indicated that the stress triaxiality at the centers of similarly-contoured axisymmetric and plane strain specimens were nearly identical. Furthermore, sectioning of companion specimens pulled to varying amounts of deformation indicated that the critical event was the sudden formation of a crack-like defect joining two large voids near the specimen center, at strain levels which were essentially the same in axisymmetric and plane strain geometries. Once this local defect formed, very little macroscopic deformation was required for fracture. Even though a plane strain to axisymmetric "ductility ratio" markedly less than unity was noted in unnotched tensile specimens, in accord with the vertex model above, this difference was not obtained for the notched geometries, as shown schematically in Fig. 1.

Apparently triaxiality effects alone can provide a sufficient explanation of the results. At the relatively low levels of triaxiality, small differences in triaxiality can lead to large differences in ductility, both as measured experimentally (5,6) and as predicted from the hole growth equations.

A more complete microstructural understanding of such experiments is currently lacking, but as Hancock noted, one implication of these tests is that, in the notched specimens at least, the deformation field between void-generating inclusions may be somewhat autonomous (relatively independent of macroscopic axial symmetry or plane strain). In this sense then, it would seem that future localization studies should focus on the heavily deformed region between the statistically determined worst inclusion pair.

Some measure of the least hole spacing can be obtained following the statistical analysis of McClintock (7), by inquiring what is the probability that all $V/\delta V$ elements in a volume V have fewer than n_1 inclusions in any element, assuming the average number of inclusions in an element to be unity from the Poisson distribution. The probability that one element will have fewer than n_1 holes is:

$$P_{\text{fewer than } n_1 \text{ in one element } \delta V} = 1 - \frac{1}{e} \sum_{n=n_1}^{\infty} \frac{1}{n!} \qquad (2)$$

For $V/\delta V$ elements,

$$P_{\text{fewer than } n_1 \text{ in each of } V/\delta V} = \left(1 - \frac{1}{e} \sum_{n=n_1}^{\infty} \frac{1}{n!} \right)^{V/\delta V} \qquad (3)$$

Equation (3) can be evaluated conveniently with a programmable calculator. For a strained volume V of 10 mm^3 and an average volume per inclusion of perhaps $(10 \, \mu m)^3$, the probability of fewer than n_1 inclusions in any element of the entire volume is 0.4 when n_1 equals 10. Since the spacing between inclusions will vary roughly as the cube root of the inclusion count, one can expect local regions in which the spacing of the inclusions is half the average spacing.

For example, consider the fracture strain as given in terms of the critical hole growth ratio ℓ_f/ℓ_o , the strain hardening exponent n , and the ratio of the two largest principal stresses to the equivalent flow stress (8):

$$\varepsilon_f = \frac{(1-n) \, \ell n \, (\ell_f/\ell_o)}{\sinh \left[(1-n) \dfrac{\sigma_1 + \sigma_2}{2\bar{\sigma}/\sqrt{3}} \right]} \qquad (4)$$

Then halving the inclusion spacing and hence the required growth ratio from 10 to 5 would reduce the fracture strain by a factor of $\ell n \, 10/\ell n \, 5 = 1.43$. (Changing the growth ratio from 5 to 2.5 would reduce the strain by a factor of 1.76). Since the inclusions very likely do occur in groups, from related causes, an even greater concentration is likely to be found. In any event, statistics can account for a significant reduction in ductility.

Needleman and Rice (2) also review models of ductile fracture due to the growth of such local statistical imperfections. They cite the work of Yamamoto (9) who used the dilatant plastic laws of Gurson (10) in conjunction with an initial imperfection consisting of a local band of material with an excess of void volume fraction as compared to the surrounding material. Although there is no precise correspondence between the assumed band angle and the porosity excess of this model and the statistical aspects of the local inclusion size, spacing and orientation, substantial reductions in ductility could be attributed to such imperfections. Again, however, localization criteria based solely on the macroscopic deformations would likely predict substantially less ductility in plane strain than axial symmetry, regardless of the stress triaxiality.

We expect these conclusions about the importance of worst defects to hold for internal crack initiation in many structural alloys with an ordinary population of inclusions. Under other loadings, or in higher purity materials, macroscopic continuum models of localization may be more appropriate. For example, we may note the localization studies of Asaro and Rice (11) on single crystals, Anand and Spitzig (12) on maraging steel, and the shear-banding in low-carbon steel observed by Cockcroft (13).

Further consideration of the evolution of deformation around voids and inclusions may be required for loadings which initiate fracture near free surfaces. As examples where such studies might be of use, we note the wolf's ear fractures under tension following torsion that were observed by Backofen, et al. (14), and the lack of correlation between ductility in tension and torsion tests found by Halford and Morrow (15) and others.

LOW-CYCLE FATIGUE CRACK INITIATION

Out-of-phase pressurization p and bending b of piping systems can cause a polygonal stress path of a type shown in Fig. 2. The drops in stress at the ends of the bending phases are associated with relaxation at a constant bend angle. Any distortion in the yield locus leads to more plastic strain than would be calculated from an isotropic hardening law, but less than would be calculated from a kinematic hardening law. Here, as in rolling contact, fatigue crack initiation is likely to depend on the worst stress-strain history on some slip system in the polycrystal. The strain histories can be surprisingly complex, as shown in Fig. 3 from calculations for steady state creep with a very high strain rate exponent, corresponding nearly to rate independent plasticity. Plastic flow relations for such calculations, taking corners on the yield locus into account, are under development. The self-consistent field method is being used to combine effects for crystals of various orientations. On individual slip planes, a combination of stress and thermal activation is assumed with energies having a multi-periodic function, as shown in Fig. 4.

$$\dot{\varepsilon} = \nu_{disloc} \exp\left[-\frac{\Delta G_o}{kT}\left(1 - \left(\frac{\tau - \tau_m}{\tau_{bm}}\right)^p\right)^q\right] \tag{5}$$

Preliminary results from this model indicate primary creep is logarithmic and secondary creep tends to be power law, with an exponent decreasing with temperature (see also Kocks, Argon, Ashby (16)). This model suggests a polarity of dislocation structure which could be revealed by cooling specimens under load, unloading them, and then looking for an elastic after-effect on reheating.*

FATIGUE CRACK GROWTH UNDER ARBITRARY HISTORIES

In turbogenerator shafts in central station power plants, the electrical load must be disconnected if there is a lightning strike on nearby power lines. For continuity of service, it is desirable to reconnect the lines as soon as possible. This can result in a random-appearing set of transient torques, approaching the yield strength of the shafts. While in many cases such torques do not seem to have caused damage, the question arises as to their effect if a shaft should contain an undetected crack. For Mode III loading on the crack, assuming non-hardening material, the crack tip displacement and the strain history in front of the crack can be found. Consider the stress history of Fig. 5, with an initial load change from Load 1 to Load 2, a second load change not quite twice as great, and a series of possible third load changes leading to three different final loads below, at, or above the initial loading. The method of finding the solution is outlined in Fig. 6. Neglecting the Bauschinger effect, the stress change due to a load change can be found from the same analysis as the initial loading, except that the yield strength is now double. When the load change on one reversal just equals the preceding load change, superposition of the stress fields just cancels out the preceding stress fields and the deformation continues as if the intermediate reversal had not existed. Any damage from plastic strains during the previous reversal does remain, however. The resulting predictions for crack growth are different depending on whether a crack tip displacement criterion is used, or a damage criterion is used based on the nested zones of residual

---

*Note added in proof: Subsequent work shows that this model necessarily leads to a non-convex hysteresis loop. Apparently the convex behavior requires a statistical distribution of dislocation resistances, and in turn a continuum of internal parameters, rather than the one parameter of Fig. 4.

plastic strain. Both criteria are different from the currently used rain flow method of Dowling, et al. (17), in that reversals rather than cycles are counted, and the history is monitored all along the crack path, rather than for the specimen as a whole. Further development and comparison with experiments are in progress.

MONOTONIC CRACK GROWTH

Recent experimental, analytical, and computational studies in elastic-plastic and fully plastic crack initiation, stable growth, and final instability have shown that under not-yet-well-defined restrictions, these phenomena can be described by a single parameter such as the J-integral or crack tip opening displacement, which serves as the amplitude of the crack tip HRR singular fields. Although these restrictions can apparently be met in face-grooved compact tension specimens of materials like A533 pressure vessel steels (18), it is also apparent that single parameter approaches cannot account for phenomena associated with a loss of HRR dominance of the crack tip fields.

The loss of the HRR field has been convincingly demonstrated (19,20) for large-scale yielding under tensile loading of plane strain, center-cracked or single edge-notched specimens, even with moderately strain-hardening materials. In such cases, intense deformation of the 45° slip planes of ideal plasticity drives the local HRR fields back toward the crack tip, squeezed against the fracture process zone.

It is typically noted in these cases that the triaxial stress elevation on the plane directly ahead of the crack tip is substantially reduced, by comparison to that given by the HRR field, while the angular distribution of crack tip plastic strain exhibits a similarly large increase near 45°, in contrast to that near 90° from the prospective crack plane in the HRR field.

For example, McMeeking and Parks (19) analyzed with finite elements a plane strain center-cracked specimen with a ratio of total crack length to specimen width of 0.5 , and a strain hardening exponent of $n = 0.1$ ($\sigma = \sigma_1 \varepsilon^n$ in the plastic range and yield strain $\sigma_o/E = 1/300$, giving $\sigma_o/\sigma_1 = 0.565$ and $\sigma_o/TS = 0.786$). They note a dramatic increase of equivalent plastic strain, $\bar{\varepsilon}_p$, along the ray 45° from the plane ahead of the crack, as fully plastic conditions are approached. The radial distance along this ray over which $\bar{\varepsilon}_p \gtrsim 0.05$ is five times larger than that which would be inferred from the HRR singularity when the plastic zone just reaches the free surface (where a normalized J is given in terms of the uncracked ligament length L by

$JE/(L\sigma_o\sigma_1) = 2.65)$. Furthermore, the maximum normal stress on the plane ahead of this crack has been reduced from approximately $3.8\sigma_o$ in small-scale yielding to about $3.1\sigma_o$ at general yield.

It may be anticipated that similar fully-plastic flow fields will interact with (and squeeze back) local HRR fields along a surface crack front in a plate or shell which is subjected to predominantly membrane tension across the crack plane. In recent elastic-plastic calculations based upon the line-spring model for surface cracks in plates, Parks (21) noted that the generalized stress state along the remaining ligament tended to that of mid-ligament tension of the plane strain single edge cracked geometry, a configuration highly susceptible to loss of HRR dominance. It may be noted that the pressure vessel industry's time-honored principle of "leak before break" is evidently associated with the loss of HRR dominance since, in leaking, surface cracks from one side of the vessel wall typically reach the other free surface along 45° planes of intense shear deformation. Some cases of the loss of dominance of singular crack tip fields have recently been reviewed by Parks (22).

Other examples of the dominance of the fully plastic, non-hardening plastic flow fields appear in doubly face-grooved plates. In aluminum-zinc specimens, the ratio of observed maximum load to standard limit load (determined from the uniaxial tensile strength and non-hardening slip line analysis) decreased as one of the notch angles was increased, apparently due to the increase in strain concentration factor in the asymmetric flow field as shown in Fig. 7 (23). A second example was noted by Marcolini (24). He found that in doubly grooved welded steel plate the transition temperature was at least 30°C above that for a similar plate with a crack in just one side, or for a Charpy specimen. This second example also shows the limitations of a one-dimensional scheme of fracture-stable design, in which the extension to fracture of a cracked member is more than the deformation supplied to it by its surroundings. Here, the through-section cracking was ductile, but subsequent lateral growth turned into an unstable cleavage mode.

Further research is necessary in order to better define the limits of applicability of single-parameter approaches, as well as to correlate fracture when no single parameter suffices. Progress will require development of plastic flow relations and local fracture models which more accurately reflect the details of the microstructural processes of deformation and fracture and their dependencies on stress and deformation history, as well as the development of practical numerical methods useful with incompressible or dilatational flow to strains of perhaps 15%. An ultimate goal should be to develop a finite element with a crack running into it, and the rule for the vector growth of the crack in terms of the nodal displacement.

CREEP CRACK GROWTH

Flow rules

While power law creep has been used for the majority of creep fracture
analyses, it should be borne in mind that the practical limitation of nominal
creep strains to the order of 1% means that primary creep will often be the
dominating mechanism. Furthermore, the far field will be well into the creep
regime, in which case the amplitude of the creep crack tip field is not governed
by an elastic singularity. This, along with the non-radial loading as a crack
sweeps by a point, means that numerical methods will be required. At the same
time, the power law analytical solutions are valuable for checking the numerical methods, as will be done here.

Consider a power law strain rate given in terms of a nominal stress σ_o,
the strain rate ε_o at the nominal stress, and the strain rate exponent m.
The power law equation under uniaxial stress is

$$\dot{\varepsilon} = \sigma/E + \dot{\varepsilon}_o (\sigma/\sigma_o)^m \tag{6}$$

The resulting stress field is given in terms of a time dependent amplitude
$C(t)$, the crack length a, a non-dimensional function I_m of the exponent
m (which varies from 2π to 3.8 as m rises from 1 to ∞), and radial
coordinates from the tip r, θ :

$$\sigma^{HRR}(r,\theta,t)/\sigma_o = \left[\frac{C(t)}{\sigma_o \dot{\varepsilon}_o a} \frac{a}{I_m r} \right]^{\frac{1}{m+1}} \sigma_{ij}(\theta;m) \tag{7}$$

For short times after an initially elastic response, Riedel and Rice (25) have
shown that $C(t)$ can be expressed in terms of the elastic stress intensity
factor K_I. "Short times" means

$$t \ll \frac{K_I^2}{(m+1)EC^*} \tag{8}$$

where $C^* = C(t \to \infty)$ under constant applied load. The creep singularity is
then approximately

$$\frac{C(t)}{\sigma_o \dot{\varepsilon}_o a} = \frac{(1-\nu^2) K_I^2}{(m+1) E\sigma_o a} (\dot{\varepsilon}_o t)^{-1} \tag{9}$$

These analytical results are used to check a numerical time integration scheme carried out explicitly, regarding the creep strains as initial strains in an elastic finite element calculation. For details, see Bassani and McClintock (26). For stability, an estimate of the allowable time step Δt_s for power law creep has been given by Cormeau (27). In terms of the maximum equivalent stress σ_{max}, at any integration point

$$\Delta t_s = \frac{4(1+\nu)}{3m} \frac{\bar{\sigma}_{max}/E}{\dot{\varepsilon}_o(\bar{\sigma}_{max}/\sigma_o)^m} \tag{10}$$

The resulting very small time steps can be increased by factors of 10 to 50 if a number of short steps, following Eq. 10, have reduced previous errors to sufficiently small values. This technique was applied to a problem with nominal tractions σ_N applied on a circular arc of radius 21a centered at the crack mouth (x = -a, y = 0). The mesh, focusing at the crack tip, has 216 eight-node isoparametric elements, giving 1380 degrees of freedom. Poisson's ratio was $\nu = 0.3$ and $\sigma_N = \sigma_o = E/2000$.

The initial elastic response gave a stress intensity factor calculated from the J integral of $K_I = 1.14$ (1.5% higher than the exact result for the half space). As shown in Fig. 8 for m = 3, the initial relaxation of the crack tip field C(t) agrees well with Riedel and Rice's short time approximation of Eq. 9. Also shown is the relaxation of the stress at the integration point closest to the crack tip (r/a = 0.002, θ = 78°) that limited Δt_s in Eq. 10. Similar results for m = 10 are shown in Fig. 9, with logarithmic scales chosen to emphasize the very large range in times required for steady state, and to show continued agreement with the Riedel-Rice approximation. The crack tip relaxation δ occurs much faster than the crack mouth relaxation Δ. Corresponding crack tip stresses and strains are shown in Fig. 10. Typical times to initiate fracture can be estimated from these results. Consider a material with a fracture process zone size ρ = 25 μm, with m = 10, $\dot{\varepsilon}_o = \dot{\varepsilon}_N = 10^{-11}s^{-1}$ (1% creep in 30 years) at $\sigma_o = \sigma_N = 5 \times 10^{-4}$ E, with an initial crack of length 5mm. If fracture occurs at a shear strain of 0.02, Fig. 10 indicates initiation at $t = 3 \times 10^{-5} t_N = 1500s$.

Crack growth

Hui and Riedel (28) have found that the amplitude of the singularity for a growing crack with m > 3 is given uniquely in terms of the crack velocity a in terms of known functions $\delta_{ij}(\theta; m)$ and $\beta(m)$, where $\beta(4) = 1.042$ and $\beta(6) = 1.237$:

$$\sigma_{ij}^{HR}(r,\theta)/\sigma_o = \beta(m) \left[\frac{\sigma_o \overset{\bullet}{a}}{E \overset{\bullet}{\varepsilon}_o r} \right]^{\frac{1}{m-1}} \hat{\sigma}_{ij}(\theta;m)$$
(11)

If the fracture criterion depends on a stress-weighted integral of strain rate from time zero, the crack velocity is related to the far-field loading. For locally-controlled cracking, however, governed by either the environment or the near tip stress and strain rate fields, the growth rate surprisingly does not depend on the far-field loading. Although this behavior has been observed in stress corrosion cracking (the so-called plateau regime, e.g. (29)), it has not been observed in creep fracture.

ACKNOWLEDGEMENTS

The stress-path and plastic flow relation parts of this work were supported by the National Science Foundation under Grant DMR-78-24185, the torsional fatigue crack growth by the Department of Energy under Contract EX-76-A-01-2295, the creep cracks by the Department of Energy under Contract EG-77-S-02-4461, the ductile nucleation by a U. K. National Research Council Grant to Glasgow University, and the part-through studies both by M. I. T. and by the Department of Energy through Sandia Laboratories and Washington University.

14

REFERENCES

(1) H. Tada, P. C. Paris, and R. Gamble, Stability Analysis of Circumferen-
     tial Cracks in Reactor Piping Systems, Office of Nuclear Reactor Regula-
     tion, NUREG/CR-0838 R1, R5, National Technical Information Service,
     Springfield, Virginia, 41 pp.

(2) A. Needleman, and J. R. Rice, "Limits to Ductility Set by Plastic Flow
     Localization", Mechanics of Sheet Metal Forming, D. P. Koistinen and
     N. M. Wang (eds.), Plenum Press, New York, (1978), pp. 237-267.

(3) J. W. Rudnicki, and J. R. Rice, "Conditions for the Localization of
     Deformation in Pressure-Sensitive Dilatant Materials", J. Mech. Phys.
     Solids 23, (1975), pp. 371-394.

(4) D. P. Clausing, "Effects of Plastic Strain State on Ductility and
     Toughness", Int. J. Fract. Mech. 6, (1970), pp. 71-85.

(5) J. W. Hancock, IUTAM Symposium on Three-Dimensional Constitutive
     Equations and Ductile Rupture, Dourdan, (June, 1980), to be published
     by North-Holland Publishing Co., Amsterdam.

(6) A. C. Mackenzie, J. W. Hancock, and D. K. Brown, "On the Influence of
     State-of-Stress on Ductile Failure Initiation in High Strength Steels",
     Engr. Fract. Mech. 9, (1977), pp. 167-188.

(7) F. A. McClintock, "On the Mechanics of Fracture from Inclusions",
     Ductility, Am. Soc. Metals, Metals Park, Ohio, (1968), pp. 255-277.

(8) F. A. McClintock, "A Criterion for Ductile Fracture by the Growth of Holes",
     J. Appl. Mech. 35, (1968), pp. 363-371.

(9) H. Yamamoto, "Conditions for Shear Localization in the Ductile Fracture
     of Void-Containing Materials", Int. J. Fract. 14, (1978), pp. 347-365.

(10) A. L. Gurson, "Plastic Flow and Fracture Behavior of Ductile Materials
     Incorporating Void Nucleation, Growth and Interaction", Ph.D Thesis,
     Brown University, (1975).

(11) R. J. Asaro, and J. R. Rice, "Strain Localization in Ductile Single
     Crystals", J. Mech. Phys. Solids 25, (1977), pp. 309-338.

(12) L. Anand, and W. A. Spitzig, "Initiation of Localized Shear Bands in
     Plane Strain", J. Mech. Phys. Solids 28, (1980), pp. 113-128.

(13) M. G. Cockcroft, "Ductile Fracture in Cold Working Operations", Ductility,
     Am. Soc. Metals, Metals Park, Ohio, (1968), pp. 199-225.

(14) W. A. Backofen, A. J. Shaler, and B. B. Hundy, "Mechanical Anisotropy
     in Copper", Trans. Am. Soc. Metals 46, (1954), pp. 655-675.

(15) G. R. Halford, and J. Morrow, "Low Cycle Fatigue in Torsion", Proc. ASTM
     62, (1962), pp. 695-709.

(16) U. F. Kocks, A. S. Argon, and M. F. Ashby, Thermodynamics and Kinetics
     of Slip, Pergamon Press, New York, (1975), 288 pp.

(17) N. E. Dowling, W. E. Brose, and W. Wilson, "Notched Member Fatigue Life
     Predictions by the Local Strain Approach", Fatigue Under Complex Loading -
     Analysis & Experiments, SAE AE-6, Warrendale, PA, (1977), pp. 55-84.

(18) C. F. Shih, H. G. DeLorenzi, and W. R. Andrews, "Studies on Crack Initia-
     tion and Stable Crack Growth", Elastic-Plastic Fracture, ASTM STP 668,
     J. D. Landes et al. (eds.), ASTM, Philadelphia, (1979), pp. 65-120.

(19) R. M. McMeeking, and D. M. Parks, "On Criteria for J-Dominance of Crack

Tip Fields in Large Scale Yielding", Elastic-Plastic Fracture, ASTM STP 668, J. D. Landes et al. (eds.), ASTM, Philadelphia, (1979), pp. 175-194.

(20) C. F. Shih, and M. D. German, "Requirements for a One-Parameter Characterization of Crack Tip Fields by the HRR Singularity", to appear Int. J. Fract.

(21) D. M. Parks,"The Inelastic Line-Spring: Estimates of Elastic-Plastic Fracture Mechanics Parameters for Surface-Cracked Plates and Shells," ASME Paper 80-C2/PVP-109 (1980),in press, Trans. ASME, J. Press. Vess. Tech.

(22) D. M. Parks, "The Dominance of the Crack Tip Fields of Inelastic Continuum Mechanics", Numerical Methods in Fracture Mechanics, Second Int. Conference at the University College, D. R. J. Owen and A. R. Luxmoore (eds.), Pineridge Press, Swansea, (1980), pp. 239-260.

(23) F. A. McClintock, "On Notch Sensitivity", Welding J. Res. Suppl. 26, (1961), pp. 202-208.

(24) R. A. Marcolini, "Notch Toughness Requirements for Welded Plate, and A Proposed Test", Ocean Engineer's Thesis, M. I. T., (1970).

(25) H. Riedel, and J. R. Rice, Tensile Cracks in Creeping Solids, Brown Univ., Report E (11-1) 3084-64, submitted for publication ASTM, (1979).

(26) J. L. Bassani, and F. A. McClintock, "Creep Relaxation of Stress around a Crack Tip", accepted for publication Int. J. Solids and Struct., (1981).

(27) I. Cormeau, "Numerical Stability in Quasi-Static Elasto Visco Plasticity", Int. J. for Num. Meth. Eng. 9, (1975), pp. 109-127.

(28) H. Hui, and H. Riedel, "The Asymptotic Stress and Strain Field Near the Tip of a Growing Crack Under Creep Conditions", submitted to Int. J. Fract., (1980).

(29) F. A. McClintock, "Continuum Description of Cracks at Macro and Micro Levels and Interaction of Crack Morphology with Environment", Stress Corrosion Cracking and Hydrogen Embrittlement of Iron Base Alloys, J. Hochmann and R. W. Staehle (eds.), National Association Corrosion Engrs., Houston, (1977), pp. 455-472.

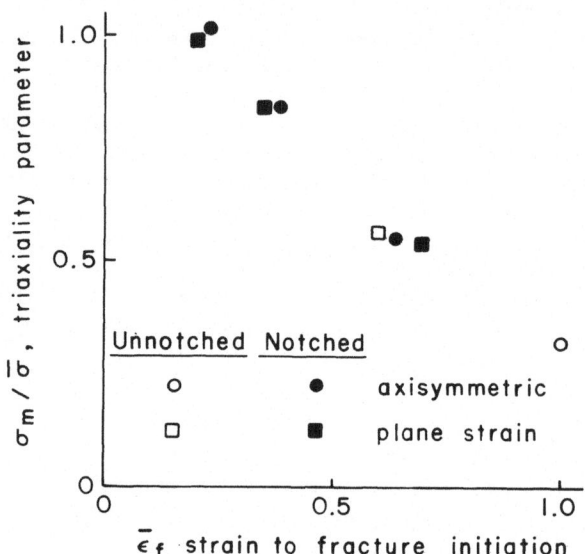

Fig.1 Effect of triaxiality and planarity on ductility.
After Hancock (5)

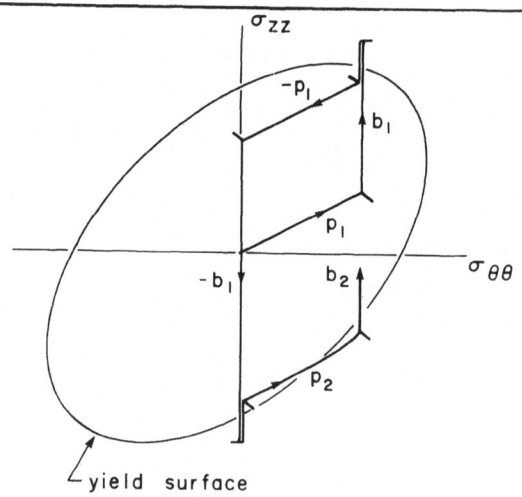

Fig. 2 Stress in a pipe for out-of-phase pressurization p,
bending b, depressurizing -p, reverse bending -b, etc.

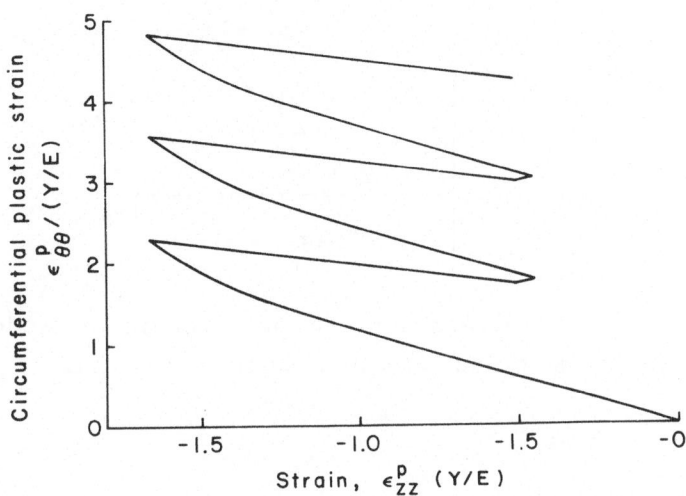

Fig. 3 Plastic strain under out-of-phase pressurization and bending

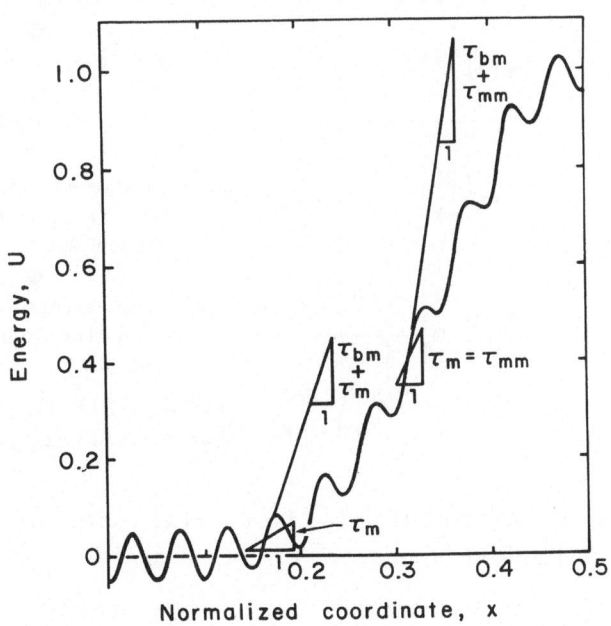

Fig.4 Potential energy function for a dislocation against Bauschinger resistance τ_m, Eq. 5

18

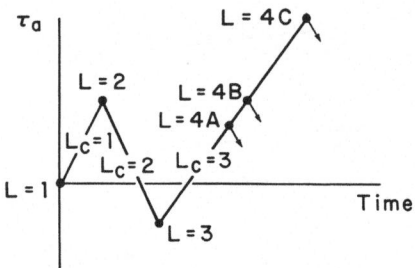

Fig. 5 Arbitrary stress history on a crack in Mode III
shear with the 3rd load change to A, B or C

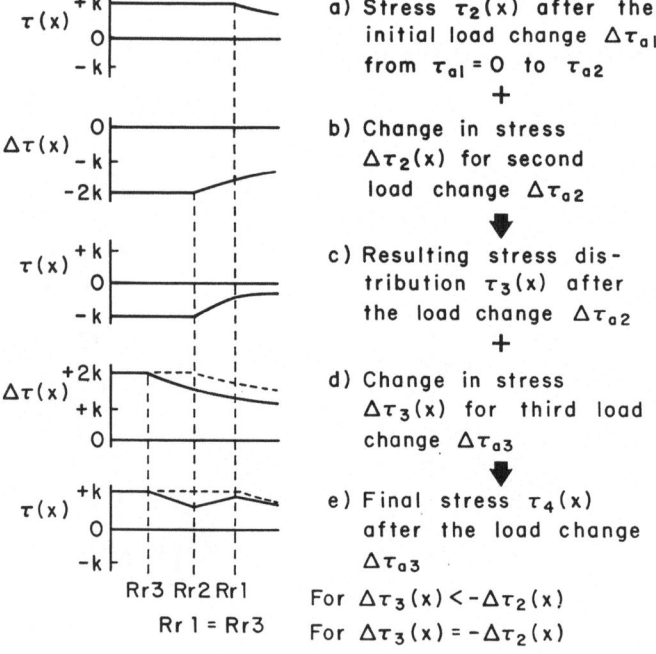

a) Stress $\tau_2(x)$ after the
initial load change $\Delta\tau_{a1}$
from $\tau_{a1} = 0$ to τ_{a2}

+

b) Change in stress
$\Delta\tau_2(x)$ for second
load change $\Delta\tau_{a2}$

c) Resulting stress dis-
tribution $\tau_3(x)$ after
the load change $\Delta\tau_{a2}$

+

d) Change in stress
$\Delta\tau_3(x)$ for third load
change $\Delta\tau_{a3}$

e) Final stress $\tau_4(x)$
after the load change
$\Delta\tau_{a3}$

For $\Delta\tau_3(x) < -\Delta\tau_2(x)$
For $\Delta\tau_3(x) = -\Delta\tau_2(x)$

Fig. 6 Development of stress $\tau(x)$ with arbitrary
loading sequence $\tau_{a1} = 0 \rightarrow \tau_{a2} \rightarrow \tau_{a3}$

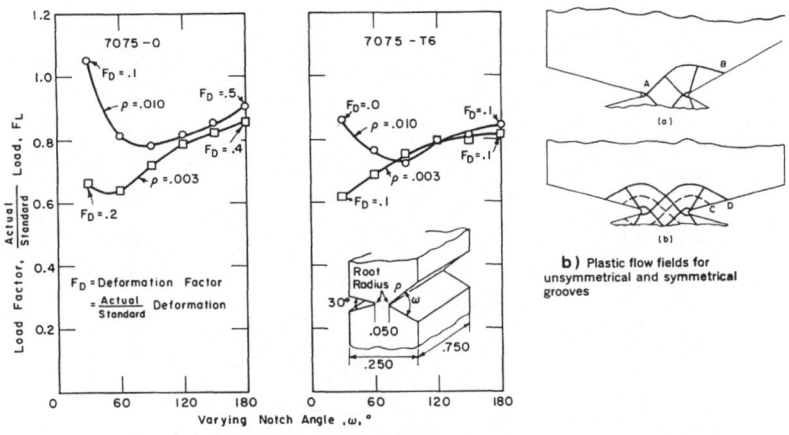

a) Load and deformation factors for grooved 7075 aluminum zinc alloy specimens

Fig. 7 Reduction in ductility with increasing notch angle (23)

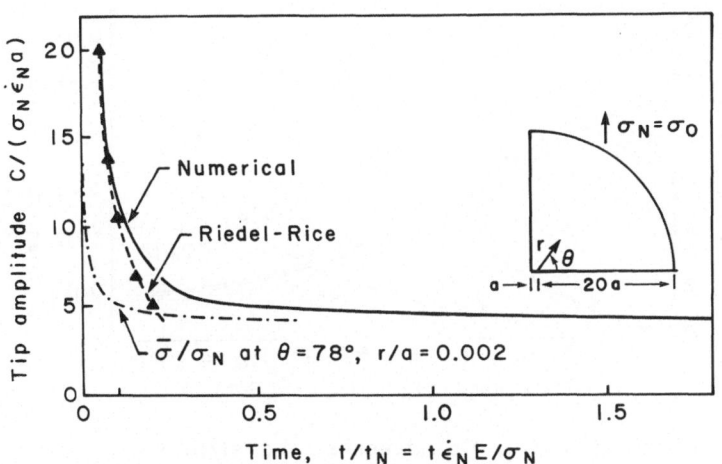

Fig. 8 Transient crack tip fields for power law creep with exponent m = 3 (Eqs. 7 and 9)

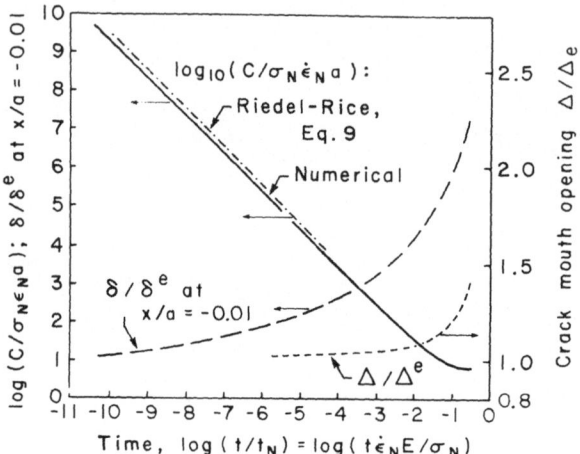

Fig. 9 Normalized tip amplitude $C/\sigma_N \dot{\epsilon}_N a$, crack tip opening δ at $x/a = -0.01$ and crack mouth opening Δ for exponent $m = 10$

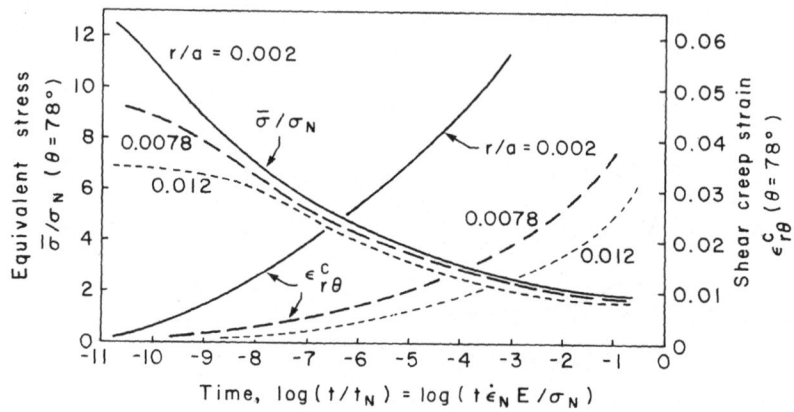

Fig. 10 Transient crack tip stresses and strains $m = 10$, $\sigma_N/E = 0.0005$

II. J-INTEGRAL — THEORETICAL CONSIDERATIONS

EXTENSION OF J–INTEGRAL AND ITS APPLICATION
TO THE ESTIMATION OF STRUCTURAL STRENGTH

M. Sakata
Professor of Mechanical Engineering

S. Aoki
Associate Professor of Mechanical Engineering

K. Kishimoto
Research Associate of Mechanical Engineering
Tokyo Institute of Technology, Ookayama, Meguro-ku, Tokyo 152, Japan

ABSTRACT

The path independent integral, $\hat{J}$, has been presented as the energy release rate during crack extension. This integral is an extension of the J-integral proposed by Rice and includes the existence of the fracture process region and the effects of plastic deformations, body forces, thermal strains, and inertia of material. In order to show the usefulness of the $\hat{J}$-integral to fracture problems, numerical values of the $\hat{J}$-integral in the presence of body forces, thermal strains, inertia effects, or preloadings that bring about large scale yielding are evaluated. The finite element method is used and illustrative examples are presented.

INTRODUCTION

In the fracture of real materials, plastic deformation occurs at the crack-tip. The linear elastic fracture mechanics is effective if the plastic zone is limited in size, compared with the dimensions of the crack and structures. Elastic–plastic treatment is necessary if the plastic zone is of considerable size. The path independent J-integral has been proposed by Rice [1], and it provides a means of extending fracture mechanics concepts from linear elastic behavior to plastic behavior [2]. The J-integral is path independent for linear or nonlinear elastic stress–strain laws, but it loses its physical significance as a crack driving force when deformation is not reversible. To be path independent for the Prandtl–Reuss relation, the J-integral needs to be extended.

Broberg [3,4] shows that the fracture process region exists near the crack-tip and plays an important role in real fracture. The fracture process region is thought to be a small region where fracture processes, such as initiation of

voids or micro-cracks and their growth and coalescence, take place, and
continuum mechanics do not work. However, global energy balance has to be
satisfied, even if the fracture process region exists, and release of energy
may play a central role in problems of the initiation of crack growth.

In this paper, the $\hat{J}$-integral, which is an extension of Rice's J-integral
and is path independent for the Prandtl-Reuss relation, is described, taking
into account the existence of the fracture process region and the effects of
plastic deformations, body forces, thermal strains, and inertia of material.
Numerical values of the $\hat{J}$-integral of cracked bodies subjected to centrifugal
force, dynamic loadings, thermal shock, or preloadings that cause large scale
yieldings are evaluated by using the finite element method. The results are
shown as illustrative examples.

PATH INDEPENDENT $\hat{J}$-INTEGRAL AS ENERGY RELEASE RATE

Energy balance during crack extension

We consider a plate containing a crack, as shown in Fig. 1, and assume
that the crack-tip moves from the initial location 0 to the final location 0',
with the distance 00' infinitesimal. $0-X_1,X_2$ is the fixed frame and $0'-x_1,x_2$
the moving frame whose origin 0' is coincident with the tip of the extending
crack. The direction of X_2 and x_2 is perpendicular to the crack surfaces.
Following Broberg, we assume that fracture occurs in a fracture process region
or an end region denoted by A_{end}. As illustrated in Fig. 1, Γ_{end} is the contour
surrounding A_{end}; Γ is any contour surrounding the end region and crack; Γ_s
curves along the crack surfaces and A, the region surrounded by Γ_{end}, Γ and
Γ_s curves.

The energy balance equation of material in A is written as

$$\int_{\Gamma+\Gamma_s} T_i(du_i/d\ell)d\Gamma + \iint_A F_i(du_i/d\ell)dA = \iint_A \rho\ddot{u}_i(du_i/d\ell)dA + \iint_A \sigma_{ij}(d\varepsilon_{ij}/d\ell)dA + \hat{J} \quad (1)$$

where T_i is surface traction, u_i displacement, F_i body force, ρ density, σ_{ij}
stress tensor, ε_{ij} strain tensor, ℓ crack length, $d/d\ell$ differentiation with
respect to length, and $\hat{J}$ energy flow into the fracture process region A_{end}.

By using the Gauss theorem, Eq. (1) reduces to

$$\hat{J} = \int_{\Gamma_{end}} T_i(du_i/d\ell)d\Gamma \quad (2)$$

With reference to Fig. 1, transformation between the fixed and moving frames
is given by

$$x_1 = X_1 \cos\theta_0 + X_2 \sin\theta_0 - l, \quad x_2 = -X_1 \sin\theta_0 + X_2 \cos\theta_0 \tag{3}$$

and the displacements are represented in terms of the fixed frame $0\text{-}X_1, X_2$ as

$$u_i(x_1, x_2, l) = u_i(X_1 \cos\theta_0 + X_2 \sin\theta_0 - l, -X_1 \sin\theta_0 + X_2 \cos\theta_0, \ l) \tag{4}$$

whence we obtain the following relation:

$$du_i/dl = \partial u_i/\partial l + (\partial u_i/\partial x_1)(\partial x_1/\partial l) = \partial u_i/\partial l - \partial u_i/\partial x_1 \tag{5}$$

Substitution of Eq. (5) into Eq. (2) yields

$$\hat{J} = \int_{\Gamma_{end}} T_i(\partial u_i/\partial l - \partial u_i/\partial x_1) d\Gamma \tag{6}$$

Since in deriving Eq. (6), no restriction has been imposed on the stress-strain relation of the material, the $\hat{J}$-integral is thought to be an energy release rate of an arbitrary material during crack extension and has a physical significance as a crack driving force.

Fracture criterion

The fracture process occuring in the fracture process region will be different in different materials and environmental conditions. For example, the initiation of voids or micro-cracks, and their coalescence will take place in metallic materials. We assume that the fracture process region is autonomous in the sense that it does not depend upon the geometry of body or crack as well as load conditions [3,4]. We assume that the deformation field in the process region is intrinsic to the material and that the relation $\partial u_i/\partial l = 0$ holds in the region. We obtain from Eq. (6)

$$\hat{J} = -\int_{\Gamma_{end}} T_i(\partial u_i/\partial x_1) d\Gamma \tag{7}$$

Let $\hat{J}_c$ denote the rate of energy that is dissipated in the fracture process region due to the crack growth of unit area; then, fracture criterion in terms of energy balance is written as

$$\hat{J} = \hat{J}_c \tag{8}$$

We again point out that in deriving Eq. (6) or (7), the stress-strain relation has been arbitrary so that Eq. (8) may be employed as a fracture criterion for a wide range of materials.

$\hat{J}$ as path independent integral

We obtain from Eqs. (3) and (4)

$$\partial u_i/\partial x_1 = \cos\theta_0(\partial u_i/\partial X_1) + \sin\theta_0(\partial u_i/\partial X_2) \tag{9}$$

and Eq. (7) is written as

$$\hat{J} = -\int_{\Gamma_{end}} T_i\{\cos\theta_0(\partial u_i/\partial X_1) + \sin\theta_0(\partial u_i/\partial X_2)\}d\Gamma \tag{10}$$

We decompose $\hat{J}$ into the components $\hat{J}_1$ and $\hat{J}_2$ as

$$\hat{J} = \hat{J}_1\cos\theta_0 + \hat{J}_2\sin\theta_0 \tag{11}$$

where $\hat{J}_k (k=1,2)$ is defined by

$$\hat{J}_k = -\int_{\Gamma_{end}} T_i(\partial u_i/\partial X_k)d\Gamma, \quad (k=1,2) \tag{12}$$

We obtain the path independent representation as

$$\hat{J}_k = \iint_A \{(\rho\ddot{u}_i - F_i)(\partial u_i/\partial X_k) + \sigma_{ij}(\partial\varepsilon_{ij}/\partial X_k)\}dA - \int_{\Gamma + \Gamma_s} T_i(\partial u_i/\partial X_k)d\Gamma \tag{13}$$

We decompose strain tensor ε_{ij} into elastic strain components ε_{ij}^e and eigen strain components ε_{ij}^* and write the result as

$$\varepsilon_{ij} = \varepsilon_{ij}^e + \varepsilon_{ij}^* \tag{14}$$

where the eigen strains are inelastic, stress-free strains which are typically exemplified by thermal strains or plastic components of strains.

The elastic strain energy density function, $W_e(\varepsilon_{ij}^e)$, is defined by

$$\partial W_e(\varepsilon_{ij}^e)/\partial\varepsilon_{ij}^e = \sigma_{ij} \tag{15}$$

Introducing Eq. (15) into Eq. (13) leads to

$$\hat{J}_k = \int_{\Gamma + \Gamma_s - \Gamma_{end}} W_e n_k d\Gamma - \int_{\Gamma + \Gamma_s} T_i(\partial u_i/\partial X_k)d\Gamma + \iint_A \{\sigma_{ij}(\partial\varepsilon_{ij}^*/\partial X_k) + (\rho\ddot{u}_i - F_i)(\partial u_i/\partial X_k)\}dA \tag{16}$$

We also consider the case where the area of the fracture process region is diminishing and the contour integral along Γ_{end} is equal to zero. Equation (16) reduces to

$$\hat{J}_k = \int_{\Gamma + \Gamma_s} \{W_e n_k - T_i(\partial u_i/\partial X_k)\}d\Gamma + \iint_A \{\sigma_{ij}(\partial\varepsilon_{ij}^*/\partial X_k) + (\rho\ddot{u}_i - F_i)(\partial u_i/\partial X_k)\}dA \tag{17}$$

We further assume: the fracture region is ignored; the crack extends in the direction parallel to the crack surfaces; the stress-strain relation of the material is elastic; and body forces, inertia of the material, and crack surface tractions are absent. Then, Eq. (17) reduces to

$$\hat{J} = \hat{J}_1 = \int_\Gamma \{W_e n_1 - T_i (\partial u_i / \partial X_1)\} d\Gamma \tag{18}$$

which coincides with Rice's J-integral.

APPLICATION OF J-INTEGRAL TO FRACTURE PROBLEM

J-integral of rotating disk and prediction of fracture initiation

In the case where body force exists, we have

$$J = \hat{J}_1 = \int_\Gamma \{W_e n_1 - T_i (\partial u_i / \partial X_1)\} d\Gamma - \iint_A F_i (\partial u_i / \partial X_1) dA \tag{19}$$

which has been used in analyzing the cracked rotating disk [5,6].

Figure 2 shows the finite element grid of a cracked rotating disk and the integrating paths for J. The numerical results of J-values vs. the rotating speeds are shown in Fig. 3. Plastic zone shapes of the disk are depicted schematically in Fig. 4. Spin tests were carried out, and J_{Ic} for a cracked rotating disk of rotor-forging steel was determined experimentally and was compared with that obtained by compact specimens. Figure 5 shows the R-curve obtained by spin tests. We conclude that prediction based on the compact tests agrees with initiation of crack growth in rotating disks.

$\hat{J}$-integral and stress intensity factors in the elastodynamic problem

Derivation of a formula for determining the dynamic stress intensity factor from J_k [7] is summarized. By the path independence of $\hat{J}$, we may take Γ to be an infinitesimal rectangle, height 2δ and width 2ε, as proposed by Freund [8]. Only singular terms of σ_{ij} and $u_{i,j}$ contribute to $\hat{J}_k$ so that we obtain

$$J_k = \lim_{\varepsilon \to 0} \lim_{\delta \to 0} \int_{-\varepsilon}^{+\varepsilon} [\{W_e(X_1, \delta, t) - W_e(X_1, -\delta, t)\} \delta_{k2} - \sigma_{i2} \{\partial u_i(X_1, \delta, t) / \partial X_k$$
$$- \partial u_i(X_1, -\delta, t) / \partial X_k\}] dX_1 \tag{20}$$

where δ_{ij} denotes Kronecker's delta. Substituting the dynamic solution [9] for σ_{ij} and u_i in the vicinity of the crack-tip and using the formula [8]

$$\lim_{\varepsilon \to 0} \int_{-\varepsilon}^{\varepsilon} \frac{H(\xi)}{\xi^{1/2}} \frac{H(-\xi)}{-\xi^{1/2}} d\xi = \pi/2 \tag{21}$$

where $H(\xi)$ denotes the unit-step function, we obtain for a stationary crack

$$\hat{J}_1 = \{K_I^2(t) + K_{II}^2(t)\}(\kappa+1)/(8\mu) + K_{III}^2(t)/(2\mu)$$
$$\}$$ (22)

$$\hat{J}_2 = -K_I(t)K_{II}(t)(\kappa+1)/(4\mu)$$

where K_I, K_{II}, and K_{III} denote the stress intensity factors for mode I, II, and III, respectevely. We decompose $\hat{J}_1$ as

$$\hat{J}_1 = \hat{J}_1^I + \hat{J}_1^{II} + \hat{J}_1^{III}$$ (23)

where the superscripts denote the deformation modes. We have the formula

$$\hat{J}_1^I = (\kappa+1)K_I^2(t)/(8\mu), \quad \hat{J}_1^{II} = (\kappa+1)K_{II}^2(t)/(8\mu), \quad \hat{J}_1^{III} = K_{III}^2(t)/(2\mu)$$ (24)

Finite element computation has been carried out for a cracked plate, shown in Fig. 6, where the tensile stresses are applied along the upper and lower boundaries at the time $t=0$ with the unit-step function time dependence. The constant strain elements are used in the computation. Figure 7 shows the grid pattern of the first quadrant and the paths over which $\hat{J}$-integrals are computed. For time integration, the Newmark-β method ($\beta=1/6$) has been used with the time step of 0.2 μs.

The normalized dynamic stress intensity factor $K_I(t)/(\sigma_0\sqrt{\pi a_0})$ is shown in Fig. 8, where the analytical solution for a finite crack in an infinite plate by Thau and Lu [10] is given by the solid curve, and the result of the finite element method using a singular element [11] is given by the dashed curve. The symbol ↓ in the figure represents the time-of-arrival of the reflected stress waves at the crack-tip from the upper and lower boundaries. A computation for an oblique crack problem has been carried out, using the finite element grid and integrating paths, as shown in Fig. 9. The time variation of the dynamic stress intensity factors is shown in Fig. 10, where the symbols L_1 and L_2 denote the time-of-arrival at the crack-tip of the reflected stress waves from the upper and lower boundaries, and from the left and right boundaries, respectively.

$\hat{J}$-integral in thermal stress field

Let α_{ij} denote the coefficient of thermal expansion and θ the temperature increment from the natural state. Thermal strains are represented by

$$\varepsilon_{ij}^* = \alpha_{ij}\,\theta$$ (25)

and Eq. (17) leads to

$$\hat{J}_1 = \int_\Gamma \{W_e n_1 - T_i(\partial u_i/\partial X_1)\}d\Gamma + \iint_A \sigma_{ij}\,\alpha_{ij}(\partial\theta/\partial X_1)dA$$ (26)

For isotropic materials under a uniform temperature field, we have

$$\hat{J}_1 = \int_{\Gamma} \{W_e n_1 - T_i (\partial u_i / \partial X_1)\} d\Gamma \tag{27}$$

Note that when the temperature is uniform, but different from that of the normal state, the elastic strain energy function does not coincide with the strain energy function, i. e.,

$$W_e = \int_0^{\varepsilon_{ij}^e} \sigma_{ij} d\varepsilon_{ij}^e \neq W = \int_0^{\varepsilon_{ij}} \sigma_{ij} d\varepsilon_{ij} \tag{28}$$

and $\hat{J}$-integral does not conform with J.

The $\hat{J}$-integral of a center cracked plate, as shown in Fig. 11, and having a temperature gradient is considered. The left and right boundaries of the plate that has the initial, uniform temperature of 0° C are suddenly subjected to the constant temperature of -100° C. The numerical values of $\hat{J}$ are computed by the finite element method. The time variation of $\hat{J}$ is shown in Fig. 12.

The $\hat{J}$-integral of a cracked cylinder, as shown in Fig. 13, is considered. The temperature of the cylinder is initially uniform (0° C). The inner wall is suddenly subjected to the constant temperature (-100° C) with the outer wall perfectly heat-insulated. The time variation of $\hat{J}$ of the cracked cylinder is shown in Fig. 14. Note that the maximum value is reached in a short time after the application of thermal loading [12].

$\hat{J}$-integral of the preloaded specimen

We consider the case that the eigen strains are composed of plastic strains, i. e., $\varepsilon_{ij}^* = \varepsilon_{ij}^p$. Equation (17) leads to

$$\hat{J}_1 = \int_{\Gamma} W_e dX_2 - \int_{\Gamma} T_i (\partial u_i / \partial X_1) d\Gamma + \iint_A \sigma_{ij} (\partial \varepsilon_{ij}^p / \partial X_1) dA \tag{29}$$

or

$$\hat{J}_1 = \iint_A \sigma_{ij} (\partial \varepsilon_{ij} / \partial X_1) dA - \int_{\Gamma} T_i (\partial u_i / \partial X_1) d\Gamma \tag{30}$$

where $\varepsilon_{ij} = \varepsilon_{ij}^e + \varepsilon_{ij}^p$, and inertia forces are neglected.

The $\hat{J}$-integral of a center cracked plate subjected to the remotely uniform stress, σ_o, as shown in Fig. 15, is considered, and numerical values of $\hat{J}$ are computed by the finite element method. The remote stress vs. the load-point displacement curve is diagrammatically shown in Fig. 15, where the part OABC indicates the preloading cycle and CDE the reloading stage. The remote stress at B is referred to as the preload, σ_{opre}. Figure 16 shows the $\hat{J}$ vs. the remote stress of the specimens that were preloaded to the level indicated in the figure. The remote stress is normalized by the yield stress of the material. When a gradually increasing load is applied, the numerical value of $\hat{J}$ almost coincides with that of J. If, following the preload, the load is reduced from its current value, $\hat{J}$ decreases almost linearly and takes a negative value owing to the compressive residual stresses produced by yielding at the crack-tip. As the load in reloading stage is brought near to its original value, the curve bends sharply over near B and the part DE becomes almost a continuation of OAB. It is noted that the present analysis is valid only for a stationary crack without growing [13].

REFERENCES

(1) J. R. Rice, "A Path Independent Integral and the Approximate Analysis of Strain Concentrarion by Notches and Cracks", Trans. ASME, J. Appl. Mech. 90, (1968), pp. 379-386.

(2) J. A. Begley and J. D. Landes, "The J Integral as a Fracture Criterion", ASTM STP 514, Am. Soc. Test. & Mat., (1972), pp. 1-23.

(3) K. B. Broberg, "Crack-growth Criterion and Non-Linear Fracture Mechanics", J. Mech. Phys. Solids, 19, (1971), pp. 407-418.

(4) K. B. Broberg, "Energy Method in Statics and Dynamics of Fracture", J. Japan Soc. Streng. Fract. Mat., 10-2, (1975), pp. 33-45.

(5) M. Sakata, S. Aoki and K. Ishii, "J-integral Analysis for Rotating Disk", Proc. Int. Conf. Fract. Mech. Tech., 1, ed. by G. C. Sih and C. L. Chow, Sijthoff and Noordhoff Pub., (1977), pp. 515-523.

(6) M. Sakata, et al., "J-integral Approach to Fracture of Rotating Disk", Trans. ASME, J. Engng. Mat. Tech., 100, (1978), pp. 128-133.

(7) K. Kishimoto, S. Aoki and M. Sakata, "Dynamic Stress Intensity Factors Using $\hat{J}$-integral and Finite Element Method", Eng. Fract. Mech., (to be published).

(8) L. B. Freund, "Energy Flux into the Tip of an Extending Crack in an Elastic Solid", J. Elasticity, 2, (1972), pp. 341-349.

(9) F. Nilsson, "A Note on the Stress Singularity at a Non-Uniformly Moving Crack Tip", J. Elasticity, 4, (1974), pp. 73-75; L. B. Freund and R. J. Clifton, "On the Uniqueness of Plane Elastodynamic Solutions for Running Cracks", J. Elasticity, 4, (1974), pp. 293-299; J. D. Achenbach and Z. P. Bažant, "Elastodynamic Near-Tip Stress and Displacement Fields for Rapidly Propagating Cracks in Orthotropic Materials", Trans. ASME, J. Appl. Mech., 42, (1975), pp. 183-189.

(10) S. A. Thau and T. H. Lu, "Transient Stress Intensity Factors for a Finite Crack in an Elastic Solid Caused by a Dilatational Wave", Int. J. Solids Struct., 7, (1971), pp. 731-750.

(11) S. Aoki, et al., "Elastodynamic Analysis of Crack by Finite Element Method Using Singular Element", Int. J. Fract., 14-1, (1978), pp. 59-68.

(12) S. Aoki, et al., "$\hat{J}$-integral in Temperature Fields", Proc. Fall Meeting of JSME, No. 780-12, (1978-10), pp. 178-180, (in Japanese).

(13) S. Aoki, et al., "Finite Element Analysis of $\hat{J}$-integral (Elasto-Plastic Problem)". Proc. Fall Meeting of JSME, No. 790-12, (1979-10), pp. 206-208, (in Japanese).

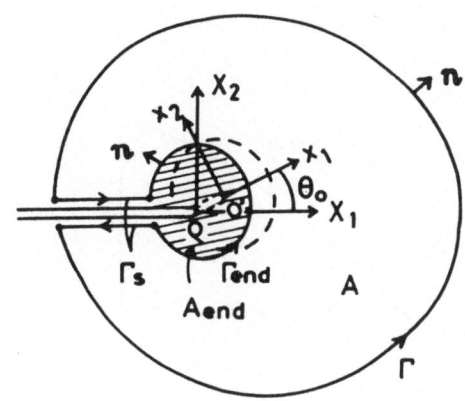

Fig. 1   Configuration of a crack-tip; A_{end}:fracture process region, Γ_{end}: boundary of A_{end}, Γ: arbitrary curve surrounding crack, Γ_s: curves along crack surface, θ_0: direction of infinitesimal crack extension.

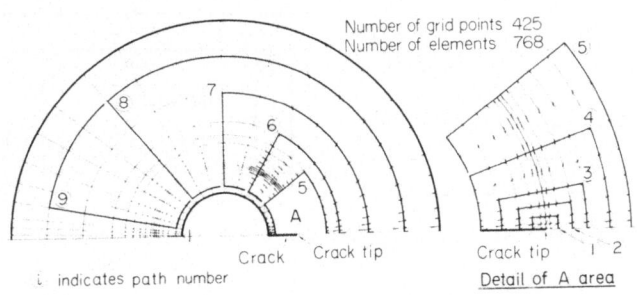

Fig. 2   Grid pattern and paths for J-integral

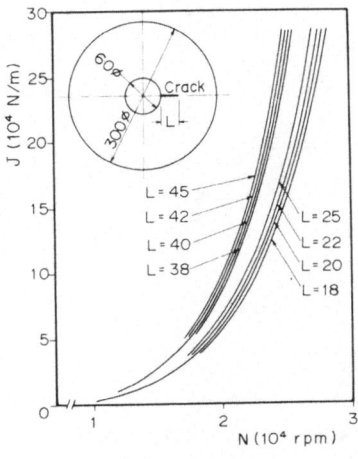

Fig. 3 J vs. rotating speed of disk, having cracks of various length

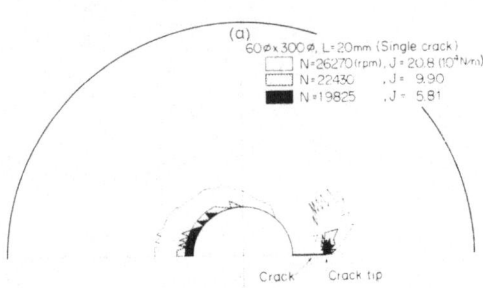

Fig. 4   Plastic zone shapes of disk

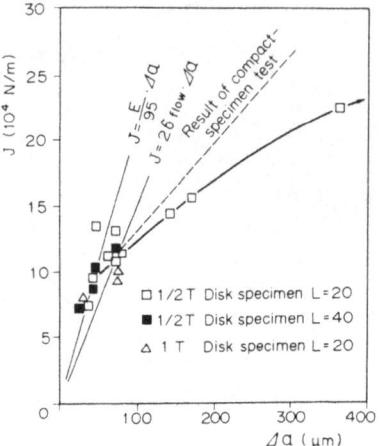

Fig. 5 J vs. Δa resistance curve obtained by spin tests

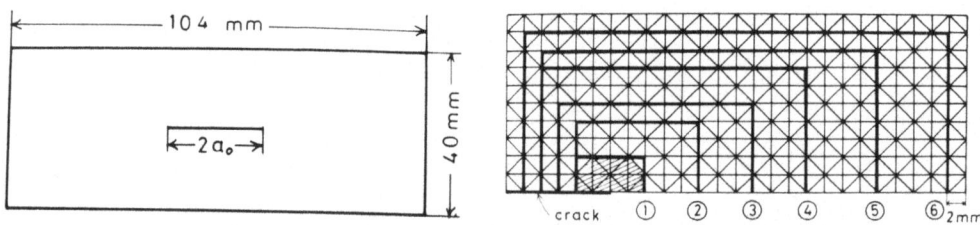

Fig. 6 Rectangular plate with crack

Fig. 7 Finite element grid pattern
(297 nodes, 520 elements)

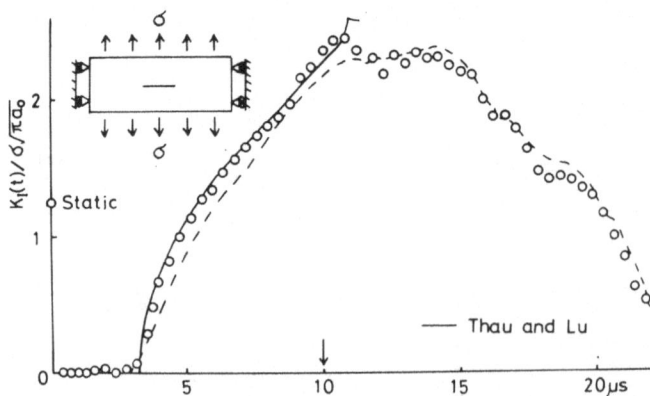

Fig. 8 Time variation of normalized dynamic stress intensity factor

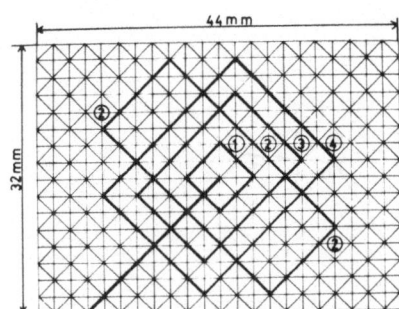

Fig. 9 Finite element grid pattern (399 nodes, 704 elements) and paths for integration

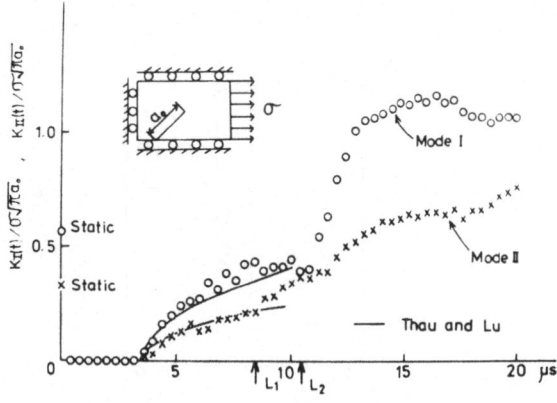

Fig. 10 Time variation of normalized dynamic stress intensity factor (mixed mode)

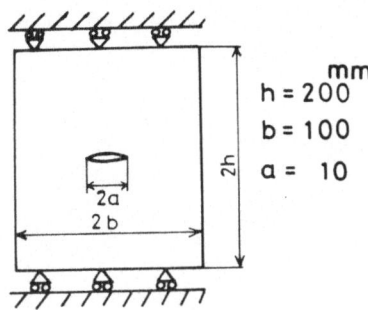

Fig. 11 Rectangular plate with central crack

$$h = 200 \quad mm$$
$$b = 100$$
$$a = 10$$

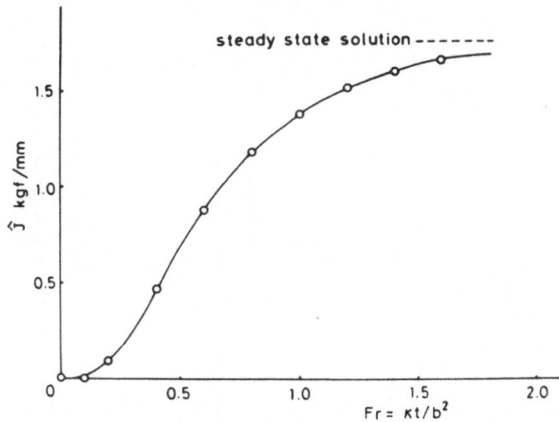

Fig. 12 Time variation of $\hat{J}$ under thermal loading

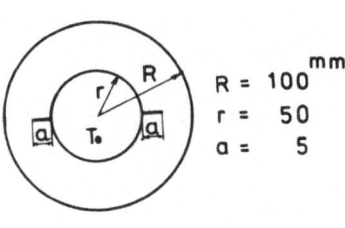

$$R = 100 \quad mm$$
$$r = 50$$
$$a = 5$$

Fig. 13 Cracked cylinder

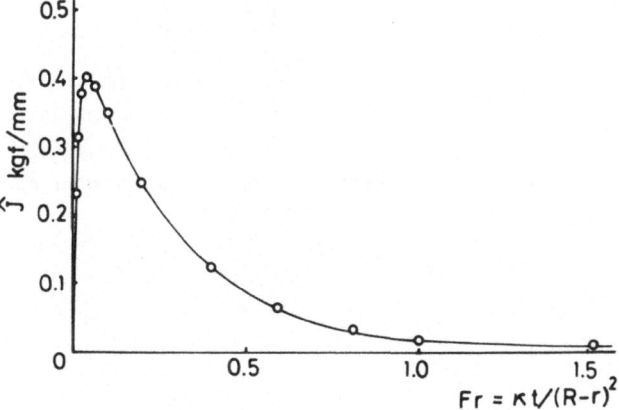

Fig. 14 Time variation of $\hat{J}$ in cracked cylinder

34

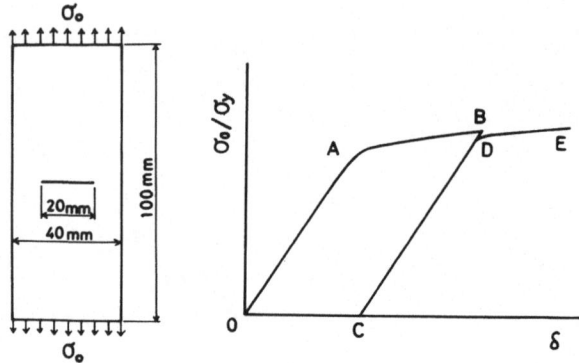

Fig. 15 Configuration of plate and remote stress vs. load-point displacement

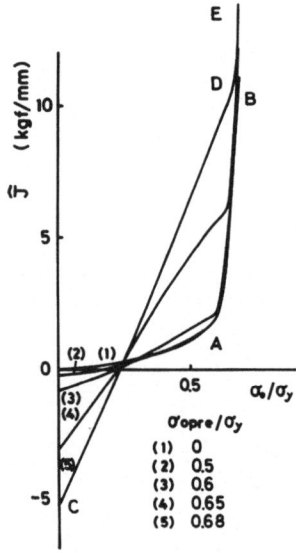

Fig. 16 $\hat{J}$ vs. remote stress in preloaded specimen

J-INTEGRAL EVALUATION OF A CRACK IN A REACTOR VESSEL.

Hiroshi Miyamoto, Masanori Kikuchi,
Department of Mechanical Engineering, Faculty of Science and Technology,
Science University of Tokyo, Yamazaki, Noda, Chiba(278), Japan

Yasuhide Sakaguchi and Takenori Shindo
Kure Works, Babcock Hitachi Co. Ltd., 6-9, Takaramachi, Kure,
Hiroshima(737), Japan

ABSTRACT

The J-integral concept is extended to several practical problems. First, the three-dimensional J-integral is introduced, and a semi-elliptical surface crack problem is analyzed. Several shapes of crack are treated, and the behaviours of crack extension are discussed. Second, a new definition of the J-integral in multi-phase material is proposed. The new J-integral has path independent charecteristics. Several examples are analyzed.

1. INTRODUCTION

In the present work, the J-integral is extended and applied to cracking in a pressure vessel of a reactor. The J-integral is used widely in fracture mechanics, but several shortcomings must be overcome for its use in the safety design of reactors:

(1) Since the crack in a reactor has three dimensional shape, the evaluation of a three-dimensional J-integral is required.

(2) To assess the safety in the welded portion, a J-integral in multi-phase material must be evaluated.

(3) Since the crack exists in a vessel having asymmetric free surfaces, the effect of this asymmetricity must be evaluated.

The J-integral proposed by Rice assumes that the crack exists in a two-dimensional isotropic and homogeneous plate and is loaded symmetrically. Rice's J-integral is applied to these problems with difficulty, so the J-integral must be extended.

This extension of the concept is carried out by using the energy momentum tensors proposed by Eshelby. By using them, the J-integral is shown as the force acting on elastic singularities (including crack tips) or elastic inhomogeneities. Then, the J-integral is the force vector having three components J_x, J_y, and J_z, and the definition of these equations in three dimensional spce is easily obtained.

In this paper, section 2 deals with the expressions of three components of the J vector. In section 3, a three-dimensional J-integral of the crack existing in a pressure vessel is evaluated, and in section 4, a new definition of the J-integral in multi-phase materials is proposed.

2.EXPRESSION OF THREE COMPONENTS OF THE J VECTOR OF THREE-DIMENSIONAL CRACKS.

The J-integral expressions in the following equations are shown by Eshelby [1] as the force acting on crack tips.

$$F_\ell = \int_\Gamma (W\delta_{j\ell} - \sigma_{ij} u_i,_\ell) ds_j = \int_\Gamma P_{j\ell} ds_j. \tag{1}$$

Where ℓ is the direction of force vector, Γ denotes the arbitrary surfaces surrounding the crack, and $P_{j\ell}$ is called the energy momentum tensor. Eq.(1) can be applied to the three-dimensional crack problem as follows.

Consider the section of a three-dimensional crack front, as shown in Fig.1. The crack front coincides with the z axis, and the surfaces surrounding the crack front are composed of S1-S2-S3. When $\ell=x$, eq.(1) becomes

$$F_x = \int_{S1+S2+S3} P_{jx} ds_j$$

$$= B\int_\Gamma (Wdy - \sigma_{ij} n_j u_i,_x ds) - \int_{S2} \sigma_{iz} u_i,_x ds + \int_{S3} \sigma_{iz} u_i,_x ds. \tag{2}$$

As F_x depends on the width B in Fig.1, so the J_x value at point o is obtained by

$$J_x = \lim_{B \to 0} \frac{1}{B} F_x = \int_\Gamma (Wdy - \sigma_{ij} u_i,_x ds_j) - \int_S (\sigma_{iz} u_i,_x),_z dS. \tag{3}$$

In the case of the two-dimensional problem, it is easily shown that eq.(3) coincides with the J-integral expression shown by Rice.

J_y and J_z are obtained by the same procedures:

$$J_y = \int_{\Gamma+\Gamma'} (-Wdx - \sigma_{ij} u_i,_y ds_j) - \int_S (\sigma_{iz} u_i,_y),_z dS, \tag{4}$$

$$J_z = -\int_\Gamma \sigma_{ij} u_i,_z ds_j + \int_S (W - \sigma_{iz} u_i,_z),_z dS, \tag{5}$$

where Γ' is the path along crack surfaces.

These values are the three components of the J vector, and each of these values is the force acting on the crack front along x, y, and z axes. When a symmetrical two-dimensional crack is considered, only the J_x value needs to be evaluated, but for the three-dimensional case, the other two components of the J vector must be considered. Moreover, if the crack surfaces are pressurized, an additional term has to be considered:

$$J_{\ell ad.} = -\int_{\Gamma'} (\sigma_{ij} u_i,_\ell) ds_j. \tag{6}$$

3. J VECTORS OF A SEMI-ELLIPTICAL SURFACE CRACK IN A PRESSURE VESSEL.

Fig.2 shows a semi-elliptical surface crack in a pressure vessel. The crack exists on the inner surface of the vessel. Fig.3 shows the dimensions and coordinates of the pressure vessel. As finite element analyses are carried out by using cylindrical coordinates, the J_r and J_z components are evaluated.

J vectors are evaluated at the six segments along the crack front, as shown in Fig.4. The pressure p acting on the inner surfaces of the vessel is assumed to be p=9.807 MN/m^2.

Four crack shapes are used by changing the value of a/c (a/c=0.33, 0.5, 0.7, and 0.9). The crack, in which a/c=0.33, is the shape recommended by ASME. For each shape of crack, two analyses are carried out: one in which the crack surfaces are unpressurized and the other in which they are pressurized. All of these analyses are carried out in elastic states. Three paths are used for the evaluation of the J vectors, and the results are shown as the average values of the three paths.

Fig.5-7 show J_r, J_z, and $\bar{J}$, the magnitude of the J vector. As a/c becomes large, J values become large. Although this is reasonable, it is interesting to notice that J_z vlaues increase rapidly at the inner surface of the vessel and J_r values increase slowly at the bottom of the crack. As the results show, $\bar{J}$ values increase rapidly at the inner surface with increasing a/c value. The maximum point of $\bar{J}$ is the bottom of the crack for small a/c values, but when a/c=0.9, the inner surface of the vessel becomes the maximum point. And the minimum point of $\bar{J}$ moves from the inner surface to the bottom of the crack as a/c value increases. It is the third segment when a/c=0.7 and the fifth segment when a/c=0.9.

It is especially interesting that when a/c increases from 0.7 to 0.9, the J_z value at the inner surface becomes larger than the J_r values at the bottom of the crack, and $\bar{J}$ values also change with the same tendencies. Therefore, The semi-elliptical surface crack in the pressure vessel grows, keeping a/c≈0.8, if J_{IC} is constant along the crack front. These results agree well with the experimental results that the surface crack extends, keeping 0.8≤a/c≤1.1[2].

Fig.8 shows the directions of the J vector when crack surfaces are unpressurized. Corresponding to the results mentioned above, the directions of the J vector change little at the bottom of the crack, but at the inner surface, they turn to the z axis as a/c value increases.

Table I shows the effect of the pressure on crack surfaces. For all a/c values, $\bar{J}$ increases about 10-14% by considering the pressure on crack surfaces, and they increase a little with increasing a/c values. From these results,

K_I values increase 5-7%, due to the pressure on crack surfaces.

A.S.Kobayashi[3] obtained K_I value at the bottom of the surface crack in a pressurized cylinder in which crack surfaces were unpressurized. To compare with his results, K_I values are calculated by using $\bar{J}$ values at the bottom of the crack with the following equation.

$$K_I = \sqrt{\bar{J}E}/(1-\nu^2). \tag{7}$$

The results are shown in Table II. The differences between the two values is 4.2% when $a/c=0.33$, and it becomes smaller as the a/c value increases.

4. EVALUATION OF J-INTEGRAL IN MULTI-PHASE MATERIALS.

In general, when a crack exists in multi-phase materials, for example, a welding structure, the deformation becomes asymmetric, and the J_y value does not become zero. So J_x and J_y values may have to be considered together. Moreover, the J values of this crack become path dependent as far as the path crosses the phase boundary, because the discontinuities of stresses and strains occur at the phase boundary. In this section, new definitions of J vector are introduced in order to remove these discontinuities.

In Fig.9(a), a notch exists perpendicular to the phase boundary; then, J_x and J_y are defined by the following equations.

$$J_x = \int_{\Gamma_1}(W\delta_{jx}-\sigma_{ij}u_{i},_x)ds_j - \int_{\Gamma_2}(-W+\sigma_{yy}u_y,_y)dy, \tag{8}$$

$$J_y = \int_{\Gamma_1}(W\delta_{jy}-\sigma_{ij}u_{i},_y)ds_j, \tag{9}$$

where Γ_1 is the path from point A to B surrounding the notch tip and Γ_2 is the closed path surrounding the phase boundary in Γ_1. In this case, the expression of J_y integral does not alter, due to the existence of the phase boundary, because $C_{ijkm},_y$ becomes zero in every region surrounded by Γ_1, where C_{ijkm} are the elastic constants.

When a notch exists along the phase boundary, as shown in Fig.9(b), it is defined as

$$J_x = \int_{\Gamma_1}(W\delta_{jx}-\sigma_{ij}u_{i},_x)ds_j, \tag{10}$$

$$J_y = \int_{\Gamma_1}(W\delta_{jy}-\sigma_{ij}u_{i},_y)ds_j + \int_{\Gamma_2}(-W+\sigma_{xx}u_x,_x)dx. \tag{11}$$

In thid case, the expression of the J_x integral is not altered by the existence of the phase boundary.

These values, defined in eqs.(8)-(11), have path independent characteristics. Examples, shown in Fig.9(a) and 9(b), are analyzed and the results are shown in Figs. 10 and 11. In these figures, x marks identify the results when additional terms of eqs.(8) and (11) are neglected and ● marks show that they are considered. Through these results, the fact that J values defined by eqs.(8)-(11) have path independent characteristics is verified. In general, when the phase boundary inclines to the crack or notch, an appropriate combination of these two cases can be used.

A notch in a three-storied lamella is analyzed, as in Fig.12. In this case, J_y is always zero because the deformation is symmetrical to x axis, and only J_x must be considered. Fig.13 shows the results. These are average values of 8 paths. As a/b increases, J_x values of a notch become larger when the notch exists in hard lamella (E_2/E_1=0.5), and in soft lamella (E_2/E_1=2.0), they become small. These results show that when a notch exists in hard material surrounded by soft material, the notch tends to grow as it grows, but in the reverse case, the notch tends to stop as it grows.

5. SUMMARY

The results are summerized as follows:

To apply the J-integral to the safety design of a reactor, the J-integral concept is extended to cover several problems.

1) A three-dimensional J-integral of a surface crack in a pressure vessel is evaluated and the results show that

   a) the crack grows, keeping a/c≈0.8;

   b) J values increase about 10-14% by considering the pressure acting on crack surfaces, and

   c) K_I values calculated by these analyses agree well with the results by A.S.Kobayashi.

2) A new definition of the J-integral in multi-phase materials is proposed and the path independent cahracteristics of this new J-integral is verified.

REFERENCES

(1) Eshelby, J.D., "The Continuum Theory of Lattice Defects", Prog. Solid State Physics, vol.3, (1956), pp.79-144

(2) Takano, T., and Okamura, H., "Fatigue Crack Growth from the Surface Crack and its Analyses", 2nd Symposium of HPI (1978), pp.94-98. (in Japanese)

(3) Kobayashi, A.S., " A Simple Procedure for Estimating Stress Intensity Factor in Region of High Stress Gradient", Proc. of the Japan-U.S. Seminar, (1973), pp.127-143.

Table I. Incremental ratios of $\bar{J}$ by
considering the pressure on crack surfaces.

$$\frac{(\bar{J}_{pr.} - \bar{J}_{unpr.})}{\bar{J}_{unpr.}} \times 100.$$

No. of	a/c			
layer.	0.33	0.5	0.7	0.9
1	10.8	10.9	12.5	13.4
2	10.8	11.0	12.4	13.3
3	10.7	11.0	13.1	13.0
4	10.7	11.0	12.3	13.0
5	11.0	11.2	12.4	13.1
6	11.7	11.5	12.6	13.2

Table II. Comparison of K_I values with the
results by A.S.Kobayashi. (K_I:MN/m$^{3\ 2}$)

a/c	0.33	0.5	0.7	0.9
K_I by A.S.Kobayashi	2.169	2.318	2.488	2.591
$K_I = \sqrt{\bar{J}E/(1-\nu^2)}$	2.077	2.369	2.535	2.585
Difference (%)	−4.2	2.2	2.0	0.3

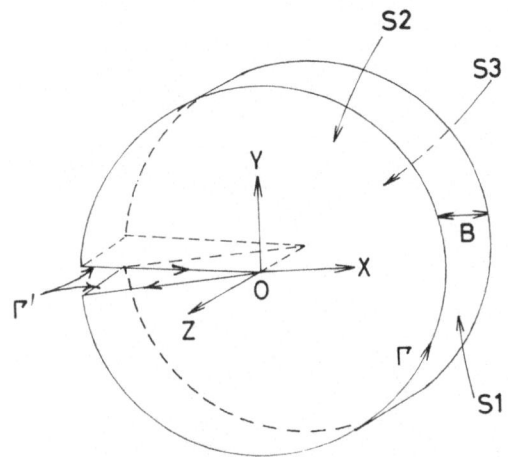

Fig.1 The section of three
dimensional crack.

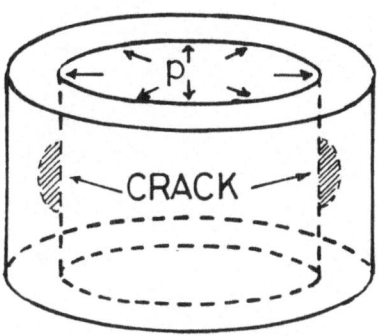

Fig.2 The semi-elliptical surface
crack in a pressure vessel.

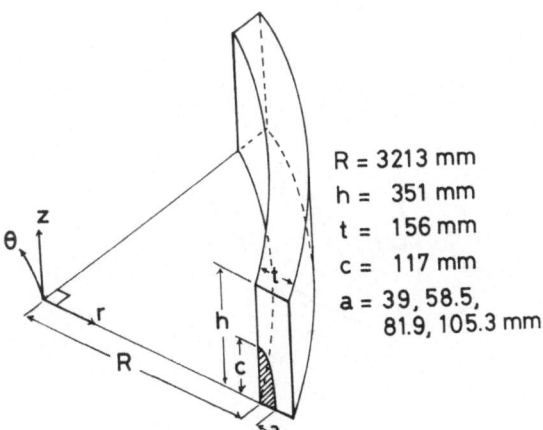

R = 3213 mm
h = 351 mm
t = 156 mm
c = 117 mm
a = 39, 58.5,
    81.9, 105.3 mm

Fig.3 Dimensions and coordinates
for the analyses.

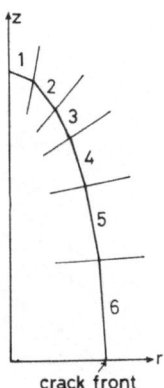

crack front

Fig.4 Seven segments of the
crack front to evaluate
J vectors.

42

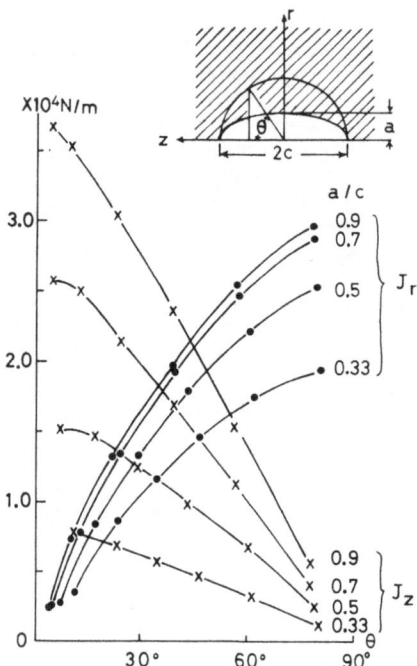

Fig.5 J_r and J_z values when crack
surfaces are unpressurized.

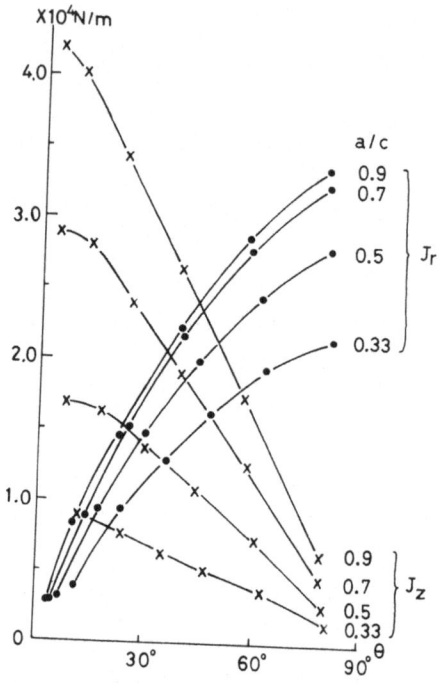

Fig.6 J_r and J_z values when crack
surfaces are pressurized.

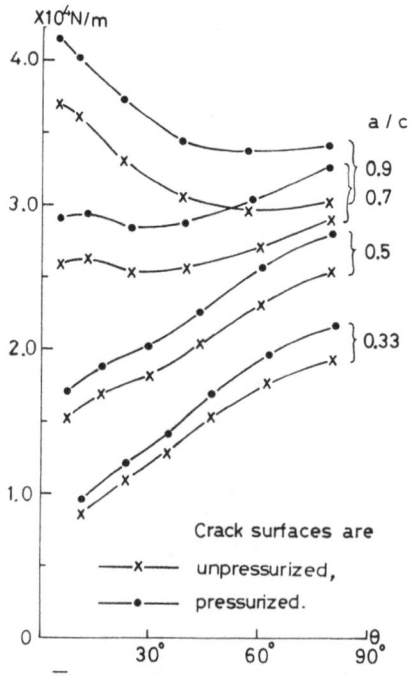

Fig.7 $\overline{J}$ values for each crack shape
and load condition.

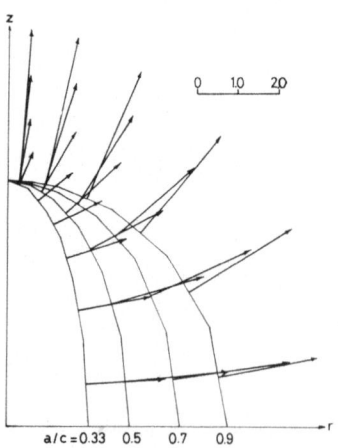

Fig.8 J vectors for four crack
shapes. (Unpressurized.)

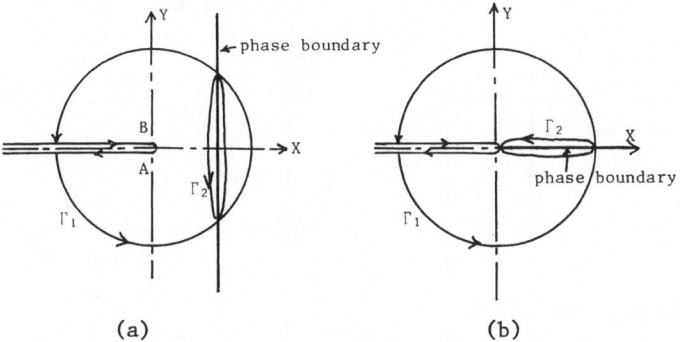

Fig.9   A notch in multi-phase material.
(a) A notch perpendicular to the phase boundary.
(b) A notch parallel to the phase boundary.

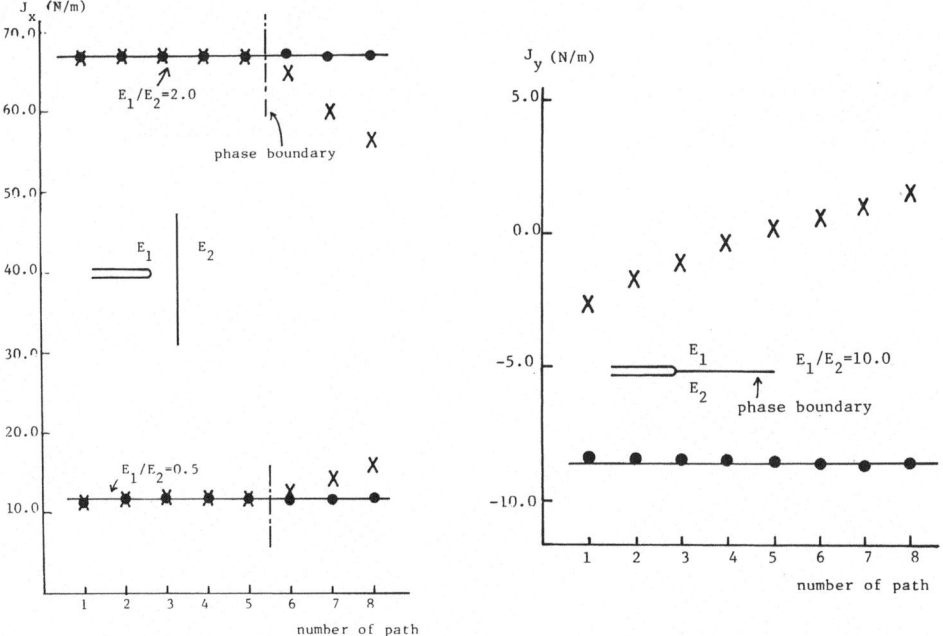

Fig.11 J_y values of a notch
parallel to the phase boundary.

Fig.10 J_x values of a notch perpendicular
to the phase boundary.
(E_1=205947 MN/m^2)

44

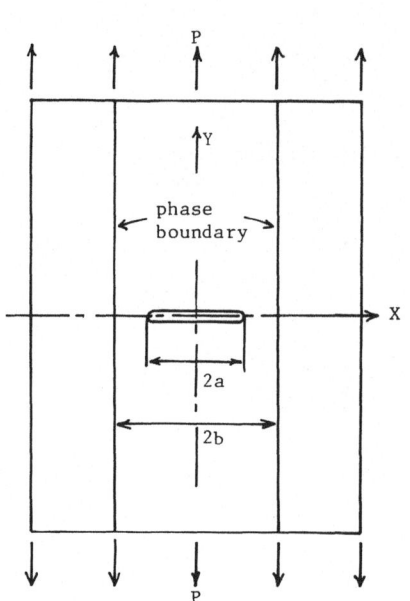

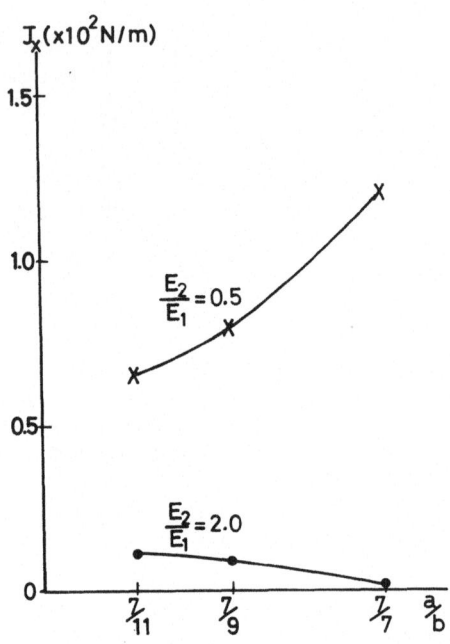

Fig.12 A notch in three storied
lamella.
$(p=9.807 \ MN/m^2)$

Fig.13 J_x vlaues of a notch in three
storied lamella.

AN ENGINEERING METHODOLOGY FOR FRACTURE ANALYSES OF TOUGH MATERIALS

C. F. Shih
General Electric Company
Corporate Research and Development
Schenectady, New York 12301, USA

ABSTRACT

The paper summarizes progress made in two research programs sponsored by the Electric Power Research Institute (EPRI), to identify viable parameters for characterizing crack initiation and continued extension, and to develop an engineering/design methodology, based on these parameters, for the assessment of crack growth and instability in engineering structures which are stressed beyond the regime of applicability of linear elastic fracture mechanics. The ultimate goal in the development of such a methodology is to establish a rational basis for analyzing the effect of flaws (postulated or detected) on the safety margins of pressure boundary components of light water-cooled type nuclear steam supply systems. The methodology can also be employed for structural integrity analyses of other engineering structures.

Extensive experimental and analytical investigations undertaken to evaluate potential criteria for crack initiation and growth, and the selection of the final criteria for analyzing crack growth and instability in flawed structures are summarized; the experimental and analytical results obtained to date suggest that parameters based on the J-integral and the crack tip opening displacement δ are the most promising. This is not surprising since from a theoretical basis, the two approaches are similar if certain conditions are met.

An engineering/design approach for the assessment of crack growth and instability in flawed structures is outlined. The approach exploits the consequences of J-controlled crack growth - the J resistance (J_R) curve is a material property, and crack driving forces can be determined from deformation plasticity analyses. Crack driving forces for the complete range of elastic-plastic deformation are obtained from a simple estimation scheme. The basic elements of the estimation scheme are the linear elastic solutions and the fully plastic solutions for the relevant crack configuration - the latter solutions are catalogued in a plastic fracture handbook. Crack driving force diagrams together with the J_R curve are employed to construct stability diagrams and predict the load-deformation and crack growth behavior of several crack geometries. The predictions are in good agreement with experimental data and full-blown numerical crack growth calculations.

1. INTRODUCTION

Nuclear steam supply system structural integrity is presently assured by designs that adhere to the ASME Pressure Vessel Code and various regulatory guides. These requirements for structural integrity are based on linear elastic fracture mechanics, which assumes that the material behave for the most part elastically, with little ductility. Consequently, the designs are conservative. Recent work in elastic-plastic fracture mechanics, has demonstrated that more realistic measures of actual design margins can be obtained through the use of these elastic-plastic analyses [1]. So far, however, the methodology for plastic or ductile fracture mechanics has required rather elaborate analysis by finite element or finite difference techniques. These techniques, while potentially providing much accurate information regarding the fracture behavior, are also very expensive and difficult to use in routine design applications and safety margin analyses. Consequently, it is desirable to develop simplified engineering approaches that can be used in routine assessments of structural integrity and fracture behavior of flawed structural components.

The paper summarizes progress made in two research programs [2, 3] sponsored by the Electric Power Research Institute (EPRI), to identify viable parameters for characterizing crack initiation and continued extension, and to develop an engineering/design methodology, based on these parameters, for the assessment of crack growth and instability in engineering structures which are stressed beyond the regime of applicability of linear elastic fracture mechanics. The ultimate goal in the development of such a methodology is to establish a rational improved basis for analyzing the effect of flaws (postulated or detected) on the safety margins of pressure boundary components of light water-cooled type nuclear steam supply systems under postulated accident conditions. The engineering/design methodology can also be employed for structural integrity analyses of other engineering structures.

2. THE BASIS FOR THE ENGINEERING APPROACH

The experimental results from the Heavy Section Steel Technology program, and tests on large plates at the Southwest Research Institute, showed that part-through cracks in A533B steels advanced in a flat fracture mode under essentially plane strain conditions for a considerable amount of crack extension; shear fracture developed only after the attainment of maximum load. In our investigation, attention has therefore been focused on flat fracture under essentially plane strain conditions and accompanied by extensive yielding.

Shih and co-workers carried out an integrated experimental and analytical investigation of stable crack extension under fully plastic conditions [1 - 7]. They employed side grooved compact specimens of A533B steels (tested at 93°C) and observed that the leading edge of the crack remained straight with crack extension. The fracture surfaces were macroscopically flat, and there were no lateral contractions along the crack front, i.e., crack extension had occurred under plane-strain conditions. Crack extensions were measured by the unloading compliance technique, and there were generally in good agreement with selected measurements made by heat tinting. The J-resistance (J_R) curve thus measured is unambiguously defined.

In the first phase of the numerical investigations, the behavior of the potential fracture criteria were observed by "forcing" the crack growth in the finite element model to follow the experimentally measured load-line displacement (LLD) vs. crack extension ($a - a_o$) data for the particular test specimen. This was accomplished by shifting the crack-tip node in the numerical model so that the simulated crack growth followed the experimental LLD vs. $a - a_o$ relationship. When the remaining element length (in the path of crack extension) reached a critical fraction of the overall element size, the crack-tip node was released and further crack extension was simulated by shifting the next crack-tip node. A number of parameters were computed at each increment of crack extension - these include the J-integral, the crack opening displacement (COD), crack opening angles (COA), and several variants of the Griffith energy release rate. In the second phase of the investigation, the numerical process is reversed; the acceptable or viable parameters are employed to control the crack extension, for e.g., crack initiation occurs at the critical value of J and further crack growth is governed by the J_R curve.

An evaluation of the analytical and experimental results led to the conclusion that J and δ are the most viable parameters for characterizing crack initiation and growth. A detailed discussion of the potential fracture criteria evaluated in the study, the strategy employed in the evaluation, procedures for measuring the J-integral, COD and crack extension, detailed examination of the analytical and experimental results, and finally an assessment of these fracture parameters that were found viable are given in an EPRI Special Report, and published papers [1 - 7].

The following main conclusions emerged from our investigations:

1. Macroscopically flat fracture surfaces with a straight leading edge can be produced by employing side grooves on test specimens. In the case of A533B steels, side groove depths of 25 percent of the specimen thickness are recommended, since they promote an essentially uniform plane strain constraint

along the crack front while producing minimal effect on specimen compliance and stress intensity factor. Side grooves also led to a sharper and unambiguous measurement of crack initiation and extension - features which are essential to the accurate determination of J_{IC} and J-resistance curves.

2. Experimental data support the characterization of some amount of crack growth by the J or δ resistance curve; these curves appear to be independent of specimen size and extent of plastic deformation, if plane-strain conditions prevail in the crack tip region and some other minimal requirements are met. In fact, the variation in the resistance curves between specimens can often be attributed to mixed mode crack extension. For the smaller sized specimens, the central portion of the crack front advances ahead and eventually the trailing edges fail in shear under conditions closer to plane stress. In side grooved specimens or thicker specimens, the extending crack front remains straight, and the fracture surfaces are macroscopically flat. Correspondingly, the J-resistance curves appear to approach a limiting lower bound curve (or plane-strain curve) with increasing specimen dimensions. These observations suggest that a portion of the J or δ resistance curve is a material property, if plane-strain conditions and some other requirements are met. The lower bound plane-strain J_R curve can be obtained from smaller specimens when side grooves are employed.

3. J-resistance data from tests with center-cracked panels (CCP) are generally higher than the corresponding data from similar sized compact specimens (CS). With increasing CCP dimensions, the slope of the J-resistance data decreases and the resistance curves seem to approach the plane strain curves from compact specimen. Observations based on available test data suggest that significantly larger CCP are required to generate valid plane strain resistance curve. Thus the lower bound plane-strain J_R curve is best obtained from compact specimen or similar bend specimens.

4. Analytical investigations strongly suggest that J and COD can be employed as characterizing parameters for crack initiation and growth. The numerical studies based on an incremental theory of plasticity revealed that J is path independent everywhere except very near the crack tip for some amount of crack growth. For specimens subjected to remote bending, e.g. compact specimen, the path independence of J is observed for growth up to 6% of the original uncracked ligament. It is noted that the slope of the J-resistance curve (dJ_R/da) is approximately constant for a relatively short interval of crack extension, while the slope of the COD-resistance curve $(d\delta_R/da$, also called the average or nominal crack opening angle) and the local crack opening angle (COA) remain practically constant over the range of crack extension explored in our experimental and analytical investigations. The J characterizing parameter approach

is valid for limited range of crack growth; the range of validity depends on strain hardening properties, crack configurations. In particular for tough materials like A533B steels at the upper shelf, the J-based criteria is valid for crack growth up to 6% of the original uncracked ligament in test specimens subjected primarily to bending, e.g. compact specimen. The COD-based criteria appear to be valid for larger amounts of crack growth.

5. The tearing modulus ($T_J = (E/\sigma_o^2) dJ/da$) proposed by Paris and co-workers [8] as a measure of material resistance to stable growth is constant for relatively small amounts of growth. Our investigations suggest that a tearing modulus based on the COA ($T_\delta = (E/\sigma_o) d\delta/da$) is an attractive alternative. The latter modulus is measurable or can be deduced directly and appears to be con- over a larger range of stable growth. Fracture toughness associated with crack initiation is measured by J_{IC} or δ_{IC} while the material resistance to crack growth is measured by T_J or T_δ. The two parameter characterization of fracture properties by J_{IC} and T_J or δ_{IC} and T_δ are analogous to the characterization of material deformation properties by the yield stress and the strain hardening exponent. These observations provide some of the basis for the treatment of crack growth and instability by an R curve approach based on J [8, 9]. Similarly, an R curve approach based on COD can be developed along the lines of the J approach [3, 15].

6. Detailed finite element stationary crack calculations showed that the HRR singularity [10, 11] do dominate the crack tip fields in strain hardening materials. However the size R of the region governed by the HRR field is strongly dependent on crack configuration. At the same level of applied J, R is larger in bend configurations than in similar sized tensile configurations [12, 13]. These results imply that the size requirement for the validity of the J approach for crack initiation and growth will depend on crack configuration or type of remote loading. In fact the size requirements for the tensile configuration are much more stringent than those for bend configurations. This conclusion is in accord with the experimental and numerical crack growth results discussed above.

7. Further analytical and numerical studies clarified the requirements for J as a characterizing parameter for crack growth or J-controlled growth. In addition to the size requirement described in the preceding paragraph, the ω parameter (defined by $\frac{C}{J} \frac{dJ}{da}$) introduced by Hutchinson and Paris [9] must be sufficiently large. Studies by Shih, Dean and German [14, 15] suggest ω values of about 10 and 80 will suffice to ensure J-controlled growth in bend and tensile configurations respectively. Under large scale yielding conditions, large T_{mat} or ψ (defined by $\frac{1}{\sigma_o} \frac{dJ}{da}$) will also promote J-controlled growth.

8. A unique relationship between J and an appropriately defined δ_t can be

obtained directly from the asymptotic crack tip fields [10, 11]. Results from detailed finite element studies of strain hardening materials for several crack configurations support the relation between J and δ_t. The slope of the J-resistance curve and the crack opening angle appear to be similarly related [16]. These results suggest that an R curve approach, based on COD, for treating crack growth and instability is consistent with the J-controlled growth approach promulgated by Paris, et al.

3. ELEMENTS OF THE ENGINEERING APPROACH

The theoretical justification and the conditions for J-controlled growth and the analyses of the stability of crack growth are adequately discussed in [8, 9, 14, 15]. Here the consequences of J-controlled crack growth and its implication on the engineering fracture methodology based on J-controlled growth are summarized.

Under conditions of J-controlled crack growth, the J_R curve obtained from fully yielded specimens will be the same as the J_R curve obtained from specimens with limited yielding (or small scale yielding) as long as plane-strain or plane stress conditions prevail in both situations. The J_R curve will also be independent of crack configurations. In other words, the J_R curve is a material property.

Secondly, stable crack extension and crack instability under large scale plasticity can be analyzed by the resistance curve approach based on J, which is a generalization of Irwin's resistance curve approach for small scale yielding based on the elastic stress intensity factor K [17]. In fact, the R curve approach based on J will be applicable to crack analyses for the complete range of elastic and elastic-plastic deformation. This approach has been employed to examine the stability of several crack configurations [3, 8, 9].

Lastly, the J-integral crack driving force can be determined from analyses using the deformation theory of plasticity. The formulas, obtained by Rice, Paris, and Merkle [18] for deeply cracked specimens and the estimation schemes developed by Shih and Shih and Hutchinson [19, 20] for more general situations, could be employed to estimate the J-driving force without recourse to detailed numerical calculations associated with incremental plasticity theory.

The above factors led to the evolvement of an engineering approach for examining crack growth and stability of flawed structures [3, 15]. The essential elements of the engineering approach are summarized below.

Fully plastic solutions of the type discussed in [19, 20] will be generated and catalogued in a Plastic Fracture Handbook [3]. Solutions will be obtained for typical test specimen geometries and certain common structural configurations (thick-walled cylinders with circumferential and longitudinal cracks).

These specimens will be treated as either plane strain, plane stress, or axially symmetric models, as appropriate. Certain more difficult configurations concerning the nozzle corner flaw will also be studied and documented. The latter configurations are three-dimensional but will be approximated by "equivalent" two-dimensional models with the appropriate boundary conditions, and the equivalent fully plastic two-dimensional results will be documented in the handbook.

Estimation scheme discussed in [19, 20] has been further refined to ensure a smooth transition in the behavior of the estimated solutions from the contained plasticity to the fully plastic state [3, 22]. The fully plastic crack solutions will be employed together with linear elastic solutions [21] to produce relatively simple formulas for quantities such as the J-integral, crack opening displacement and other relevant parameters for the complete range of elastic-plastic deformation. The formulas, which have been shown to be very accurate in the problems thus far examined will be further developed for complex crack configurations.

An engineering methodology for predicting crack growth and fracture instability in flawed structures is being developed [3, 15, 22]. The methodology is in essence a resistance curve approach based on the J-integral and/or the crack opening displacement and is similar to the procedure employed by Paris, Hutchinson and co-workers. The material J_R curve will be given by the statistical mean of all crack growth data from specimens which meet the conditions for J-controlled growth, i.e. primarily from compact specimens. The J-integral crack driving force diagrams determined from the estimation scheme together with the appropriate experimentally measured J_R curve are the essential components of the stability analyses. The methodology assumes that J-controlled growth is applicable. If such is not the case, the methodology will give conservative estimates of the load carrying capacity and the fracture resistance of the structure.

4. SOME APPLICATIONS OF THE ENGINEERING APPROACH

The predictions of the engineering approach are compared with test measurements obtained from carefully controlled experiments and full numerical calculations based on a J_2 incremental plasticity theory and observed crack growth behavior. For the particular test specimen under consideration, the J-integral crack driving force diagram is constructed using the estimation scheme. Figure 1 shows the crack driving force for two limiting situations - the solid line indicates the variation of J with applied load held fixed and the dashed line corresponds to displacement Δ_L being held fixed. The experimentally measured J_R curve indicated by the solid line is superimposed on the

52

diagram at the crack length appropriate to the specimen being examined. Equilibrium of crack growth requires that the applied J equal the material resistance J_R. Thus the solid line (load) and dashed line (displacement) that intersect at a particular point on the J_R curve give the respective values of P and Δ_L which are in equilibrium at that crack length. By repeating the process at different points along the J_R curve, the complete load-deformation behavior is obtained.

The load-deflection behavior obtained by the above procedure, the measured load-deflection record for the A533B steel 4T compact specimen with an initial crack length of 4.615 inches, and the finite element crack growth calculations for the same configuration based on J_2 flow theory of plasticity, are shown in Figure 2. The agreement between all three results is very good, in fact the estimated curve follows completely the trend of the experimental data and the full numerical calculations. Details concerning the experimental data and the crack growth calculations based on flow theory are given in [2, 4, 5].

In one application related to the leak rate analyses of pipes, the mouth opening area (MOA) of a through crack in a pipe subjected to internal fluid pressure (i.e. axial load only) were determined by the engineering fracture approach. The estimated MOA were 10% to 30% larger than the measurements made by Battelle [23] - values obtained from the Dugdale or Goodier and Field models significantly underestimated the test data. The results are summarized in Figure 3. Details of the analyses are given in [2].

ACKNOWLEDGMENT

The author acknowledges the many contributions made by Dr. V. Kumar and Mrs. M. D. German to this work. This work was sponsored by the Electric Power Research Institute (EPRI), Palo Alto, California. The encouragements given by Dr. D. F. Mowbray of General Electric Company and Drs. R. L. Jones and T. U. Marston of EPRI is gratefully acknowledged.

REFERENCES

[1] EPRI Ductile Fracture Research Review Document, ed. T. Marston, EPRI NP-701-SR, Special Report, February 1978.

[2] C. F. Shih, et al, "Methodology for Plastic Fracture," 1st through 10th Quarterly Reports to Electric Power Research Institute, Inc., Contract No. RP601-2, General Electric Company, Schenectady, New York, September 1976 - July 1979.

[3] C. F. Shih and V. Kumar, "Estimation Technique for the Prediction of Elastic-Plastic Fracture of Structural Components of Nuclear Systems," 1st Semiannual Report to EPRI, Contract No. RP1237-1, General Electric Company, Schenectady, New York, July 1, 1978 - January 31, 1979.

[4]   C. F. Shih, H. G. deLorenzi, and W. R. Andrews, "Studies on Crack Initia-
      tion and Stable Crack Growth," in Elastic-Plastic Fracture, ASTM Special
      Technical Publication 668, 1979, pp. 65-120.

[5]   W. R. Andrews and C. F. Shih, "Thickness and Side-Groove Effects in J-
      Resistance Curves," in Elastic-Plastic Fracture, ASTM Special Technical
      Publication 668, 1979, pp. 426-450.

[6]   C. F. Shih, H. G. deLorenzi, and M. D. German, "Crack Extension Modeling
      with Singular Quadratic Isoparametric Elements," International Journal
      of Fracture, Vol. 12, 1976, pp. 647-651.

[7]   C. F. Shih, H. G. deLorenzi, and W. R. Andrews, "Elastic Compliance and
      Stress-Intensity Factors for Side-Grooved Compact Specimens," Inter-
      national Journal of Fracture, Vol. 13, 1977, pp. 544-548.

[8]   P. C. Paris, H. Tada, A. Zahoor, and H. Ernst, "The Theory of Insta-
      bility of the Tearing Mode of Elastic-Plastic Crack Growth," in Elastic-
      Plastic Fracture, ASTM Special Technical Publication 668, 1979, pp. 5-36.

[9]   J. W. Hutchinson and P. C. Paris, "Stability Analysis of J-Controlled
      Crack Growth," in Elastic-Plastic Fracture, ASTM Special Technical
      Publication 668, 1979, pp. 37-64.

[10] J. W. Hutchinson, "Singular Behavior at End of Tensile Crack in Hardening
      Material," Journal of the Mechanics and Physics of Solids, Vol. 16, No.
      1, January 1968, pp. 13-31; also "Plastic Stress and Strain Fields at
      Crack Tip," Journal of the Mechanics and Physics of Solids, Vol. 16, No.
      4, September 1968, pp. 337-347.

[11] J. R. Rice and G. F. Rosengren, "Plane Strain Deformation Near Crack Tip
      in Power-Law Hardening Material," Journal of the Mechanics and Physics
      of Solids, Vol. 16, No. 1, January 1968, pp. 1-12.

[12] R. M. McMeeking and D. M. Parks, "On Criteria for J-Dominance of Crack
      Tip Fields in Large-Scale Yielding," in Elastic-Plastic Fracture, ASTM
      Special Technical Publication 668, 1979, pp. 175-194.

[13] C. F. Shih and M. D. German, "Requirements for a One Parameter Char-
      acterization of Crack Tip Fields by the HRR Singularity," General
      Electric Company TIS Report No. 79CRD076, 1979; also to appear in
      International Journal of Fracture.

[14] C. F. Shih, R. H. Dean, and M. D. German, "On J-Controlled Crack Growth:
      Evidence, Requirements and Applications," General Electric Company TIS
      Report submitted for publication.

[15] C. F. Shih, "An Engineering Approach for Examining Crack Growth and
      Stability in Flawed Structures," presented at the OECD-CSNI Specialist
      Meeting at Washington University, St. Louis, Missouri, September 1979,
      submitted for publication.

[16] C. F. Shih, "Relationships Between the J-Integral and the Crack Opening
      Displacement for Stationary and Extending Cracks," General Electric
      Company TIS Report No. 79CRD075, April 1979; submitted for publication.

[17] Fracture Toughness Evaluation by R-Curve Methods, ASTM Special Technical
      Publication 527, 1973, p. 118.

[18] J. R. Rice, P. C. Paris, and J. G. Merkle, "Some Further Results of J-
      Integral Analysis and Estimates," Progress in Flaw Growth and Fracture
      Toughness Testing, ASTM Special Technical Publication 536, 1973, pp.
      231-245.

54

[19] C. F. Shih, "J-Integral Estimates for Strain Hardening Materials in
      Antiplane Shear Using Fully Plastic Solutions," in Mechanics of Crack
      Growth, ASTM Special Technical Publication 590, 1976, pp. 3-22.

[20] C. F. Shih and J. W. Hutchinson, "Fully Plastic Solutions and Large-
      Scale Yielding Estimates for Plane Stress Crack Problems," Transactions
      of the ASME: Journal of Engineering Materials and Technology, Series H,
      Vol. 98, No. 4, October 1976, pp. 289-295.

[21] H. Tada, P. C. Paris, and G. R. Irwin, The Stress Analysis of Cracks
      Handbook, Del Research Corporation, Hellertown, Pennsylvania, 1973.

[22] V. Kumar and C. F. Shih, "Fully Plastic Crack Solutions, Estimation
      Scheme, and Stability Analyses for Compact Specimen," presented at ASTM
      Symposium on Fracture Mechanics, St. Louis, Missouri, May 1979; sub-
      mitted for publication in ASTM Special Technical Publication.

[23] M. F. Kanninen, et al, "Mechanical Fracture Predictions for Sensitized
      Stainless Steel Piping with Circumferential Cracks," EPRI NP-192, Project
      585-1, Final Report, September 1976.

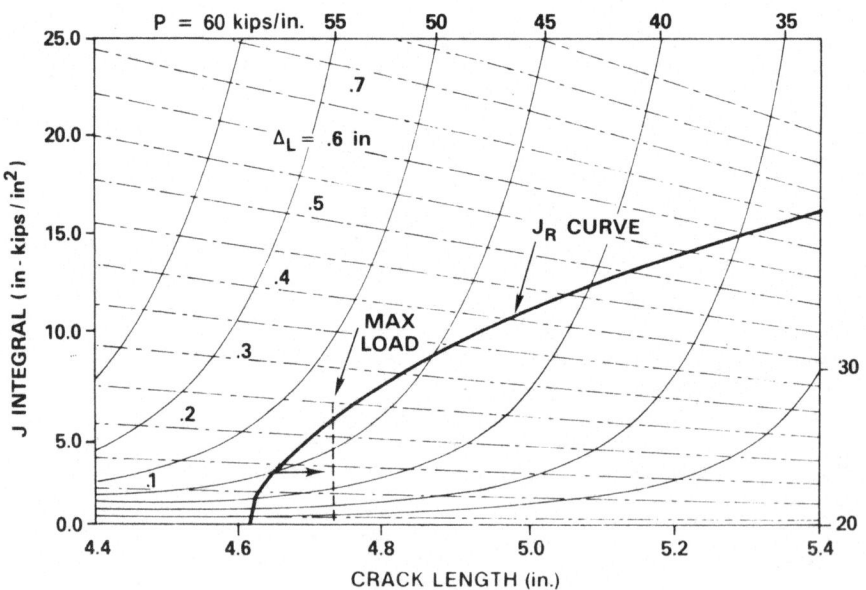

FIGURE 1. A typical J-integral crack driving force diagram for 4T compact
           specimen - A533B steel.

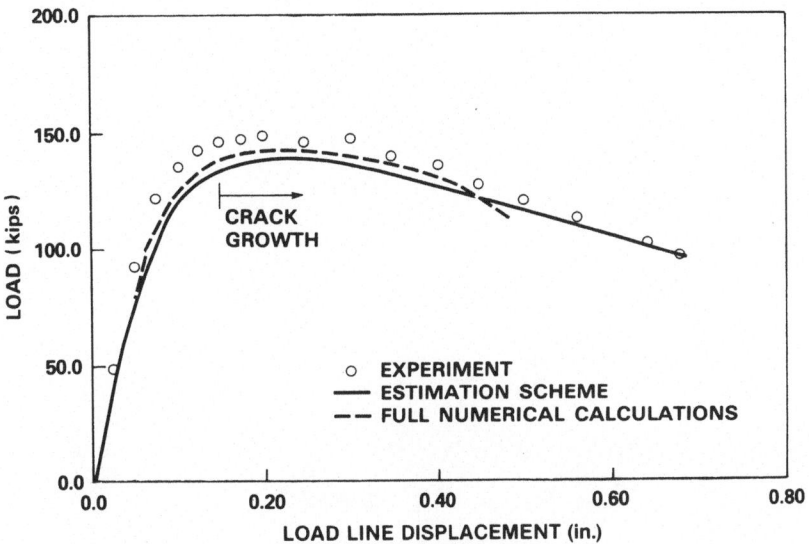

FIGURE 2. Comparison of predicted and experimentally measured load–displacement relationship for 4T A533B compact specimen. Results from full numerical calculations based on J_2 flow theory of plasticity are also included.

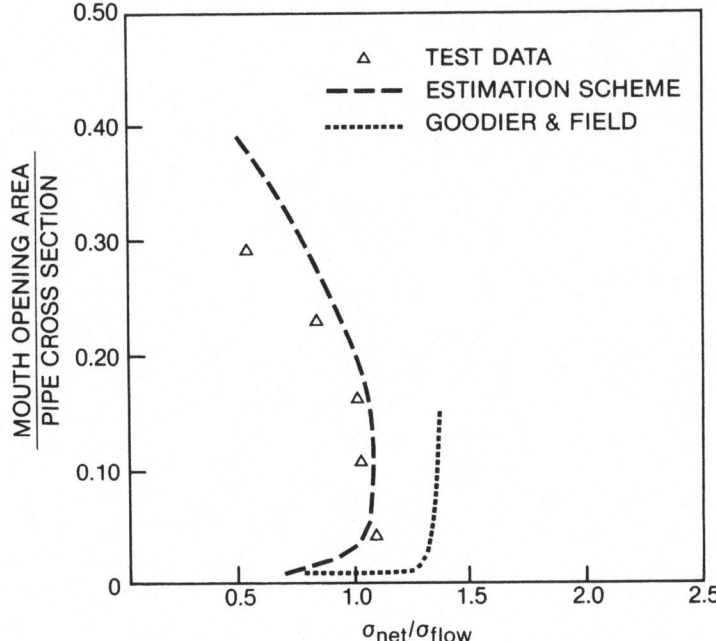

FIGURE 3. Predicted mouth opening area. Battelle test data and Goodier and Field solution.

TOWARD THE DEVELOPMENT OF
PLASTIC FRACTURE MECHANICS CAPABILITIES FOR
ASSESSING THE RELIABILITY OF
ENERGY RELATED STRUCTURES

M. F. Kanninen

Manager, Fracture Mechanics Projects Office
Battelle, Columbus, Ohio   43201 USA

ABSTRACT

Research to develop a plastic fracture mechanics methodology for conditions where a significant amount of stable crack growth precedes fracture instability is described. Several different candidate plastic fracture criteria were systematically evaluated by an integrated program of experimentation and finite element analyses. It was learned that the J-resistance curve and a constant value of the local crack tip opening angle may be the most effective choices. In addition, a combination of the two appears to retain the advantages of each without suffering from their limitations.

INTRODUCTION

The implementation of virtually all energy producing and utilizing schemes is constrained--sometimes severely--by the properties of the available structural materials. Because design considerations are usually related to the stiffness, strength, and weight of a structure, the properties that are involved in material selection are the elastic moduli, the yield stress, and the density. However, in recent years, a wide-spread recognition of two key facts has enlarged this set. First, all materials either contain intrinsic crack-like flaws or are exposed to conditions during fabrication and service that introduce such flaws. Second, the conventional mechanical properties of a material do not characterize its ability to resist the enlargement of a crack. Accordingly, fracture mechanics has evolved to provide quantitative assessments of the possibility that the failure of a structure can occur from the growth of a crack. This discipline has introduced another material property--the fracture toughness--into the structural design and evaluation process.

The discipline of fracture mechanics treats any failure process which can emanate from crack growth. In principle, this is true regardless of the constitutive relation obeyed by the material, the geometry of the crack and of the body containing it, and the type and rate of the loads applied to the body. The only essential requirement is that at least one crack-like defect can be identified for analysis purposes. However, at present, the practical applications of fracture mechanics are largely limited to quasi-static conditions and to materials that behave in an essentially linear elastic manner. The application of fracture mechanics to ductile and tough materials—particularly where significant amounts of crack tip plastic deformation and stable crack growth can precede general instability—cannot now be routinely accomplished in a rigorous manner.

There are a number of specific research areas involved in the application of fracture mechanics to energy related structures using tough ductile materials. These include:

1. crack initiation and stable growth accompanied by large scale plastic deformation and local inhomogeniety

2. dynamic unstable crack propagation and arrest, particularly under high loading rates

3. subcritical crack growth from cyclic loading, elevated temperature, and aggressive environments

4. three dimensional crack-structure geometries.

This paper summarizes the progress in a research program conducted by the author and his colleagues at Battelle's Columbus Laboratories which concentrates on the first of these areas [1-4]. The work was aimed at the development of a basically correct plastic fracture methodology capable of treating stable crack growth and instability in nuclear pressure vessels and piping. These materials tend to be ductile and tough, particularly when fracture is a primary concern. Because of the fundamental point of view that was taken, the findings of this program should be generally applicable to the materials used in a wide range of energy related applications.

OBJECTIVE AND BASIC APPROACH

Because inspection procedures are so intensive in nuclear systems, a great many flaws are detected. A typical example is the circumferential crack found in a heat affected zone of a girth welded Type 304 stainless steel BWR 4-inch diameter recirculation bypass line shown in Figure 1 [5]. Another is a crack that reached almost three-fourths of the way around a 10-inch primary cooling pipe fitting in the Duane Arnold nuclear plant [6]. Cracking in Feedwater nozzles, the torus system and the belt line region are other areas of concern.

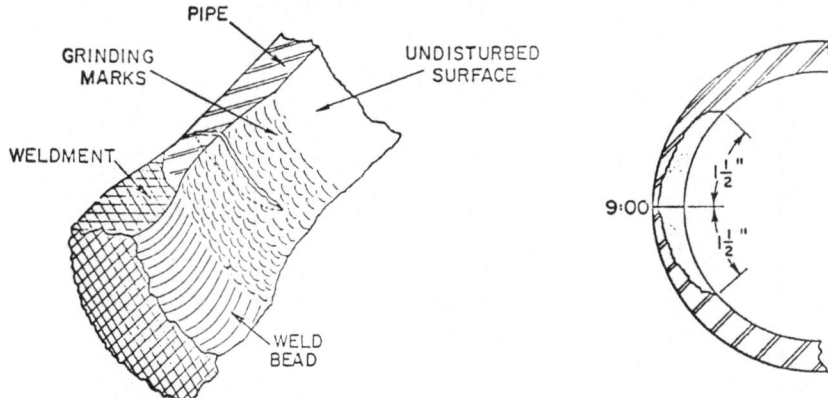

Figure 1.   Profile of Circumferential Crack Detected in 4-Inch
            Recirculation Bypass Line (Loop B) of the Quad Cities
            II Boiling Water Reactor

The impetus for the development of a plastic fracture mechanics methodology
for the flaws discovered (or hypothesized) in nuclear components comes from the
fact that linear elastic fracture mechanics (LEFM) can significantly underesti-
mate the strength of a cracked component when it is made of a tough ductile
material.   Figure 2 illustrates this with results from two part-through wall
cracked aluminum tension panels.   These results show that, if crack growth initiation (points a_1 and a_2 in Figure 2) are taken as the failure point, the actual
load carrying capacities of the panels (points b_1 and b_2) are underestimated by
a factor of three.   As can be clearly seen in Figure 2, a significant amount of
stable crack growth occurs under rising load prior to fracture instability which
accounts for the greater strength.   Clearly, to assess the added margin of
safety conferred by the stable crack growth process requires that a plastic
fracture mechanics methodology be developed for the region beyond where LEFM is
valid.

It can be seen in Figure 2 that net section yield based on the initial
cracked area (points c_1 and c_2) overestimates the fracture loads in this
instance.   There are, in addition, many circumstances in which the load-
geometry combination is complicated enough to make the application of a net section yield criterion inappropriate; e.g., seismic loading.   This further indicates that the proper approach is a plastic fracture mechanics treatment using
a crack tip crack growth criterion and its geometry-independent critical values
to determine fracture instability from a calculation of the progress of stable
crack growth.   This is generally referred to as a resistance curve approach.

One of the attractive features of LEFM is the equivalence that exists between all of the various fracture parameters.   This makes the choice of K, G,
COD, or J, in a particular application largely a matter of experimental and

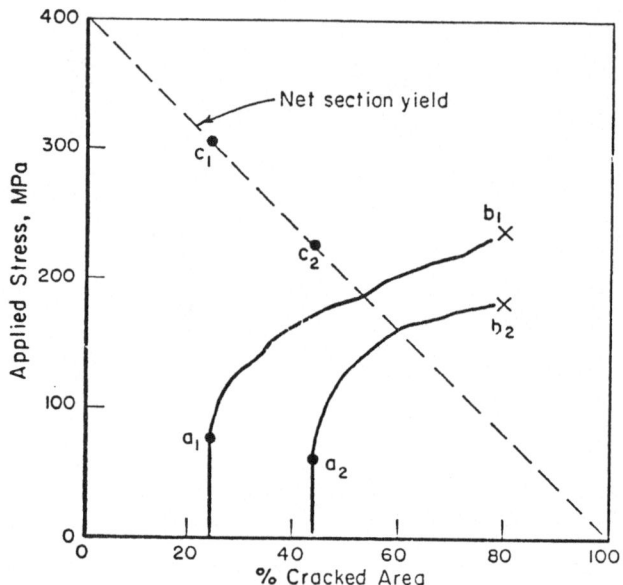

Figure 2. Nominal Applied Stress Values As a Function of
the Extent of Stable Crack Growth In Two Aluminum
Plates With Initial Part—Through Cracks

computational convenience. However, this equivalence becomes a detriment when
one wishes to identify one of the LEFM parameters as the most fundamental for
use as the basis of a plastic fracture mechanics formulation. Recognizing this
difficulty, a pragmatic approach was taken by the author and his colleagues.
The first step was to compile all of the more probable candidate criteria for
evaluation by a combination of experimentation and analysis. The plastic fracture criteria considered are shown in Table 1.

In Table 1, a delineation is made between the crack driving force parameter
and its critical value for crack growth. Taking the J-integral as a heuristic
example, stable crack growth is presumed to occur when the equality

$$J(a,\sigma,b) = J_c(\Delta a) \tag{1}$$

is satisfied. In Equation (1), J is considered to be a material-independent
function of the system containing the cracked component and, as such, J will depend upon the crack length a, the applied stress σ, and the dimensions of the
cracked body b. It is a computed quantity. The parameter, J_c, in contrast, is
supposed to be a geometry-independent material property that depends primarily
upon the extent of crack advance Δa. It should be clear that not every possible
fracture criterion will fullfill this expectation. The basic goal of the research is, therefore, to examine each of the candidate criteria displayed in
Table 1 to determine which of them is most likely to do so.

Criterion Symbol	Identification	Designation of Critical Value
G	LEFM energy release rate	R
J	J-integral	J_c
T	Tearing modulus	T_{mat}
G	Generalized energy release rate	R
COA	Average crack opening angle	$(COA)_c$
CTOA	Crack tip crack opening angle	$(CTOA)_c$
F	Finite element crack tip Node force	F_c
G_o	Elastic-plastic energy	G_{oc}
G_Z	Computational Process Zone Energy release rate	G_{Zc}

Table 1. Candidate Fracture Criteria for the Basis of
a Plastic Fracture Methodology

ANALYSIS PROCEDURE AND RESULTS

The research to develop an effective plastic fracture mechanics methodology
consisted of a three-stage program of integrated experimentation and computation. The objective was to single out the more effective candidate criteria
from the compilation given in Table 1. First, laboratory test pieces of A533B
steel and two "toughness-scaled" aluminum alloys were tested to obtain data on
crack growth initiation, stable growth, and instability. Second, "generacion-
phase" analyses were performed in which the observed applied stress versus
stable crack growth data were used as input to a finite element model to generate critical values of each candidate criterion. Third, "application-phase"
analyses were made using each criterion to predict fracture for the several
different geometries and loading conditions for which experimental results were
available.

The generation-phase results indicated that, with the exception of the LEFM
parameters, all of the candidate criteria have some merit. But, at the same
time, they all exhibited distinct deficiencies. Hence, to a surprising degree,
these results cannot eliminate any of the candidates. They instead suggest that
the most effective parameters might be those offering the greatest computational
convenience coupled with the least conceptual weaknesses. In this light, the

62

finite element node force and the various energy release rate parameters can be eliminated. This leaves J, COA, and CTOA as the leading candidates.

The use of the J-resistance curve in plastic fracture mechanics owes much of its popularity to the work of Shih et al [7-8] and to Paris, et al [9]. Its great advantage is that, because it is based on deformation plasticity, the value of J depends only on the current loads and crack length, independent of prior load history and stable crack growth. As Hutchinson and Paris [10] have shown, the plastic unloading which accompanies crack growth will not disqualify a deformation plasticity approach provided the extent of stable growth is limited. Consequently, a geometry-independent J-resistance curve can be expected. At the same time, it is clear that this region of validity is inexorably lost with continued crack growth. Hence, for conditions under which the amount of growth that can be allowed is too limited, either the J-resistance curve must be evaluated specifically for the geometry of interest or an alternative procedure found.

The CTOA criterion has been extensively examined in the work of Kanninen, et al [1-4] and Shih, et al [7-8]. The most attractive feature of this criterion is the observation that stable growth seems to occur with a constant value of the CTOA. This is illustrated in Figure 3 which shows a sequence of crack profiles in a center-cracked Type 304 stainless steel tension panel. This

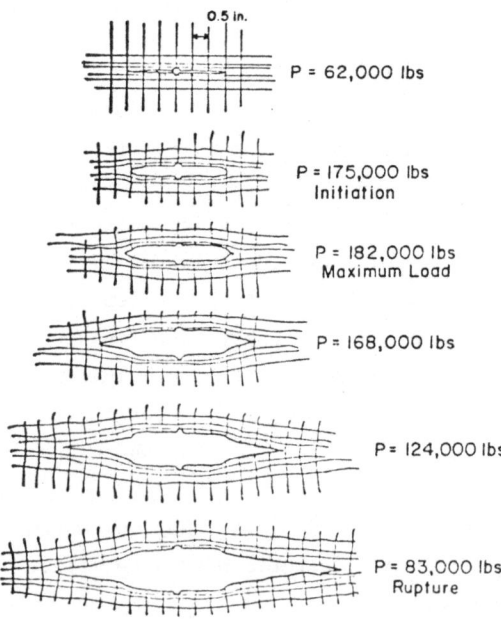

Figure 3.   Stable Crack Growth in a Type 304 Stainless Steel Center Cracked Tension Panel—Initial Crack Length of 3 Inches, Panel Width 12 Inches.

observation has also been found to be true in both A533B steel and in aluminum. This obviously makes CTOA very convenient for crack growth computations, as already shown in an application by Emery, et al [11]. However, CTOA also has its disadvantages. First, close examination via a generation phase computation reveals that there is actually a transient period at the onset of growth in which CTOA varies somewhat. Second, the direct measurement of this quantity is very difficult. Third, its precise definition in a finite element or other analysis method is somewhat arbitrary.

A parameter related to the CTOA is the COA--the average crack opening angle--which is the ratio of the COD at the initial crack tip to the total amount of stable crack growth. The COA resistance parameter exhibits similar behavior to the CTOA but is much easier to define and to measure. However, its connection to the events taking place at the crack tip is somewhat nebulous.

In an effort to distinguish among the leading plastic fracture criteria, the application-phase calculations shown in Figure 4 were performed. These used the J-resistance curve and, separately, constant critical values of CTOA and COD--all as determined from compact tension test results on 2219 aluminum-- to predict the load-crack growth behavior of a wide center cracked panel of the same material. Note that initiation was specified to occur in all cases at the same J_{Ic} value.

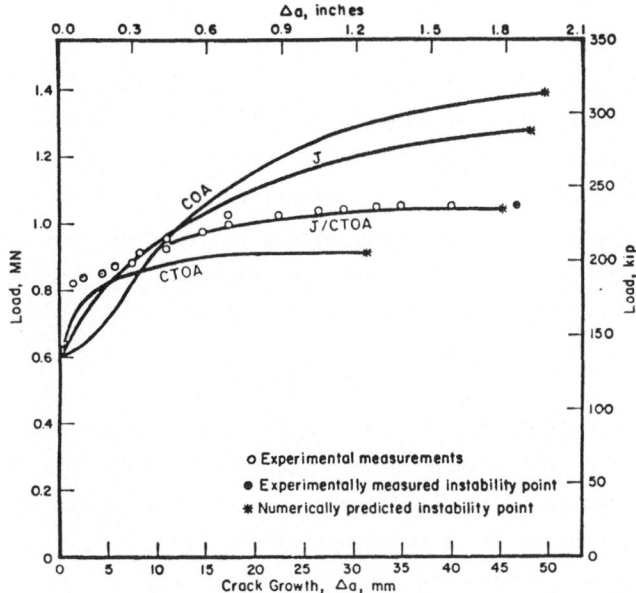

Figure 4. Critical Comparison of Plastic Fracture Criteria By Application Phase Analyses of a Wide Center-Cracked Panel of 2219 Aluminum

It can be seen from the results of Figure 4 that the predictions using J and CTOA are both reasonable and somewhat better than that using COA. However, the combination of J and CTOA designated as J/CTOA in Figure 4 is clearly the best. In the combined approach, the J-resistance curve is used for initiation and some amount of stable crack growth. An internally computed CTOA value is used thereafter. This procedure seems most promising as it takes full advantage of the J-resistance curve approach without suffering from its limitations with regard to small amounts of crack growth. It also uses the observed constancy of CTOA while circumventing the complications of the initial transient in this parameter. Yet, the combined parameter approach does not require any further experimental measurements because the critical CTOA to be used for extensive growth is one that is calculated during the early stages of crack growth.

CONCLUSIONS

The results of the research reviewed in this paper show that a number of different fracture criteria can be used as the basis of a resistance curve approach to predict stable crack growth and instability in tough ductile materials. Of all the candidate criteria examined, J and CTOA appear to be the most useful. Nevertheless, both have significant advantages and disadvantages:

- J is computationally convenient, but the J-resistance curve exhibits geometry-dependence after a limited amount of stable crack growth.
- CTOA exhibits a constant resistance value (after an initial transient) for substantial stable crack growth but its experimental determination is difficult.

A combination approach in which the initiation and some amount of crack growth is governed by the J-resistance curve with a computed critical CTOA value used thereafter appears to be a highly promising technique for accurately predicting extended amounts of stable growth prior to fracture.

ACKNOWLEDGEMENT

The results reviewed in this paper were largely drawn from research project RP601-1 conducted at Battelle's Columbus Laboratories for the Electric Power Research Institute, Palo Alto, California. The author would like to thank Ted Marston, Robin Jones, and Karl Stahlkopf of EPRI for their support and encouragement of the work. Thanks are also owed to the many of the author's present and former colleagues who participated in these projects; primarily, G. T. Hahn, D. Broek, C. W. Marschall, R. S. Stonesifer, and I. S. Abou-Sayed.

REFERENCES

(1) M. F. Kanninen, D. Broek, G. T. Hahn, C. W. Marschall, E. F. Rybicki, and G. M. Wilkowski, "Towards an Elastic-Plastic Fracture Mechanics Predictive Capability for Reactor Piping", Nuclear Engineering and Design, Vol 48 (1978), pp 117-134.

(2) G. T. Hahn, D. Broek, M. F. Kanninen, C. W. Marschall, A. R. Rosenfield, E. F. Rybicki, and R. B. Stonesifer, "Methodology for Plastic Fracture", EPRI Ductile Fracture Research Review Document, T. U. Marston, editor, Electric Power Research Institute, Palo Alto, California (1978), Section 5.

(3) M. F. Kanninen, E. F. Rybicki, R. B. Stonesifer, D. Broek, A. R. Rosenfield, C. W. Marschall, and G. T. Hahn, "Elastic-Plastic Fracture Mechanics for Two-Dimensional Stable Crack Growth and Instability Problems", Elastic-Plastic Fracture, J. D. Landes, et al, editors, ASTM STP 668 (1979), pp 121-150.

(4) M. F. Kanninen, G. T. Hahn, D. Broek, R. B. Stonesifer, C. W. Marschall, I. S. Abou-Sayed, and A. Zahoor, "The Development of a Plastic Fracture Methodology", Battelle Columbus Laboratories Report to the Electric Power Research Institute, in preparation, 1979.

(5) C. F. Cheng, W. A. Ellingson, and J. Y. Park, "Effect of Residual Stress and Microstructure on Stress-Corrosion Cracking in BWR Piping", Corrosion/76, Houston, Texas, March, 1976.

(6) Anon, "Cracks Found at Arnold, Vermont Yankee", Nuclear News, August, 1978, p. 34.

(7) C. F. Shih, W. R. Andrews, H. G. Delorenzi, R. H. Van Stone, S. Yukawa, and J. P. D. Wilkinson, "Crack Initiation and Growth Under Fully Plastic Conditions: A Methodology for Plastic Fracture", EPRI Ductile Fracture Research Review Document, op cit, Section 6.

(8) C. F. Shih, H. G. DeLorenzi, and W. R. Andrews, "Studies on Crack Initiation and Stable Crack Growth", Elastic-Plastic Fracture, op cit, pp 65-120.

(9) P. C. Paris, H. Tada, A. Zahoor, and H. Ernst, "Instability of the Tearing Mode of Elastic-Plastic Crack Growth", Elastic-Plastic Fracture, op cit, pp 5-36.

(10) J. W. Hutchinson and P. C. Paris, "The Theory of Stability Analysis of J-Controlled Crack Growth", Elastic-Plastic Fracture, op cit, pp 37-64.

(11) A. F. Emery, A. S. Kobayashi, W. J. Love, and A. Jain, "Dynamic Propagation of Circumferential Cracks in Two Pipes with Large Scale Yielding", Journal of Pressure Vessel Technology, in press, 1979.

ELASTIC-PLASTIC ANALYSES OF A THREE-POINT BEND SPECIMEN AND A FRACTURING PIPE

A. S. Kobayashi
Professor of Mechanical Engineering
University of Washington, Seattle, Washington 98195, USA

J. S. Cheng
Research Engineer
Fatigue and Fracture Mechanics, AD 34
Rockwell International
Los Angeles, California 90009

A. F. Emery
Professor of Mechanical Engineering
University of Washington
Seattle, Washington 98195 USA

S. N. Atluri
Regents Professor of Mechanics
Georgia Institute of Technology
Atlanta, Georgia 30332

W. J. Love
Professor of Mechanical Engineering
University of Washington
Seattle, Washington 98195 USA

ABSTRACT

An assumed displacement, hybrid finite element method was used to analyze the elastic-plastic plane strain state of a three point bend specimen. A shell code was used to analyze the elastic dynamic propagation of a girth crack in a stainless steel pipe subjected to uniaxial tension.

INTRODUCTION

The three point bend specimen is a widely used specimen for elastic-plastic static and dynamic characterizations of structural steel [1,2]. A compilation of the widely varying numerical results generated by nine investigators under a task effort of the ASTM Committee E24 Task Group E24.01.09 was published recently [3]. As another divergent source of data, the writer and his former associates report their elastic-plastic analysis of the same specimen with material properties similar to an A533B steel used in the ASTM round robin analysis.

In the second part of this paper, the dynamic response of a circumferential crack propagating in a net-section yielded pipe is discussed. These results were obtained through the continuing efforts by the writer and his colleagues for analyzing the dynamic fracture responses of propagating axial and girth cracks in large pipes and in particular in the presence of large plastic deformation. Highlights of the recent results obtained for a

propagating girth crack in an 18-inch stainless steel pipe subjected to uniaxial tension is given in the following.

THREE-POINT BEND SPECIMEN

Specimens

Geometry and loading condition of the three-point bend specimen together
with the finite element mesh used are shown in Figure 1. The elastic-plastic
material properties used in the analysis is shown in Figure 2. The strain
hardening characteristics of the material studied differs slightly with that
used in Reference [3] where the coefficient B_o in the Ramberg-Osgood true
stress versus true strain relation was 144×10^3 psi in Figure 2 while
$B_o = 120 \times 10^3$ psi in Reference [3]. This slight difference in the stress-
strain relation should be borne in mind when the present results are compared
with those of Reference [3].

Numerical Technique

The numerical technique used in this analysis is the assumed displacement
(small) hybrid finite element method with a $1/\sqrt{r}$ and HRR [4] singularities
embedded in the four quadrilaterial element surrounding the crack tip for
elastic and elastic-plastic analyses, respectively. Despite the relative
coarseness of the finite element break down shown in Figure 1, high numerical
accuracy is achieved by the use of additional quadric and quadratic displacement fields in the singular and non-singular elements, respectively. Von Mises
yield criterion and the flow rule were used in the plasticity analysis. Further
details of the numerical procedure can be found in References [5,6].

J integral [7] was computed by evaluating numerically the path integral
along the four rectangular paths shown in Figure 1. The smallest path is completely embedded in the singular elements. Since these paths did not coincide
with the interelement boundaries, smoothly varying stress and strain fields were
obtained and integrated by using higher order integration schemes. J was also
computed by an approximate procedure, which was suggested by Rice et al.[8],
through the use of the numerically obtained single load displacement relation.

Results

Figure 3 shows the growth of the yield zones with increasing loads. The
yield zone propagated from both the crack-tip and the point of load application
and is more extensive at the latter. Also note that at the early stage of
loading, the plastic zones extended at approximately 40°-50° to the axis of
the crack. At a load of $P = 8.06$ kips, the entire uncracked ligament yielded
and the yield zone spread to approximately 60% of the total plate width.

Figures 4 and 5 shows the log-log plots of the effective stresses and strains ahead of the crack tip in the uncracked ligament. Note the gradual transition from the elastic $1/\sqrt{r}$ singularity in the stresses and strains to the $r^{-0.1}$ and $r^{-0.9}$ singularities, respectively of the HRR model. Figure 6 shows the J calculated for the four paths at successively increasing load-point displacements (or loads). The standard deviation of J for the four paths was within 5 percent of the average J which is denoted as $\bar{J}$. Within the expected numerical accuracy, Figure 6 shows that the path independent property of J is maintained. J also appears to be proportional to the square of the load point displacement prior to a load-point displacement of $u_A \doteq 10 \times 10^{-3}$ inch and is linearly related to u_A beyond this loading.

Figure 7 shows the mean $\bar{J}$ and the approximate J computed from the load displacement relation [8] for increasing load-point displacement. Good agreement between the two Js are noted.

Figure 8 shows the ratio of $\bar{J}/K_I$ where K_I is the stress intensity factor computed from equivalent elastic analysis. The ratio of $\bar{J}/K_I$ remains close to its elastic value of 0.03 up to an applied load of $P = 24 \times 10^3$ lbs with small scale yielding present as shown in Figure 3. Beyond this load, the plasticity effects become increasing noticeable as shown in Figure 8.

Figure 9 shows the relation between various crack opening displacement (COD) and the crack mouth opening of v_B. COD was obtained directly by linearly extrapolating the crack opening profile obtained from finite element analysis. Other COD curves shown for comparison are those due to Wells [9], Hayes [10] and the CODA formula [11]. The numerical COD of Figure 9 can be represented as $\delta = \bar{J}/2.7\,\sigma_{ys}$ where σ_{ys} is the yield strength. This relation is slightly different from Hayes' correlation formula of $\delta = \bar{J}/2\,\sigma_{ys}$.

Conclusions

(1) For the three point bend specimen analyzed, the path independence of J integral is maintained within a standard deviation of ± 5 percent from the averaged $\bar{J}$. The J integral evaluated at the smallest path embedded in the singular element agrees with those evaluated outside these singular element.

(2) Rice's simple estimation of J-integral agrees well with J computed from the field values, without or with the presence of large scale yielding.

(3) COD defined as a linearly extrapolated crack opening profile agrees reasonably well with the empirical formula of CODA [11] and can be represented as $\delta = \bar{J}/2.7\,\sigma_{ys}$.

PROPAGATING CIRCUMFERENTIAL CRACK

Introduction

As mentioned previously the writer and his colleagues have analyzed axial and circumferential cracks propagating in steel pipes with [12,13] and without [14,15] coupled fluid depressurization analysis for the past five years. Despite the large number of experimental papers [16-20] published to date on this subject, little data is available for verifying these mathematical modelings of pipe failures as well as others [21-23]. Numerical experimentation, however, is continuing with the hope that these analyses will identify the critical combinations of loading conditions, pipe geometries and materials which can then be subjected to further experimental scrutiny. In the following, some recent findings on a circumferential crack propagating in a field of large-scale plastic yielding is presented.

Pipe, Loading and Material Property

A 316 stainless steel pipe with an average radius of R = 8.1 in. (20.6 cm) and wall thickness 2h = 1.78 in. (4.52 cm) was considered. The half cylinder, as shown in Figure 10, was taken to be three radius long with an outgoing stress wave condition prescribed at one end and a plane of symmetry at the x = 0 plane. A pre-existing through crack of 1/20th the circumference, or half crack length a = 3.82 in. was located at the x = 0 plane of the half cylinder. The entire cylinder was pre-loaded under uniform axial tension with the crack being welded together to transmit the applied tensile stress. At time t = +0, the two crack surfaces were released and the resultant dynamic stress redistribution opened the crack until the prescribed fracture criterion was satisfied and the crack propagated. The cylinder was not pressurized in this analysis since the added hoop stress only decreased slightly the axial stress level at yielding and did not materially affect the crack propagation.

The materials properties for 316 stainless steel including the postulated fracture toughness K_{IC}, are shown in Table 1.

TABLE 1

MATERIAL PROPERTIES OF 316 STAINLESS STEEL

	Modulus of Elasticity E ksi	Poisson Ratio ν	Yield Strength σ_{ys} ksi	Fracture Toughness K_{IC} ksi$\sqrt{in}$	Crack Opening Angle (COA) $d\delta/da$
316 SS	29×10^3	0.3	82	300*	0.19

---

* Postulated plane strain fracture toughness.

Conspicuously missing in the above Table is a dynamic ductile fracture criterion necessary to propagate the crack. Such criterion, which is not available to date, is necessary to estimate a more realistic crack propagation velocity in the presence of large scale yielding surrounding the crack tip. In view of the urgency for dynamic ductile fracture analysis of a fracturing pipe, a relatively new static ductile fracture criterion of critical crack tip opening angle (CTOA), which is advanced among others by Shih et al. [24] and Kanninen et al. [25], was used in this dynamic analysis. Although this concept of critical CTOA has not been tested under conditions of dynamic crack propagation, the use of this quasi-static ductile fracture criterion in dynamic fracture does not appear to be grossly unjustified in view of the post-mortem observations of extensive ductile tearing of the fracture surfaces.

Numerical Technique

Figure 11 shows the finite difference mesh of a one quarter cylinder segment which was considered in the analysis due to symmetries of the problem. This quarter cylinder was divided into uniform finite difference mesh sizes of $\Delta\theta = 9°$ and $\Delta\theta/a\Delta\theta = 1$.

The dynamic shell code used in this study has been described amply in prior papers [12-15] and thus will not be repeated here. Although small strains together with the additional restriction of plane section remaining plane were assumed, large deflections and rotations of the shell element were accounted for in this algorithm. Piola-Kirchoff stresses, successively updated Lagrangian-strains and von-Mises yield criterion together with the associated incremental flow rule involving incremental plastic strains were used in this analysis. Strain hardening and dynamic plasticity effects, however, were not considered.

Results

Figure 12 shows the extent of large scale plastic yielding for three axial loads of 65, 70 and 80 ksi. In all cases, the crack started to propagate after net-section yielding and was obviously governed by the large plastic strains surrounding the propagating crack tip.

Figure 13 shows the crack tip position versus time for the postulated CTOA criterion of $d\delta/da = 0.19$ plus two additional values of $d\delta/da = 0.10$ and 0.30. The significant differences in propagation histories for each critical CTOA values indicate the large influence of the crack tip deformation field in dynamic ductile fracrure problems. Also shown for comparative purpose is the crack tip release mechanism used in References [12-15] where discrete crack tip motion was used to model continuous crack propagation. The new algorithm, which attempts to model continuous crack tip motion and which is thus more realistic, results in slower crack tip motion.

Figure 14 shows the crack velocity versus time relations for the three CTOA values of $d\delta/da$ = 0.10, 0.19 and 0.30. The higher terminal velocity for the lower $d\delta/da$ = 0.10 as well as the higher terminal velocity for the old numerical algorithm are noted.

Discussions

The reductions in terminal crack velocities under large scale yielding substantially increased the estimated total crack transit time along the pipe circumference. The larger differences in crack velocities illustrates the importance of establishing an accurate dynamic ductile fracture criterion if reliable fracture dynamic analysis is to be conducted on ductile materials which are commonly used pressure vessel material of today. The significantly slower crack speed and the larger crack opening areas will no longer allow the analyst to ignore the depressurization due to fluid loss. Thus the fluid-structure interaction effect must be reconsidered in such dynamic fracture of ductile pipes.

ACKNOWLEDGEMENT

Results reported in this paper have been obtained through research con-tracts supported by the US Air Force AFOSR-75-2478 and the Electric Power Research Institute EPRI RP 231-1.

REFERENCES

(1) "Standard Method of Test for Plane-Strain Fracture Toughness of Metallic Materials, E-399-74," Book of ASTM Standards, Part 31, ASTM, Philadelphia, 1974.

(2) D.R. Ireland, "Critical Review of Instrumented Impact Testing," Proc. of Int'l Conf. Dynamic Fracture Toughness, The Welding Institute, Cambridge, July 1976, pp. 47-62.

(3) W.K. Wilson and J.R. Osias, "A Comparison of Finite Element Solutions for an Elastic-Plastic Crack Problem," Int'l Journ. of Fracture, Vol. 14, 1978, pp. R95-R108.

(4) J.R. Rice, "Mathematical Analysis in the Mechanics of Fracture," Fracture (edited by H. Liebowitz), Vol. 2, Chap. 3, Academic Press, 1968, pp. 273-277.

(5) J.S. Cheng, "Assumed Displacement Hybrid Finite Element Method for a Finite Strain Analysis and Elastic-Plastic Fracture Mechanics," Ph.D. Thesis submitted to University of Washington, 1976.

(6) J.S. Cheng, S.N. Atluri and A.S. Kobayashi, "Assumed Displacement Hybrid Finite Element Method for Non-linear Elastic and Elastic-Plastic Analyses," A Collection of Technical Papers on Structures and Materials, Vol. A, AIAA/ASME 18th Structures, Structured Dynamics and Material Conf., San Diego, California, March 21-23, 1977, pp. 279-289.

(7)   J.R. Rice, "A Path Independent Integral and the Approximate Analysis of
      Strain Concentration by Notches and Cracks," Journ of Applied Mechanics,
      Trans. of ASME, Vol. 38, Series E, No. 2, June 1968, pp. 379-386.

(8)   Bucci, R.J., Paris, P.C., Landes, J.D. and Rice. J.R., J Integral
      Estimation Procedure, ASTM STP 514, 1972, pp. 40-69.

(9)   A.A. Wells, "Application of Fracture Mechanics at and Beyond General
      Yielding," British Welding Research Report, Nov. 1963, pp. 563-570.

(10) D.J. Hayes and C.E. Turner, "An Application of Finite Element Techniques
      to Post-Yield Analysis of Proposed Standard Three-Point Bend Fracture
      Test Pieces," Int'l Journ of Fracture, Vol. 10, No. 1, March 1974,
      pp. 17-32.

(11) "Fracture Toughness Testing of Metallic Materials, Part II, COD Testing,"
      COD Application Panel of the Navy Department Advisory Committee on
      Structural Steel, UK, December 1970, p. 54.

(12) Emery, A.F., Love, W.J., and Kobayashi, A.F., "Fracture in Straight Pipes
      under Large Deflection Conditions — Part 1, Structural Deformation,"
      ASME Journal of Pressure Vessel Technology, Vol. 99, Series J, No. 1,
      February 1977, pp. 122-127, and Part II, "Pipe Pressures," pp. 128-136.

(13) Emery, A.F., Kobayashi, A.S., and Love, W.J., "Pipe Stress Intensity
      Factors and Coupled Depressurization and Dynamic Crack Propagation," EPRI
      RP231-1, Annual Report No. 4, 1977, Chapter 3.

(14) Emery, A.F., Kobayashi, A.S., and Love, W.J., "Fracture Dynamics of
      a Propagating Crack in a Pressurized Ductile Cylinder," Trans. of 4th
      International Conference on Structural Mechanics in Reactor Technology,
      Vol. F, Paper F-7, 1977.

(15) Emery, A.F., Kobayashi, A.S., and Love, W.J., "An Analysis of the Propa-
      gation of a Brittle Circumferential Crack in a Pipe Subjected to Axial
      Stress." ASME Preprint 78-PVP-101, 1978.

(16) Maxey, W.A., Kiefner, J.F., Eiber, R.J., and Duffy, A.R., "Ductile
      Fracture Initiation, Propagation and Arrest in Cylindrical Vessels,"
      Fracture Toughness, ASTM STP 514, September 1972, pp. 70-81.

(17) Shoemaker, A.K., McCartney, R.F., "Displacement Consideration for a
      Ductile Propagating Fracture in a Line Pipe," ASME Journal of Engineering
      Material Technology 96, 1974, pp. 318-322.

(18) Darlaston, B.J.L., and Harrison, R.P., "Ductile Failure of Thin Walled
      Pipes with Defects Under Combinations of Internal Pressure and Bending,"
      Proceedings of the Third International Conference on Pressure Vessel
      Technology, Part II Materials and Fabrication, ASME, 1977, pp. 669-676.

(19) Watanabe, M., Mukai, Y., Kageo, S., and Fujihara, S., "Mechanical
      Behavior on Bursting of Longitudinally and Circumferentially Notched AISI
      304 Stainless Steel Pipes by Hydraulic and Explosion Tests," ibid loc cit,
      pp. 667-683.

(20) Nora, Y., Fukuda, M., Nozaki, N., Koga, T., and Takeuchi, I., "Study on
      Resistivity of Various Types of Steels Against Propagating Shear Fracture
      by Modified West Jefferson Type Burst Test," ASME Preprint 78-PVP-71, 1978.

(21) Kanninen, M.F., Lampoth, S.F., and Popelar, "Steady State Crack Propaga-
      tion in Pressurized Pipelines without Backfills," ASME Journal in
      Pressure Vessel Technology, Vol. 98, Feb. 1976, pp. 56-65.

74

(22) Freund, L.B., Parks, D.M., Rice, J.R., "Running Ductile Fracture in a Pressurized Line Pipe," Mechanics of Crack Growth, ASTM STP 590, 1976, pp. 243-262.

(23) Parks, D.M. and Freund, L.B., "On the Gasdynamics of Running Ductile Fracture in a Pressurized Line Pipe," ASME Journal of Pressure Vessel Technology, Vol. 100, February 1978, pp. 13-17.

(24) Shih, C.F., Delorenzi, H.G., and Andrews, W.R., "Studies on Crack Initiation and Stable Crack Growth," Elastic-Plastic Fracture (edited by J.D. Landes, J.A. Begley and G.A. Clarke), ASTM STP 668, 1979, pp. 65-120.

(25) Kanninen, M.F., Rybicki, E.F., Stonesiter, R.B., Brock, D., Rosenfield, A.F., Marschall, C.W. and Halm, G.T., "Elastic-Plastic Fracture Mechanics for Two-Dimensional Stable Crack Growth and Instability Problems," ibid loc cit., pp. 121-150.

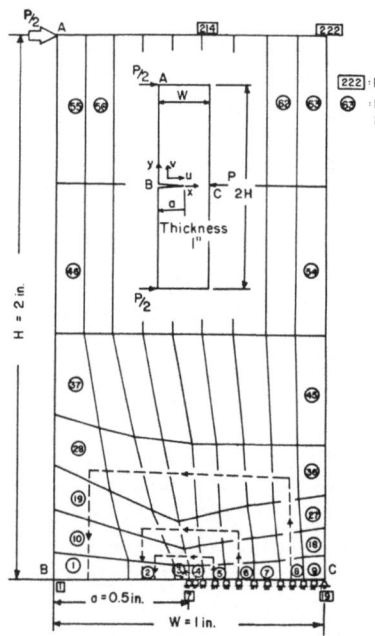

Figure 1. MODEL OF THREE-POINT BEND SPECIMEN
(shown in insert) AND PATHS OF J INTEGRAL

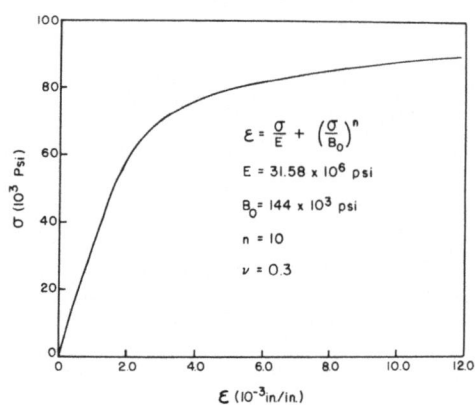

$$\varepsilon = \frac{\sigma}{E} + \left(\frac{\sigma}{B_0}\right)^n$$

$E = 31.58 \times 10^6$ psi

$B_0 = 144 \times 10^3$ psi

$n = 10$

$\nu = 0.3$

Figure 2. STRESS-STRAIN CURVE

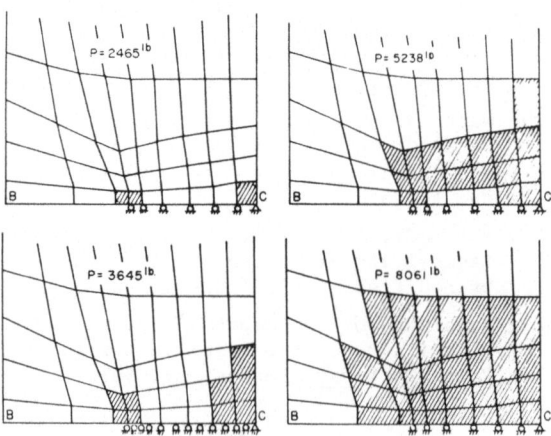

Figure 3. PLASTIC YIELD ZONES AT VARIOUS LOAD LEVELS

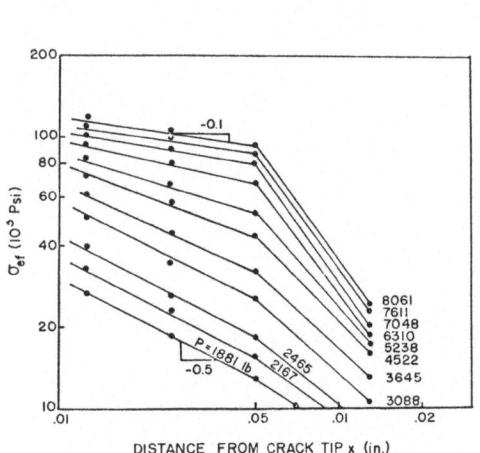

Figure 4. DISTRIBUTION OF EFFECTIVE STRESS
AHEAD OF CRACK (Log-Log Scale)

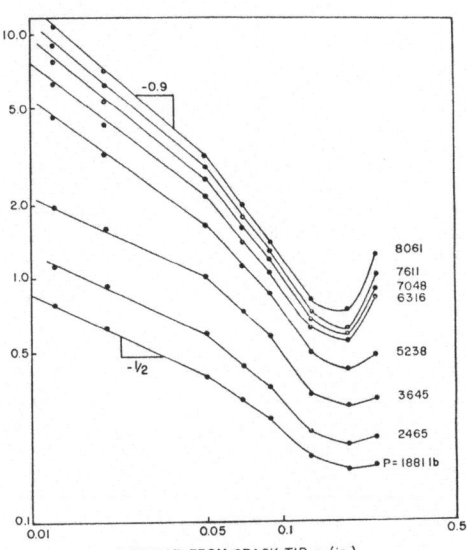

Figure 5. DISTRIBUTION OF EFFECTIVE TOTAL STRAIN
AHEAD OF CRACK (Log-Log Scale)

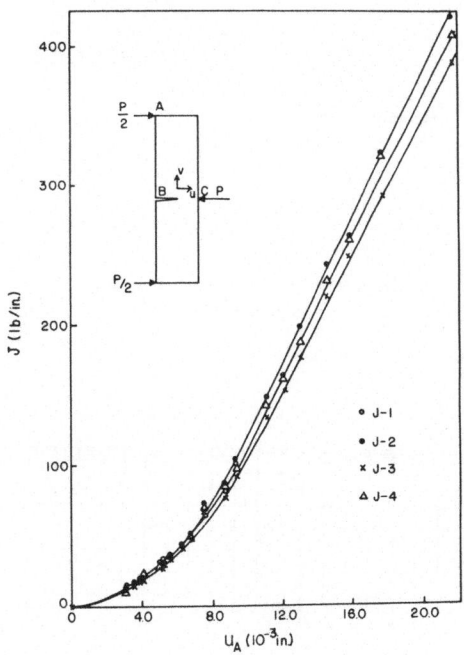

Figure 6. J INTEGRALS AT VARIOUS PATHS

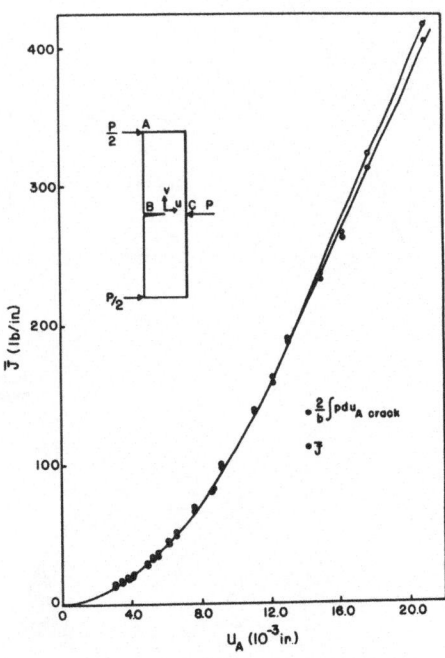

Figure 7. J INTEGRAL ESTIMATION AT VARIOUS
LOAD LEVELS

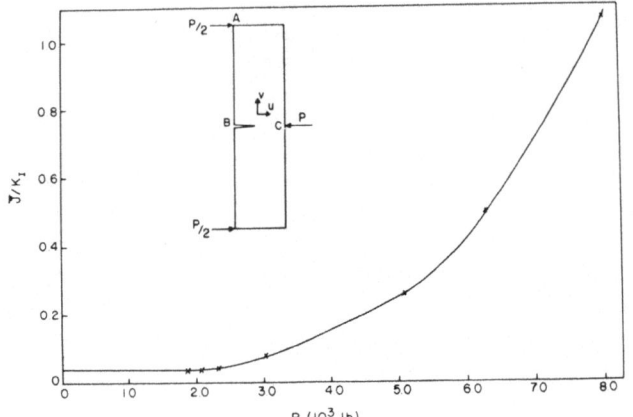

Figure 8. $\overline{J}/K_I$ AT VARIOUS LOAD LEVELS

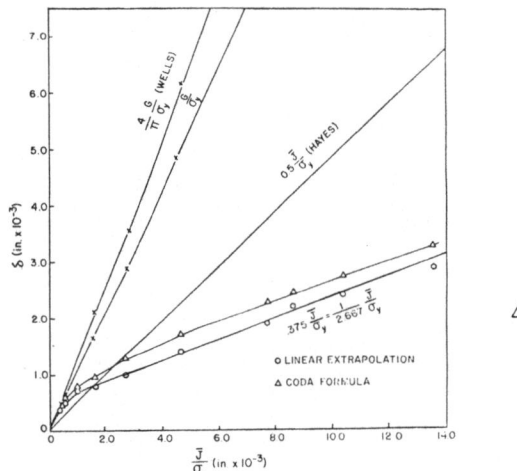

Figure 9. COMPARISONS OF VARIOUS COD

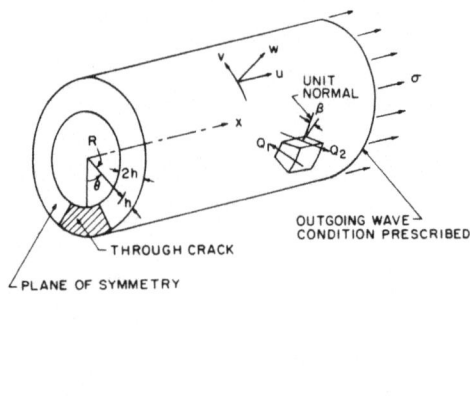

Figure 10. FRACTURING CYLINDRICAL SHELL.

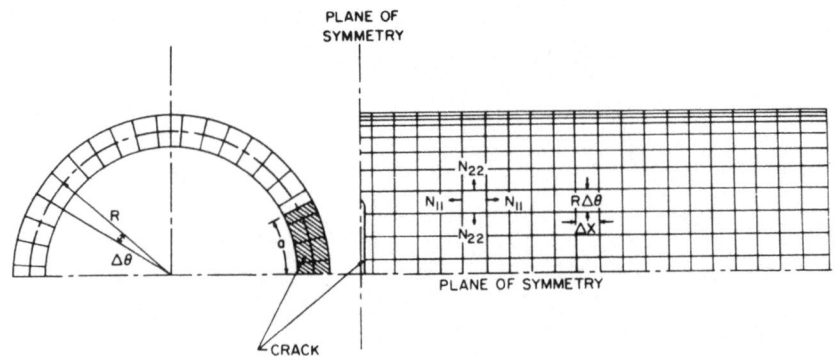

Figure 11. FINITE DIFFERENCE MESH.

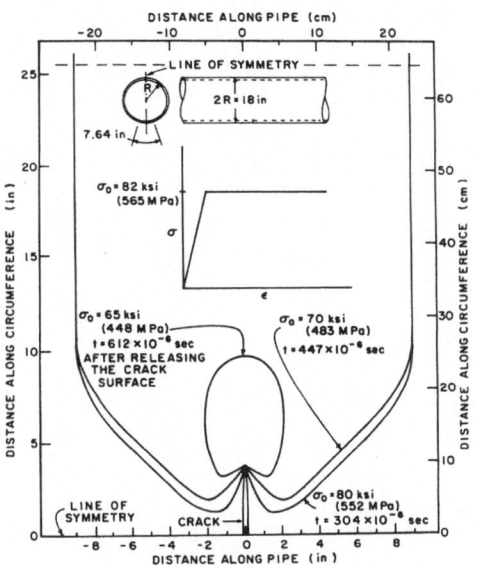

Figure 12. PLASTIC YIELD ZONES ALONG THE PIPE PERIPHERY (EXPANDED VIEW).

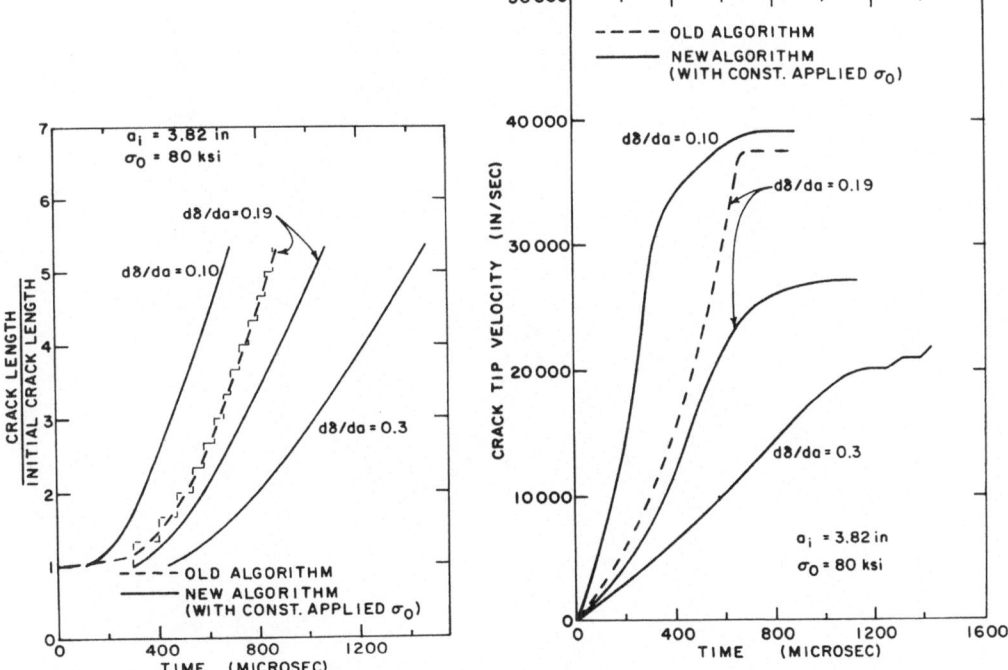

Figure 13. CRACK EXTENSION WITH TIME

Figure 14. CRACK VELOCITY CHANGE WITH TIME.

ANALYSIS OF DUCTILE FRACTURE BEHAVIOR OF TYPE 304 STAINLESS STEEL

Shinji Sakata   Kunio Hasegawa
Researcher   Researcher
Tasuku Shimizu   Shigeru Shida
Senior Researcher   Department Manager
Mechanical Engineering Research Laboratory
Hitachi, Ltd.
3-1-1 Saiwai-cho, Hitachi-shi, Ibaraki-ken, 317, Japan

ABSTRACT

This report describes the plastic deformation behavior of notched plates of type 304 stainless steel and the ductile fracture criteria for such plates. Elastic-plastic finite element analyses were carried out to determine the relation between crack opening displacement and applied load. The results of the analyses agreed rather well with those obtained in experiments over a relatively small strain, but there were large differences were apparent between the calculated and experimental values in the large strain in which cracks initiate around notches. A simple equation was proposed to predict the load at the onset of penetration of the part-through notched plates.

INTRODUCTION

It is reported that intergranular stress corrosion cracks (IGSCC) have been found under some particular conditions in the heat affected zones (HAZ) of butt-welds in type 304 stainless steel pipes of some light water reactor (LWR) plants. Several institutions and fabricators have been investigating countermeasures for at least four years. Several remedies have been developed and already applied in some plants. Hence, it has been possible to substantially reduce pipe line problems caused by IGSCC. The occurrence of pipe failure is unlikely under service conditions, because only an extraordinarily high load could trigger a catastrophic pipe fracture at the site of cracks. The fracture analysis of a pipe with a flaw is very difficult because of material ductility which causes a multiaxial state of stress in the vicinity of a flaw in a pipe. Recently, several fracture criteria have been investigated; various parameters, such as COD, J-value, and net section collapse failure [1], have been developed.

In this report, ductile fracture prediction criteria, based on an elastic-plastic analyses with a large deformation and experiments with plates having artificial center flows are discussed.

EXPERIMENTAL METHOD

An example of the test specimens used in the experiments is shown in Fig. 1. Specimens were made of type 304 stainless steel. The chemical compositions and mechanical properties of the steel are shown in Table 1. A center notch was fabricated with an electrical discharge machine. Notch depth varies from 2.2 mm to 5.5 mm and specimen thickness is 5.5 mm. The length of a notch is 30 mm, which is 30 % of the plate width. Notch width is 0.2 mm and the tip radius is 0.1 mm. A differential transformer was installed on the specimen in order to measure the displacement between two gage marks which were located at a distance of 100 mm from the center of a notch in both directions along a load axis. The relations between applied loads and COD were obtained with a clip gage and a series of photographs of the notch, taken intermittently during loading. All experiments were conducted at room temperature using a standard static tensile testing machine.

ELASTIC-PLASTIC ANALYSIS

To obtain quantitative values for the stress and strains around the tip of a notch and a ductile fracture criterion based on the above stresses and strains, elastic-plastic analysis, which take large deformation effects into account, were conducted with the finite element program HI.EPIC. This program was developed by Hitachi for the inelastic analysis of structures. An 8-node quadrilateral isoparametric finite element was used for 2-D analysis and a 20-node 3-D isoparametric one was used for 3-D analysis.

The material constants used are shown in Table 2. A bilinear approximation relation was used to represent the stress - strain relation in the calculations. The yield criterion is taken from von Mises's theory and the strain hardening rule from kinematic hardening theory. Yielding follows the flow rule.

A two dimensional plane stress model was used for plates with through thickness center notches; three dimensional models were used for plates with part - through notches. The cost of using the computer for the 3-D elastic-plastic analysis was extremely high. To lower this cost, more effective models must be designed. The boundary conditions were determined by the symmetry conditions, and a load was applied at the unnotched end of a model. The constraint conditions were such that the displacement at the end of a model caused constraint all along the boundary.

RESULTS OF EXPERIMENTS AND CALCULATIONS

Through-thickness notched specimen

Figure 2 shows the relation between COD and applied load for a through-thickness notched plate under tension. The maximum load in the figure is the one at which a crack initiates at the tip of the notch. The experimental and calculated results agree rather well with each other up to 145 KN, but they differ by 40 % at the maximum load because the bilinear stress-strain relation, which approximates an actual relation in the strain range of 5 % to 10 %, was used in the calculations.

Quantitatively, the calculated notch configuration was very similar to the experimental one, but as mentioned above, the critical strain or critical COD of crack initiation could not be defined. Many investigators have researched the J-integral as a possible parameter of ductile fracture criterion. A simplified J-value calculation was used to estimate the fracture toughness of center notched specimens. The stress-strain relation is represented by Ramberg - Osgood as

$$\varepsilon^T = B\sigma^{1/n} \tag{1}$$

The values 9.36×10^{-23} and 0.12 were obtained for the material constants B and n, respectively, from an experimental curve. The value J is expressed as in Eq. (2) for a crack in an infinite plate.

$$J = 2\pi a[W^e + f(n)W^p] \tag{2}$$

where   2a : crack length

$$W^e = \sigma_o^2/2E$$
$$W^p = \sigma_o\varepsilon_o^p/(n+1) \tag{3}$$
$$f(n) = (n+1)[3.85(1-n)/\sqrt{n} + \pi n]/2$$

The crack initiation load was $\sigma_o/\sigma_y = 1.53$ or $\sigma_o = 300$ MPa; then Eq. (2) gives $J = 1800$ mm MPa. Reference [2] says $J = 780$ mm MPa for type 304 stainless steel. This difference is caused by the follows: the crack initiation load was obtained from observations with the unaided eye of the surface of the plate specimen and the specimen is not thick enough to find a critical J-value. But if the J-value is assumed to be 713 mm MPa, then the predicted stress for a test specimen is $\sigma_o = 272$ MPa, which is very close to the experimental result.

Part-through notched specimen

Figure 3 shows the calculated deformation of a part-through notched specimen with a 4.4 mm deep flaw, The out-of-plane deformation is significant in the vicinity of the notch. The relations between COD and applied load are plotted in Fig. 4. As described in Fig. 2, the calculated COD does not agree with the experimental one in the high strain range for deep notched plate specimens. Tensile rigidity increases and COD becomes small in shallow notched plate specimens. The calculated results for specimens with a 2.2 mm deep flaw suggest this, but experiments show that the COD obtained with these specimens is larger than that obtained with specimens with a 4.4 mm deep flaw at crack initiation load. Observations with the unaided eye showed that the specimen surface around a 2.2 mm deep notch was extremely deformed, locally and in an out-of-plane direction. Figure 5 shows photographs of deformed notches under loading. The deformation pattern of specimens with a 2.2 mm deep notch might be compared to that of specimens with a 4.4 mm deep notch. At present, it is very difficult to obtain values for local deformation in a shallow notched plate, shown in Fig. 5, with the finite element method because of the enormous expense of the computing time.

Ductile fracture criterion of a notched plate

The ductile fracture of a notched plate cannot be preducted with the parameters discussed above. Local stresses and strains could not be parameters either, because, from the view point of structural design, the calculations are too difficult to do locally. Therefore, a simple ductile fracture criterion, which can predict the crack penetration load of a part-through notched plate, is discussed. Figure 6 shows displacement vs applied load for specimens with notches of various depths. The displacements were measured at a gage length of 200 mm on the plate specimens. The load indicated by A are those at which cracks penetrate through the thickness of the plate specimens. The loads indicated by B are those at which small cracks initiate at the tip of through-the-thickness flaws and proceed across the plate width. Nominal stresses at the onset of crack penetration, which are listed in Table 3, were obtained from Fig. 6. The relation [1] initial crack area vs nominal stress, which was proposed by M.F. Kanninen, is plotted in Fig. 7. The experimental data for the through-the-thickness notched specimens agreed rather well with the postulated line, but another line can be expected for part-through notched specimens from this figure. Stress distributions in the notched section were investigated in finite element calculations, the results of which are shown in Fig. 8. The stress distributions can be divided into two parts and can be expressed as a rectangular one, shown in Fig. 9. Using this figure the onset of penetration load P_p can be given by

$$P_p = \sigma_p A + \sigma_f (1-\phi) A_L \tag{4}$$

where $A = t(W-\ell)$

$\qquad$ $A_L = (t-d)\ell$

$\qquad$ σ_p : stress on area A

$\qquad$ σ_f : true stress at failure

$\qquad$ ϕ : reduction factor of area in tensile test

Equation (4) shows that the ligament stress at the onset of penetration will reach failure stress and that the ligament area will be reduced. Hence, the onset of the penetration load, which is given by the net section stress σ_p in Fig. 9, can be calculated with Eq. (4). The stress σ_p was obtained from the experimental results obtained in the present study and those obtained in the studies described in the reference. The stress σ_p were plotted as a function of the ratio of initial crack depth to plate thickness, shown in Fig. 10. The stress σ_f and the reduction factor used here were

$$\sigma_f = \begin{array}{l} \text{1683 MPa at R.T. for the present experiments} \\ \text{1200 MPa at 205 °C for Kanninen's data} \end{array}$$

$$\phi = 0.72$$

As the figure shows, the relation between σ_p and d/t is expressed by a single linear curve, and the extrapolated values at d/t = 0 and 1.0 are approximately the same as the ultimate tensile strength σ_{UTS} and the yield strength σ_{ys} at 0.2 % strain. The linear equation for σ_p is obtained as follows:

$$\sigma_p = \sigma_{UTS} + \frac{d}{t} (\sigma_{ys} - \sigma_{UTS}) \tag{5}$$

A physical interpretation of Eq. (5) is given below. An unnotched plate (that is, the case where d/t = 0) breaks when nominal stress reaches the ultimate strength of the material. Significant crack opening displacement occurs in a through-the-thickness notched plate, that is the case where d/t = 1.0, when the nominal stress reaches the yield strength of the material.

CONCLUSIONS

$\qquad$ 1. The relations between crack opening displacement and applied load were obtained for both through-the-thickness and part-through notched specimens through experiments and in finite element calculations. It turns out that calculations for shallow notched plates are very difficult to obtain because of the presence of large local deformations.

$\qquad$ 2. Ductile fracture criteria were discussed and a simple equation for predicting the load at the onset of penetration of a part-through notched plate was introduced.

REFERENCES

(1) M. F. Kanninen, D. Broek, G. T. Hahn, C. W. Marschall, E. F. Rybicki,
     and G. M. Wilkowski, "Towards an Elastic-Plastic Fracture Mechanics
     Predictive Capability for Reactor Piping", Nuclear Engineering and
     Design, 48 (1978) 117-134.

(2) W. H. Bamford and A. J. Bush, "Elastic-Plastic Fracture", ASME STP-668,
     April (1979).

Table 1 Chemical and mechanical properties

Chemical Composition, %						Mechanical Prop.		
C	Si	Mn	S	Ni	Cr	σ_{ys} (MPa)	σ_{UTS} (MPa)	El. (%)
0.05	0.58	0.95	0.004	9.01	18.22	245	567	65

σ_{ys} : Yield strength at $\varepsilon = 0.2$ %

Table 2 Material constants for finite
element analysis

Elastic modulus	E	1.94×10^5 MPa
Yield stress	σ_y	1.96×10^2 MPa
Strain hardening coefficient	H	2.45×10^3 MPa
Poisson's ratio	ν	0.26

σ_y : Yield strength at $\varepsilon = 0.1$ %

Table 3 Notch geometry and onset of penetration

Spec. No.	Initial notch		Onset of penetration	
	Depth, d (mm)	Lengh, ℓ (mm)	Nominal Stress σ_i (MPa)	COD (mm)
C-1	2.2	30	384	2.77
C-2	3.3	30	318	2.50
C-3	4.4	30	261	1.86
C-4	5.5 (through wall)	30	304*	6.10*

* at maximum load

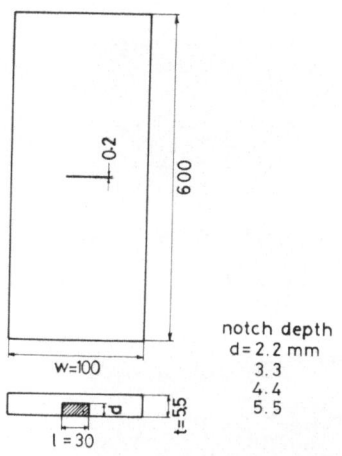

Fig. 1   Center notch tension specimen

notch depth
d = 2.2 mm
3.3
4.4
5.5

w = 100
l = 30
t = 55

600
0.2

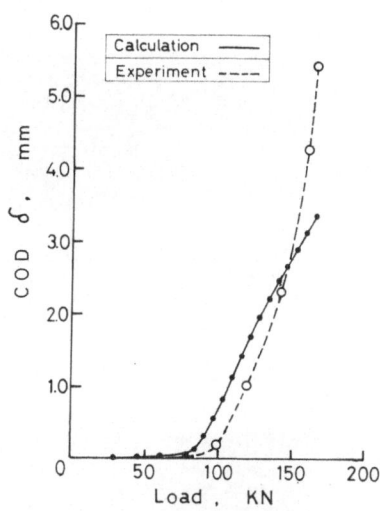

Fig. 2   Crack opening displacement vs load
in through thickness notched plate

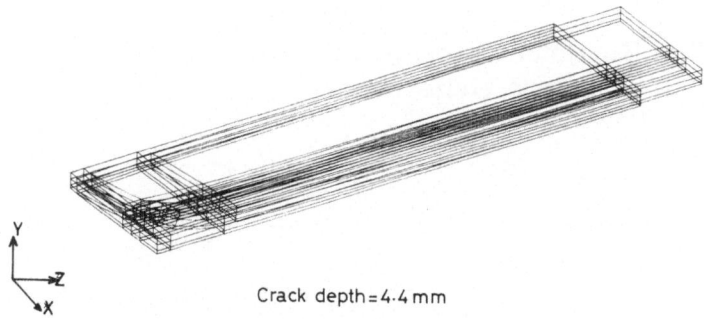

Crack depth = 4.4 mm

Fig. 3   Deformation of part-through notched plate
by finite element analysis

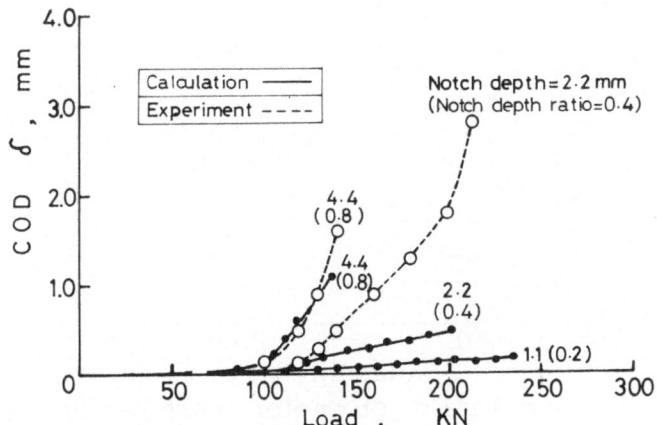

Fig. 4   Crack opening displacement vs load in
surface notched plate

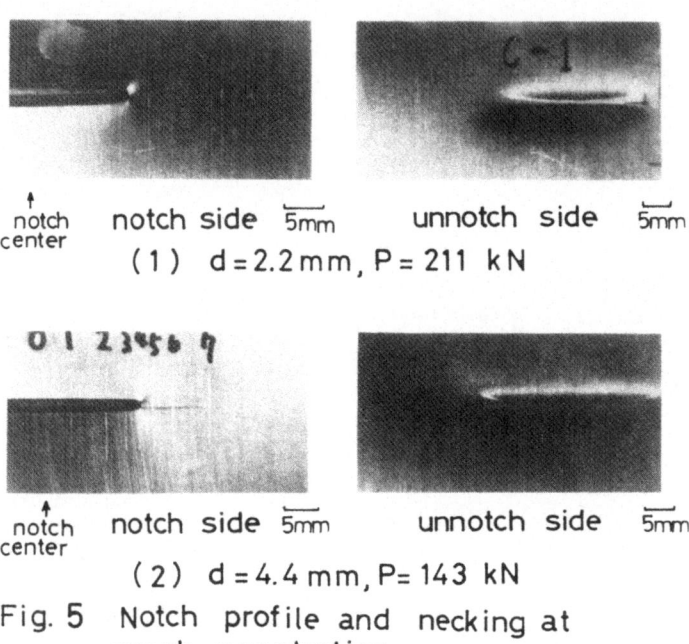

↑
notch   notch side ⌒5mm   unnotch side ⌒5mm
center
　　　　(1)　d = 2.2 mm , P = 211 kN

↑
notch   notch side ⌒5mm   unnotch side ⌒5mm
center
　　　(2)　d = 4.4 mm, P = 143 kN

Fig. 5　Notch profile and necking at
　　　　crack penetration

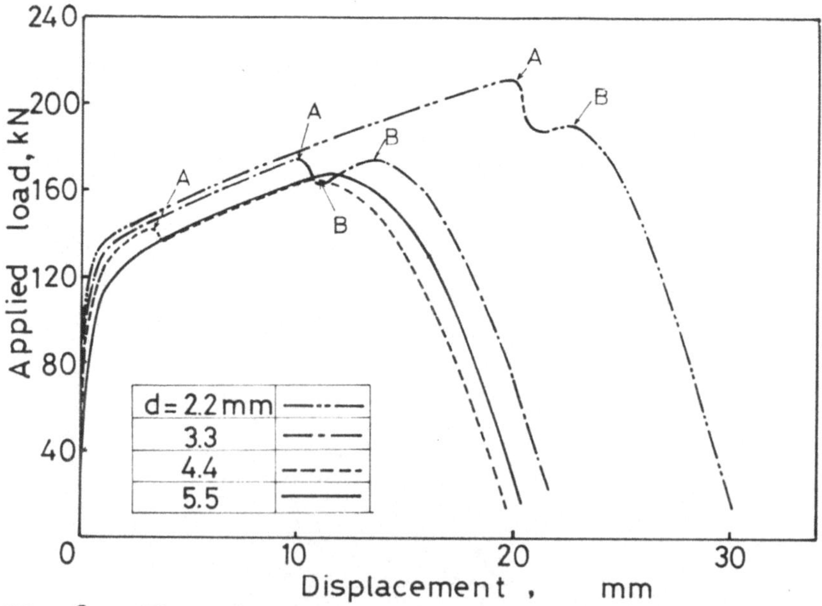

Fig. 6　Flow diagram of center notch tension specimen

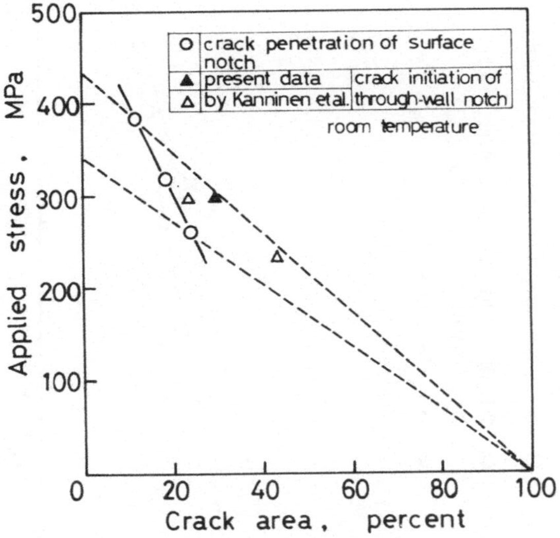

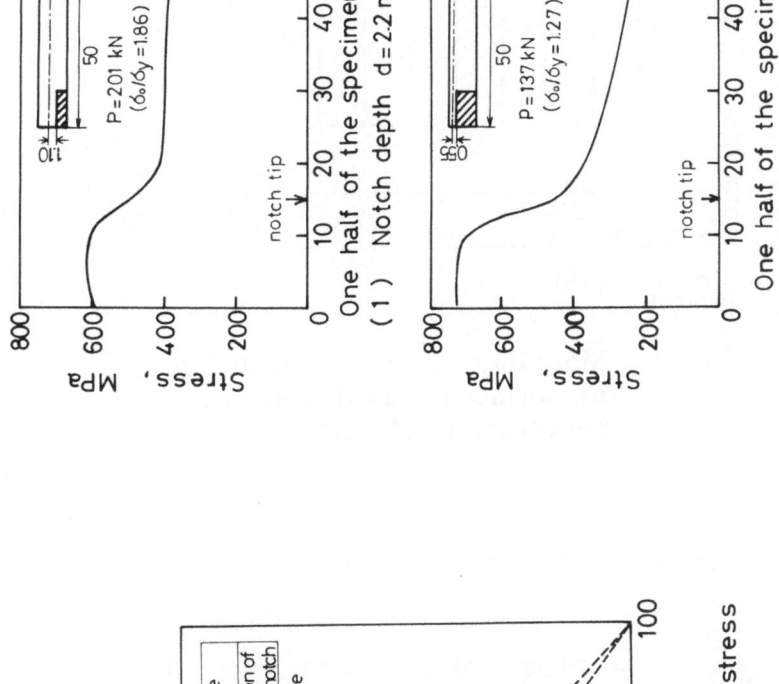

Fig.8 Longitudinal stress distribution of notch section

(1) Notch depth d=2.2 mm

P=201 kN
(σ_o/σ_y=1.86)

notch tip

One half of the specimen width

Stress, MPa

(2) Notch depth d=4.4 mm

P=137 kN
(σ_o/σ_y=1.27)

notch tip

One half of the specimen width

Stress, MPa

Fig. 7 Relationship between applied stress and cracked area

Crack area, percent

Applied stress, MPa

O	crack penetration of surface notch	crack initiation of
▲	present data	through-wall notch
△	by Kanninen et al.	room temperature

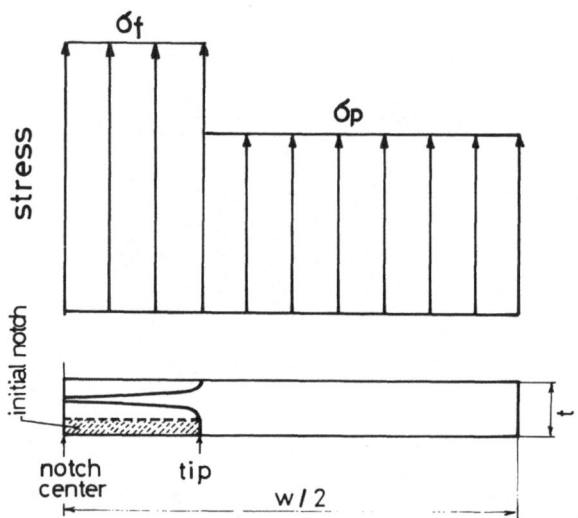

Fig. 9　Modelling of stress distribution in surface notched plate at penetration of notch

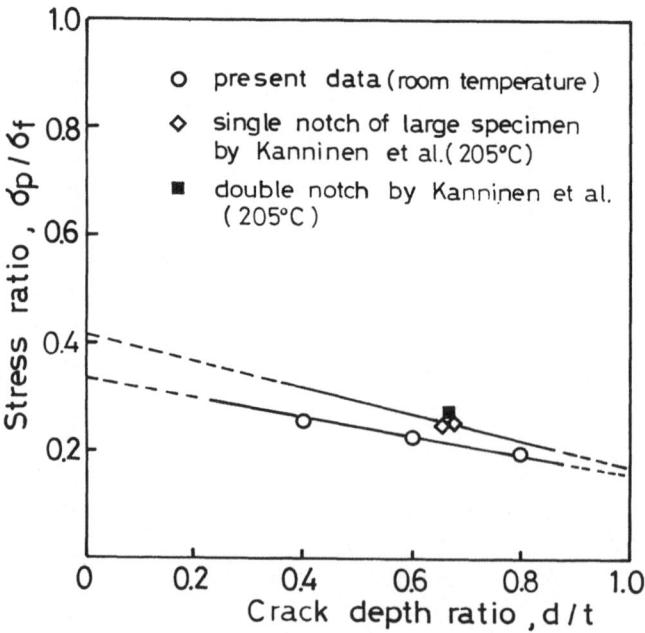

Fig. 10　Stress σ_p of unnotched area at crack penetration as a function of crack depth

EVALUATION OF J-INTEGRAL AND COD
FOR COMPACT TENSION SPECIMEN

M. Shiratori
Associate Professor of Mechanical Engineering
Yokohama National University, Tokiwadai, Hodogaya-ku, Yokohama, 240 JAPAN

T. Miyoshi
Associate Professor of Precision Machinery Engineering
University of Tokyo, Hongo, Bunkyo-ku, Tokyo, 113 JAPAN

ABSTRACT

Based upon a lip line field analysis, a unified approach has been carried our for a compact tension specimen in order to give explicit forms of the plastic constraint factor, rotational factor, crack opening displacement, and J-integral. These expressions are compared with the results of finite element analyses and have been found reasonable and useful for fracture toughness testing. Finally, the relation between J and COD has been considered.

INTRODUCTION

The compact tension (CT) specimen, as well as the three point bend bar, is one of the most important specimens in fracture toughness testing. It is recommended as one of the standard specimens by the ASTM E24 Task Group on Elastic-Plastic Fracture Criteria[1]. In the draft of the standard, an equation proposed by Merkle and Corten[2] is recommended to evaluate the J-integral from a single load vs. load-line-displacement, P vs. Δ, curve. This equation, taking account of the effect of axial force, is more reasonable for the CT specimen than Rice's equation[3], which is developed for the deep-notched bending specimen. But the effect of plastic constraint, which is essential to the plane strain deformation, is not considered in their equation. Therefore, it is desirable to modify the equation by taking account of both effects of axial force and plastic constraint at the same time. On the other hand, no recommendation is found about using the CT specimen to evaluate crack opening displacement (COD) in the draft of the British Standard[4]. The main reason may come from the fact that no explicit and reasonable formula exists for evaluating COD from the clip gage displacement.

In this paper, a unified approach has been carried out, based upon a slip line field analysis, in order to give explicit forms for the plastic constraint factor, rotational factor, COD, and J-integral. These expressions are compared with the results of finite element analyses. And finally, the relation between J and COD has been considered.

SLIP LINE FIELD ANALYSIS

Fig. 1 shows the shape of a CT specimen and a possible slip line field at general yielding. The pattern of the slip line field is the same as the one proposed by Green[5] for pure bending, but the ratio of R/b is affected by the axial force in the case of the CT specimen, where R is the radius of the rigid hinge, OO_1, and b is the ligament length. If B and P are denoted as the thickness of the specimen and the applied load, respectively, the load per unit thickness can be expressed by $T = P/B$. Equilibrating the internal stress along α-slip line, OQR', to the external forces, the authors have obtained the following expressions for the general yielding load[6,7]:

$$T_1 = \frac{T}{2kb} = \frac{r}{r_0} - 1 \tag{1}$$

where

$$\left. \begin{array}{l} r = \frac{R}{b}\sin(2\lambda - \frac{\pi}{4}) = -(\frac{W}{b} - 1) + \{(\frac{W}{b} - 1)^2 + 2r_0(\frac{W}{b} - \frac{1}{2})\}^{1/2} \\[2mm] r_0 = \frac{R_0}{b}\sin(2\lambda - \frac{\pi}{4}) = 0.370 \end{array} \right\} \tag{2}$$

$$\lambda = \frac{1}{2}\angle OO_1Q = 1.021 = 58.5° \tag{3}$$

where k is the yielding stress in shear and W is the distance from the load line to the bottom surface of the specimen. In eq. (1), the parameter T_1 is obtained by the general yielding load, T, divided by 2kb, which is the tensile yielding load of the plain specimen with width b.* Therefore, T_1 can be considered a plastic constraint factor defined for the axial force.

Since the rigid part on the right hand side of the slip line, OQS, rotates about the point O_1, the projection of this point to the symmetrical axis gives the rotational center for this specimen. Therefore, the parameter, r, can be considered the rotational factor for the rigid-plastic CT specimen by the geometrical interpretation of eq. (2). On the other hand, r_0 can be interpreted as that for pure bending because the radius of the rigid hinge, R_0, is constant so that $R_0/b = 0.389$ in the case of pure bending.

A lower bound general yielding load can be obtained by the assumption of the stress distribution, shown in Fig. 2. If a Tresca material is assumed, the resulting equation for the lower bound general yielding load is also given by eq. (1) but with $r_0 = 0.5$ instead of $r_0 = 0.370$. Table 1 shows the values of T_1 and r compared with T_{1L} and r_L, respectively, where the suffix 'L' means that

---

* For Tresca material $\sigma_y = 2k$, where σ_y is the tensile yielding stress.

these values are the results of lower bound analysis.

EVALUATION OF CRACK OPENING DISPLACEMENT

Wells's equation

Wells[8] proposes the following equations in order to evaluate COD at the crack tip, δ, from the clip gage displacement, V_g, (see Fig. 3):

$$\delta = \frac{0.45b}{0.45W+0.55a+Z} (V_g - \phi) \qquad \text{for} \quad V_g \geq 2\phi$$

$$\delta = \frac{0.45b}{0.45W+0.55a+Z} \frac{V_g^2}{4\phi} \qquad \text{for} \quad V_g < 2\phi \qquad \Biggr\} \qquad (4)$$

$$\phi = \zeta\sigma_y W(1-\nu^2)/E, \qquad \zeta = 2.34 \quad \text{for } a/W = 0.5 \qquad (5)$$

where a is the crack length and Z is the distance from the load line to the point at which the clip gage is set. It is usual to put $Z = 0$, since the clip gage is often set on the load line in the case of CT specimen.

COD for rigid-plastic body

The concept of the rotational factor is useful to derive a relation between δ and V_g for a rigid-plastic CT specimen. From a kinematical consideration, shown in Fig. 3, the relation is simply given by the following equation:

$$\delta = \frac{rb}{W+Z-(1-r)b} V_g = \frac{rb}{rW+(1-r)a+Z} V_g \qquad (6)$$

Note that $r = 0.452$ for $a/W = 0.5$ (see table 1), which is quite close to the value shown in Wells's equation.

A modification due to plastic adjustment of the stress intensity factor

Eq. (6) is correct only for the rigid-plastic body; therefore, a modification is necessary to develop an equation which is valid for a more realistic material. The modification can be done by introducing the concept of plastic adjustment of the stress intensity factor, K. The crack opening displacement, based upon the plastic adjustment of K, is given by[9]

$$\delta = \frac{1-\nu^2}{E} \frac{K^2}{\sqrt{3}\sigma_y} \qquad (7)$$

where the value of K for the CT specimen is given by

$$K = P\sqrt{a}Y/BW, \qquad Y = 29.6-185.5\xi+655.7\xi^2-1017.0\xi^3+638.9\xi^4, \qquad \xi = a/W \qquad (8)$$

If the deformation is purely elastic, the clip gage displacement is proportional to the external force; therefore, the parameter α defined by the following equation is constant.

$$\alpha = EV_g/(P/B) \qquad (9)$$

Table 2 shows the values of α against various crack lengths which are the results of a numerical analysis by Newman [10] for a standard CT specimen with $W = 50$ mm. Eliminating K and P from eq. (7) – (9),

$$\delta = \frac{(1-\nu^2)E}{\sqrt{3}\sigma_y} \left(\frac{Y}{\alpha}\right)^2 \frac{a}{W^2} V_g^2 \qquad (10)$$

This equation is assumed to hold until general yielding occurs in the specimen. The following equation will hold after general yielding:

$$\delta - \delta_0 = \frac{rb}{rW+(1-r)a+Z} (V_g - V_{g0}) \tag{11}$$

where δ_0 and V_{g0} are the values given by eqs. (7) and (9), respectively, at general yielding whose load is given by eq. (1). For example, concrete expressions of eqs. (10) and (11) for a/W = 0.5 and 0.6 are

$$\begin{aligned}
\delta &= 0.37V_g^2 \; (V_g<0.51), \quad \delta = 0.31(V_g-0.20) \quad (V_g>0.51), \quad \text{for a/W=0.5} \\
\delta &= 0.25V_g^2 \; (V_g<0.51), \quad \delta = 0.22(V_g-0.22) \quad (V_g>0.51), \quad \text{for a/W=0.6}
\end{aligned} \tag{12}$$

Examination by finite element analysis

In order to examine the validity of the formulae proposed above, the relations between δ and V_g have been obtained through an elastic-plastic finite element analysis, where the deformation theory based upon the von Mises' yield criterion with linear work hardening rate, H', has been assumed. The mechanical properties of materials are chosen to simulate those of A533B steel at room temperature, i.e.,

$E = 2.06 \times 10^5$ N/mm², $\nu = 0.3$, $\sigma_y = 480$ N/mm², and $H' = E/100$.

Isoparametric elements with eight nodes have been used in the analysis. This type of element can express a singularity of $\frac{1}{\sqrt{r}}$ by shifting the midside node to a quarter point[11]; therefore, it is utilized at the crack tip. An example of the element breakdown is shown in Fig. 4.

Fig. 5 shows the crack opening displacement, V, along the crack line. The axis of abscissa in the figure expresses the normalized coordinate where the load line and the crack tip correspond to 0.0 and 1.0, respectively. The clip gage displacement, V_g, is chosen to be the load line displacement at (a-x)/a=0. As far as the results of the finite element analysis is concerned, the real crack tip opening displacement is zero, as shown in Fig. 5, since the point of the crack tip is fixed. Therefore, an extrapolation technique, shown by dotted lines in the figure, has been used to obtain δ. This technique is reasonable because the crack tip opening displacement discussed in this paper is essentially based upon the concept of the rotational center. This hypothetical crack opening displacement, however, may differ from the real one which would be observed in the real materials in the experiment. Discussions about the real opening displacement have been presented by Miyoshi[12].

Fig. 6 shows the results for a/W = 0.5 and 0.6. In the figure, "G.Y." means the general yielding occurs at these points. Agreement is observed between the authors' or Wells's equations and the results of the finite element analysis. Though Wells's equations are excellent, they are proposed only for a/W = 0.5; the authors' equations are applicable for arbitrary crack length.

Therefore, the authors recommend using their equations in the estimation of COD from clip gage displacements.

EVALUATION OF J-INTEGRAL

Definition of J-integral and its expression for a rigid-plastic body

Rice[3, 13] has developed the following expression for the J-integral within the framework of a total strain formulation of elastic-plastic deformation:

$$J = -\frac{1}{B}\int_0^{\Delta}(\frac{\partial P}{\partial a})_{\Delta}d\Delta \tag{13}$$

For the rigid-plastic CT specimen, the general yielding load per unit thickness, T, is independent of the load-line-displacement, Δ; therefore, integration of eq. (13) and substitutuion of eqs. (1) and (2) result in

$$J = 2k\Delta(T_1+b\frac{\partial T_1}{\partial b}) = 2k\frac{1-r_0}{r_0}\frac{rb}{W-(1-r)b}\Delta \tag{14}$$

A modification of Merkle and Corten's equation

Merkle and Corten have obtained the following equation for $(\partial P/\partial a)$ in eq. (13) by neglecting elastic contribution and assuming that the amount of rigid rotation, θ, is a function of P/BT where P is the instantaneous load in the real P-Δ curve, and BT is the collapse load at general yielding:

$$-(\frac{\partial P}{\partial a})_{\Delta} = \{(\frac{\partial T}{\partial b})_{\Delta}/T\}P + \{(\frac{\partial \theta}{\partial b})_{\Delta}/(\frac{\partial \theta}{\partial \Delta})_b\}(\frac{\partial P}{\partial \Delta})_b \tag{15}$$

From a kinematical consideration about Fig. 3, θ is given by

$$\theta = \frac{\Delta}{a+rb} = \frac{\Delta}{W-(1-r)b} \tag{16}$$

By substituting eqs. (1) and (16) into eq. (15), eq. (13) becomes

$$J = \frac{1}{b}[\eta_r\int_0^{\Delta}(\frac{P}{B})d\Delta + \eta_c\int_0^{P/B}\Delta d(\frac{P}{B})] \tag{17}$$

$$\left.\begin{array}{l}\eta_r = 1 + \dfrac{1}{1-(1-r)b/W} \\[2mm] \eta_c = \dfrac{b/W}{1-(1-r)b/W}\{(1-r) - \dfrac{r-r_0}{1-(1-r)b/W}\}\end{array}\right\} \tag{18}$$

This is an expression for evaluating the J-integral from a single P-Δ curve. In eqs. (18), r is the rotational factor given by eq. (2). If $r_0 = 0.5$, which is the result of the lower bound analysis, eqs. (17) and (18) coincide with Merkle and Corten's equations. But the authors recommend use of $r_0 = 0.370$, which is the result of the slip line field analysis. Table 3 shows the values of η_r and η_c for various crack lengths. In the limit of $b/W \rightarrow 0$, eq. (18) results in $\eta_r=2.0$ and $\eta_c=0.0$; therefore, eq. (17) coincides with Rice's equation.

Examination by finite element analysis

In order to examine the validity of the proposed equations, a finite element analysis has been carried out. The method of analysis is the same as that mentioned in the previous section.

The J vs. V_g curves for a/W = 0.5 and 0.6 are shown in Fig. 7, where J_S and J_M are obtained from a single P-Δ curve resulting from the finite element analysis. The clip gage displacement is taken to be the same as the load-line-displacement here, i.e., $V_g = \Delta$. J_S corresponds to the values of the J-integral obtained by eqs. (17), (18), and (2) with $r_0 = 0.370$, which are called authors' equations, while J_M corresponds to those with $r_0 = 0.5$, which are called Merkle's equations. $\overline{J_p}$ in Fig. 7 shows the results of the path integral, due to the original definition of the J-integral[13]. An average of 10 paths has been taken to give the value of $\overline{J_p}$. The results of Fig. 7 show that the coincidence between J_S and J_p is excellent and that J_S seems to be more reasonable than J_M.

RELATION BETWEEN J-INTEGRAL AND COD

For the small scale yielding range, the relation between J and COD is given by $J = \lambda \sigma_y \delta$, where the parameter λ is reported to be constant, ranging from 1.0 for DBCS model (plane stress) to about 2 as the finite element results [15, 16]. On the other hand, no such equation has been reported for the large scale and general yielding ranges. Therefore, it may be useful to consider the relation between J and COD for these ranges. For a rigid-plastic CT specimen after general yielding, the result of the slip line field analysis and the concept of the rotational center are useful to obtain the relation between J and COD. Since $V_g = \Delta$ for Z = 0 in Fig. 3, eq. (6) results in

$$\delta = \frac{rb}{W-(1-r)b} \Delta \tag{6'}$$

This equation is substituted into eq. (14) to eliminate Δ and then

$$J = \frac{1-r_0}{r_0} 2k\delta = \lambda \sigma_y \delta \tag{19}$$

where
$$\lambda = \frac{2k}{\sigma_y} \frac{1-r_0}{r_0} = \begin{cases} 1.70 & \text{for Tresca material} \\ 1.97 & \text{for Mises material} \end{cases} \tag{20}$$

It is very difficult to obtain the relation for the large scale yielding range. The finite element method may be the only way for obtaining it. Fig. 8 shows the results of the finite element analysis for a/W = 0.5 and 0.6 for whole ranges of plasticity. The relation given by eqs. (19) and (20) is also plotted as a solid line in Fig. 8. Note that the relation is not affected by the crack length, which is predicted by eq. (19). The coincidence between eq. (19) and the results of the finite element analysis is remarkable.

CONCLUDING REMARKS

Based upon a slip line field analysis, a unified approach has been carried out for the CT specimen in order to give explicit forms of the plastic constraint factor, rotational factor, COD, and J-integral. These expressions are compared with the results of finite element analyses, and have been found reasonable and useful for fracture toughness testing. Further, a simple relation between J and COD has been obtained.

Finally, the authors would like to express their thanks to Mr. Arita and Mr. Nemoto for their help in the finite element analyses.

REFERENCES

(1) G.A. Clerk, et al., "A Procedure for the Determination of Ductile Fracture Toughness Values Using J Integral Techniques", J. of Testing and Evaluation, 7-1, (Jan. 1979), pp.49-56.

(2) J.G. Merkle and H.T. Corten, "A J-Integral Analysis for the Compact Specimen, Considering Axial Force as well as Bending Effects", J. of Pressure Vessel Technology, Trans. of ASME, J-96, (Nov., 1974), pp.286-292.

(3) J.R. Rice, et al., "Some Further Remarks of J-Integral Analysis and Estimates", Progress of Flaw Growth and Fracture Toughness Testing, ASTM STP 536, (1973), pp.231-245.

(4) British Standard Institution, "Methods for Crack Opening Displacement (C.O.D.) Testing", Draft for Development 19, (1972).

(5) A.P. Green, "The Plastic Yielding of Notched Bars due to Bending", Quart. J. of Mech. and Appl. Math., 6, (1953), pp.223-239.

(6) M. Shiratori and B. Dodd, "The General Yielding of Compact Tension Specimens", OUEL Report, Univ. of Oxford, Dept. of Engng. Sci., 12707, (1977).

(7) M. Shiratori and T. Miyoshi, "An Evaluation of J-Integral for CT Specimen Based upon Slip Line Field Analysis", Trans. of JSME, 45-389, (Jan., 1978), pp.50-56, (in Japanese).

(8) A.W. Pense and R.D. Stout, "Fracture Toughness Related Characteristics at Cryogenic Temperature", WRC, Britain, 205, (1975), p.8.

(9) H. Okamura, Introduction to Linear Fracture Mechanics, Baihukan, (1976), 226 pp., (in Japanese).

(10) J.C. Newman Jr., "Stress Analysis of the Compact Specimen Including the Effects of Pin Loading", Fracture Analysis, ASTM STP 560, (1974), pp.105-121.

(11) R.S. Barsoum, "On the Use of Isoparametric Finite Elements in Linear Fracture Mechanics", Int. J. for Num. Methods in Engng., 10, (1976), pp.25-37.

(12) T. Miyoshi, "Relation between J-Integral and Crack Opening Displacement", J. of Faculty of Engng., Univ. of Tokyo (B), 34-2, (1977), pp.371-375.

(13) R.R. Rice, "A Path Independent Integral and the Approximate Analysis of Strain Concentration by Notches and Cracks", J. of Appl. Mech., Trans. of ASME, 35, (June, 1968), pp.379-386.

(14) M. Shiratori and T. Miyoshi, "Evaluation of Constraint Factor and J-Integral for Single-Edge Notched Specimen", Proc. of ICM 3, Cambridge, England, 3, (1979), pp.425-434.

(15) J.F. Knott, Fundamentals of Fracture Mechanics, Butterworths, (1973).

(16) D. Broek, Elementary Engineering Fracture Mechanics, Noordhoff, (1974).

Table 1 Plastic constraint factor
and rotational factor

a/W	T_1	T_{1L}	T_1/T_{1L}	r	r_L
0.1	0.542	0.357	1.52	0.570	0.688
0.2	0.449	0.303	1.48	0.536	0.651
0.3	0.364	0.252	1.45	0.504	0.626
0.4	0.289	0.205	1.41	0.477	0.603
0.5	0.223	0.162	1.38	0.452	0.581
0.6	0.166	0.123	1.35	0.431	0.562
0.7	0.116	0.088	1.32	0.413	0.544
0.8	0.072	0.055	1.30	0.396	0.528
0.9	0.034	0.026	1.28	0.382	0.513

Table 2 Dimensionless clip gage displacements $\alpha = EV_g/(P/B)$

a/W	0.2	0.3	0.4	0.5	0.6	0.7	0.8
Load line (Z=0.0)	3.90	6.49	10.39	16.82	28.81	57.12	140.4
Edge of plate (Z=12.5)	7.99	11.27	16.40	24.86	40.45	76.23	182.4

Table 3 Coefficient η_r and η_c of eq. (17)

a/W	Merkle and Corten		Authors	
	η_r	η_c	η_r	η_c
0.3	2.355	0.193	2.531	0.310
0.4	2.313	0.207	2.458	0.321
0.5	2.265	0.200	2.377	0.299
0.6	2.213	0.176	2.295	0.254
0.7	2.159	0.141	2.214	0.195

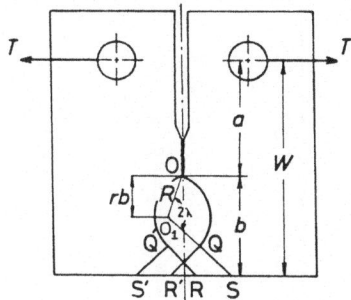

Fig. 1 Shape of the CT specimen and a possible slip line field

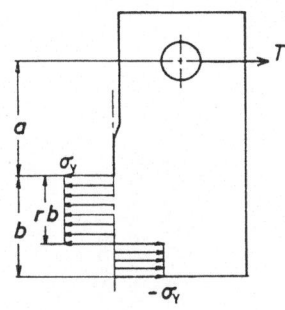

Fig. 2 Assumption of stress distribution across the ligament which gives a lower bound general yielding load

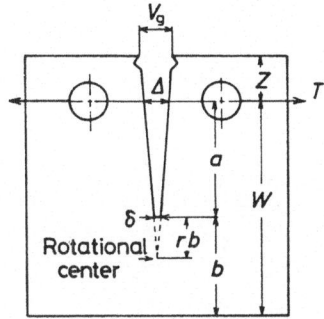

Fig. 3 A schematical diagram explaining the concept of the rotational center

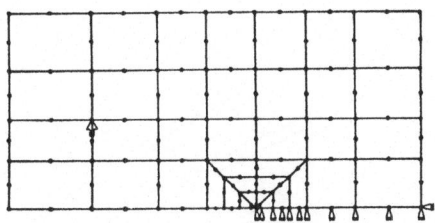

Fig. 4 Element breakdown of the half of the specimen with eight-noded isoparametric elements

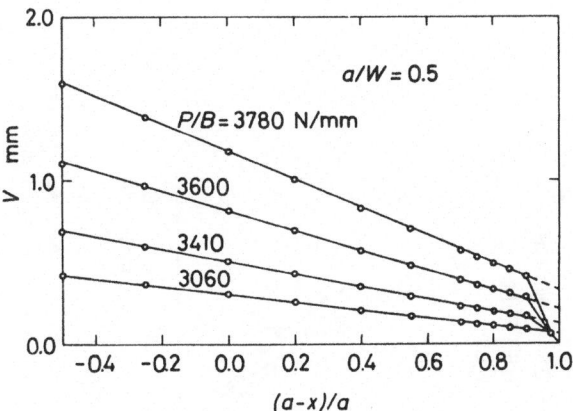

Fig. 5 Crack opening displacements; results of finite element analysis

98

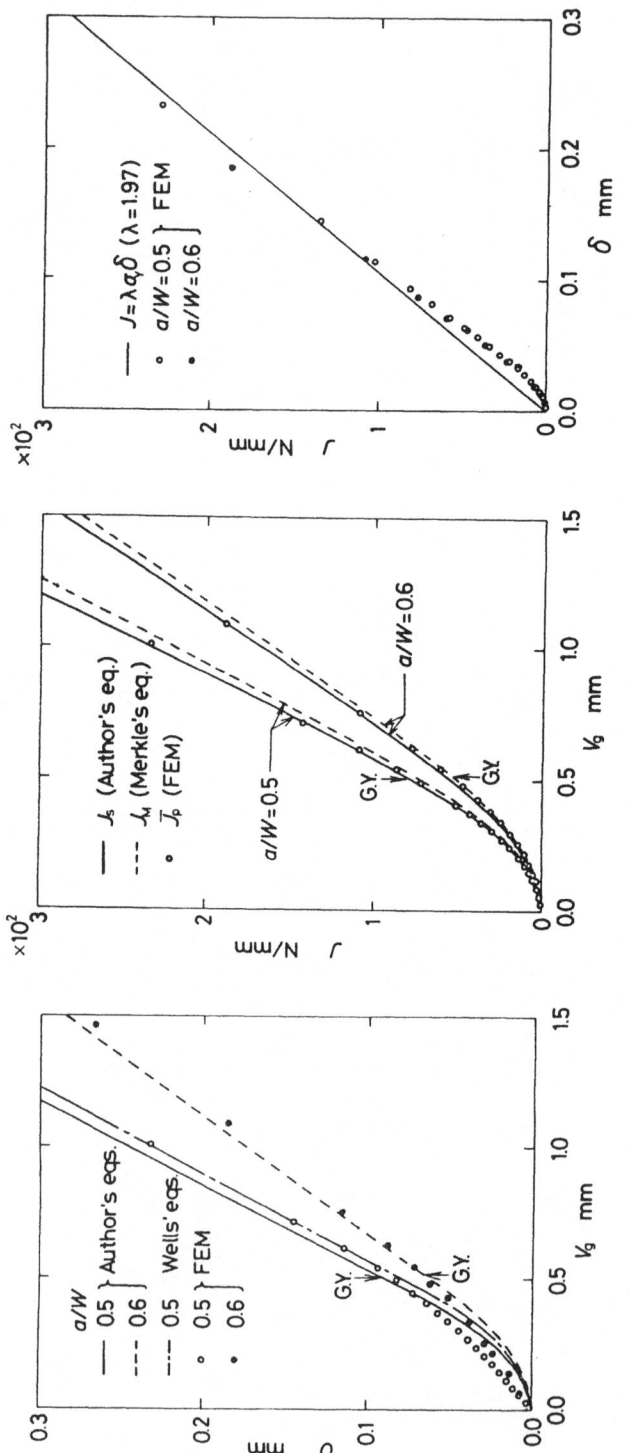

Fig. 8 J-integral vs. crack tip opening displacement

Fig. 7 J-integral vs. clip gage displacement

Fig. 6 Crack tip opening displacement vs. clip gage displacement

RELATION BETWEEN J-INTEGRAL AND
CRACK TIP OPENING DISPLACEMENT

T. Miyoshi
Associate Professor of Precision Machinery Engineering
University of Tokyo, Hongo 7-3-1, Bunkyo-ku, Tokyo, JAPAN

M. Shiratori
Associate Professor, Mechanical Engineering
Yokohama National University, Tokiwadaira 156, Hodogaya-ku,
Yokohama, JAPAN

ABSTRACT

This paper describes the analysis by the finite element method, which shows the correlation between the J-integral (J) and crack tip opening displacement (CTOD). The result is that the non-dimensional J-integral $\sim$ non-dimensional CTOD relation seems to be expressed by a single curve for various metals. This is very effective for the J_{IC} evaluation by the stretched zone size.

INTRODUCTION

There is a relation between the fracture toughness and the stretched zone size which is measured from the fractograph of the specimen [1,2]. Therefore, the fracture toughness can be evaluated from the stretched zone size if the quantitative relation between the fracture toughness and the stretched zone size is obtained. In the K_{IC} fracture toughness test, the test methods are standardized [3], and K_I values are obtained for the various specimen geometries and loading modes. So, the merit of the K_{IC} evaluation by the stretched zone size is little. In the J_{IC} fracture toughness test, the initiation of the stable crack growth must be determined, and J_{IC} value is evaluated at this point. The initiation of the stable crack growth is detected by the electric potential method [4], AE method, etc. These need special, delicate, and expensive equipment. On the other hand, the stretched zone has not changed its size after the stable crack growth, which enables the evaluation of the J_{IC} value without the detection of the stable crack growth.

If the relation between J and the stretched zone size can be expressed by a single curve for various metals, it is very effective for J_{IC} evaluation by the stretched zone size, because J_{IC} evaluation can be made from a single curve for a variety of metals. In this paper, the relation between J and the stretched zone size is analyzed by special finite element method.

RELATION BETWEEN FRACTURE TOUGHNESS AND STRETCHED ZONE SIZE

For the small scale yielding range, the stretched zone depth $2d$, i.e., the crack tip opening displacement (CTOD), is related to K_I as follows (Fig. 1):

$$\text{CTOD} = m \; \frac{K_I^2}{E \; \sigma_Y} \; , \tag{1}$$

in which E is Young's modulus, σ_Y is the yield stress of the material, and m is constant as shown in Table 1.

In the large scale yielding and the fully plastic ranges, the relation between J and CTOD may be stated:

$$J = m \; \sigma_Y \; \text{CTOD}. \tag{2}$$

The value $m = 1$ is given by the Dugdale model, but the relation between J and CTOD is generally difficult to get. Four factors [5] cause this difficulty: (1) constraint, (2) work-hardening, (3) the increase of J, (4) the lack of uniqueness in the definition of crack tip opening displacement. The relative importance of these factors depends on the degree of yielding, and they interact with each other to give a particular value of m. m ranges $1 \lessgtr m \lessgtr 3$ for virtually all studies [5].

Recently, it has been reported that J/E (E; Young's modulus) $\sim$ the stretched zone size relation is expressed by a single curve for various materials [2]. Figure 2 shows this experimental result, which does not agree with the expected variation in m.

FINITE ELEMENT ANALYSIS FOR THREE-POINT BEND BAR SPECIMEN

In order to investigate the relation between J and CTOD, a finite element analysis for the three-point bend bar is carried out. The geometry of the specimen is shown in Fig. 3, and five materials, shown in Table 2, are considered. In Table 2, A533B, SUS304, SNCM8, and Ti-6Al-4V are nuclear pressure vessel steel, austenite stainless steel, Ni-Cr-Mo-V rotor steel, and titanium alloy, respectively. The work-hardening rate, $H' = 0$, is given for this analysis. The finite element analysis is carried out for one-half of the specimen due to the symmetry. Singular fan elements [6] are used near the crack tip, and ordinary linear triangular elements are employed in the other area (Fig. 4). The displacement function of the fan elment is given by

$$\left.\begin{array}{l} u_r = \alpha_1 + \alpha_2 r + \alpha_3 \theta + \alpha_4 r\theta, \\ u_\theta = \alpha_5 + \alpha_6 r + \alpha_7 \theta + \alpha_8 r\theta. \end{array}\right\} \tag{3}$$

This element becomes a singular one at the crack tip, shown in Fig. 4, and the non-zero CTOD is obtained at the crack tip. The CTOD is calculated by the following equation:

$$\text{CTOD} = u_r(0, \pi) - u_r(0, -\pi). \tag{4}$$

On the contrary, the ordinary linear triangular element gives the zero CTOD at the crack tip. The J-integral evaluation is carried out by Rice's simple formula [7]:

$$J = \frac{2A}{b}, \tag{5}$$

in which A is the strain energy per unit thickness and b is the ligament length. The strain energy of the specimen is easily calculated by finite element analysis.

RESULTS AND DISCUSSIONS

Figure 5 shows the correlation between the J-integral and the CTOD, and the five curves differ for each other. The relation between J/σ_Y and CTOD is described in Figure 6. On the other hand, the non-dimensional J-integral $\sim$ non-dimensional CTOD relations for five materials are presented by only one curve. This is a substantial advantage for J_{IC} evaluation by the stretched zone size, because J_{IC} value is determined without the detection of the slow crack growth, once the non-dimensional $J \sim$ non-dimensional CTOD curve is obtained by analytical and experimental techniques (Fig. 7).

The theoretical background on which the one curve presentation depends is HRR singular solution [8]. For non-work-hardening material, the HRR solution is given as follows:

$$\frac{\sigma_{ij}}{\sigma_Y}(r,\theta) = \widetilde{\sigma_{ij}}(\theta), \tag{6}$$

$$\frac{\varepsilon_{ij}}{\varepsilon_Y}(r,\theta) = \frac{JE}{I\sigma_Y^2 a} \frac{1}{(r/a)} \widetilde{\varepsilon_{ij}}(\theta), \tag{7}$$

$$\frac{u_i}{a}(r,\theta) = \frac{JE}{I\sigma_Y^2 a} \widetilde{u_i}(\theta). \tag{8}$$

The alternative form of the equation (7) is given by

$$\frac{E}{\sigma_Y a}[\varepsilon_{ij}(r,\theta)\cdot r] = \frac{1}{I}\frac{JE}{\sigma_Y^2 a}\widetilde{\varepsilon}_{ij}(\theta).\tag{9}$$

CTOD can be considered as the elongation of "small tensile specimen at the crack tip," so the equation (9) is the theoretical background of the non-dimensional $J \sim$ CTOD relation. Finite element analysis shows that $1/I$ varies according to the degree of yielding, and this value is equivalent to m in the equation (2). Figure 8 shows the $J/E \sim$ CTOD relation and a single curve correlates the results of all five materials. The reason is not completely analyzed, but it will be in the near future.

REFERENCES

[1] D. Broek, "Correlation between Stretched Zone Size and Fracture Toughness", Engineering Fracture Mechanics, 6-1, (March,1974), pp. 173-181.

[2] H. Kobayashi, H. Nakamura, and H. Nakazawa, "A Relation between Crack Tip Plastic Blunting and the J-Integral", Proc. of ICM3, Vol.3, (August, 1979), pp. 529-538.

[3] W. F. Brown and J. G. Kaufman, eds., "Developments in Fracture Mechanics Test Methods Standardization". ASTM STP 632, (1977), pp. 221-240.

[4] British Standards Institution, "Crack Opening Displacement (COD) Testing", DD19, (1972).

[5] D. G. H. Latzko, "Post-Yield Fracture Mechanics", Applied Science Publishers, (1979), pp. 23-210.

[6] N. levy, P. V. Marcal, W. J. Ostergren, and J. R. Rice, "Small Scale Yielding near a Crack in Plane Strain: A Finite Element Analysis", NASA Technical Report NGL 40-002-080/1, (1969).

[7] J. R. Rice, P. C. Paris, and J. G. Merkle, "Some Further Results on J-Integral Analysis and Estimates", ASTM STP 536, (1973), pp. 231-245.

[8] J. W. Hutchinson, "Singular Behavior at the End of a Tensile Crack in a Hardening Material", J. Mech. Phys. Solids, 16 (1968), pp. 13-31.

[9] H. Liebowitz, ed., "Fracture", Academic Press, Vol.3, (1971), pp. 47-225.

Table 1 The value of m in equation (1).

	Irwin	Hahn and Rosenfield	Broek
m	0.75	0.5	0.4

Table 2 Mechanical properties of five materials.

Material		Young's Modulus (GPa)	Yield Stress (MPa)	E/σ_Y
A533B	—▼—	206	481	429
SUS304	—△—	181	235	770
SNCM8	—●—	206	1471	140
SNCM8	—○—	206	981	210
Ti-6Al-4V	—□—	103	941	109

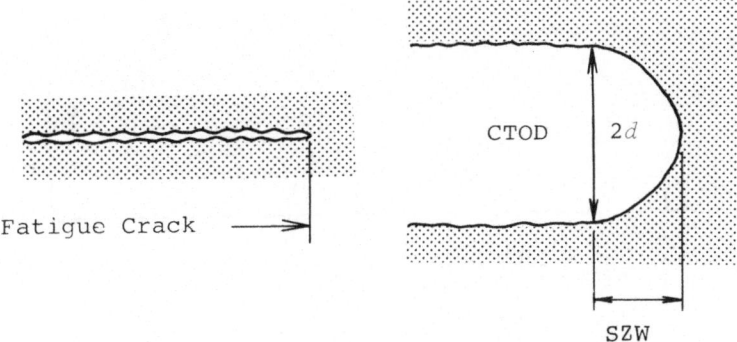

Fig.1 Crack tip opening displacement (CTOD).

104

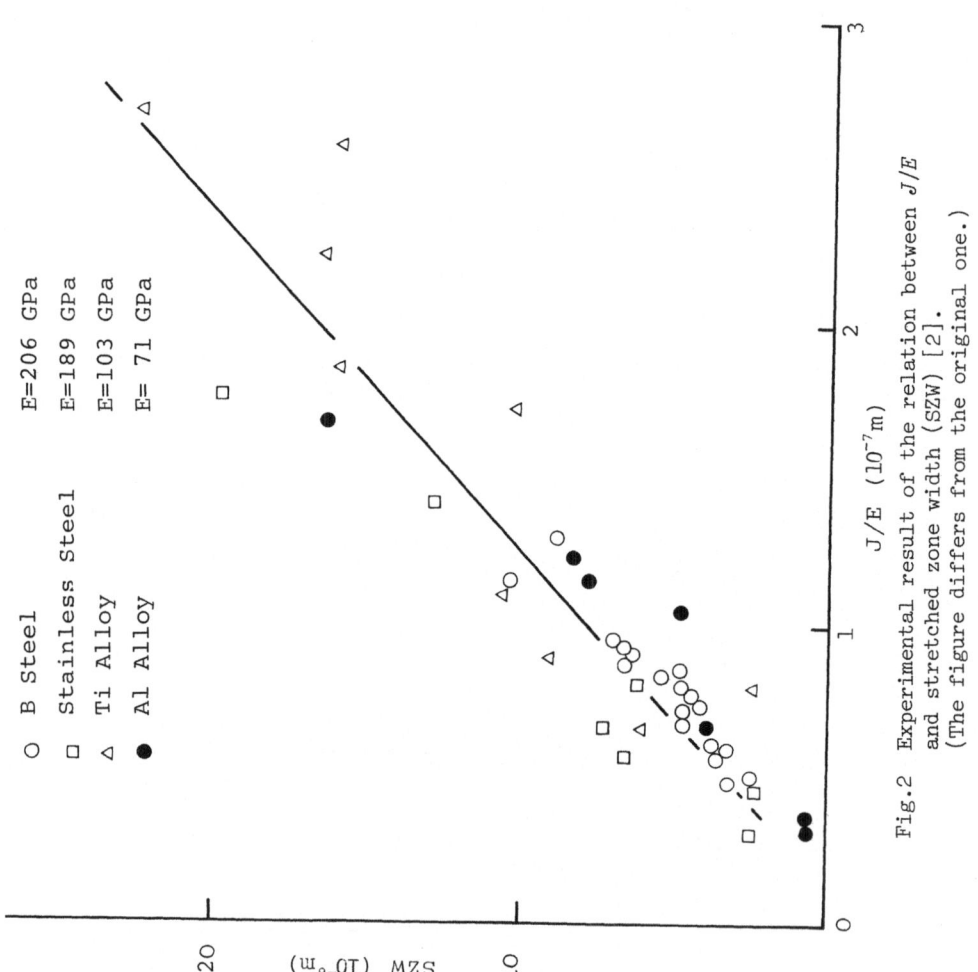

Fig.2 Experimental result of the relation between J/E
and stretched zone width (SZW) [2].
(The figure differs from the original one.)

P
b
W
a
B
S
D

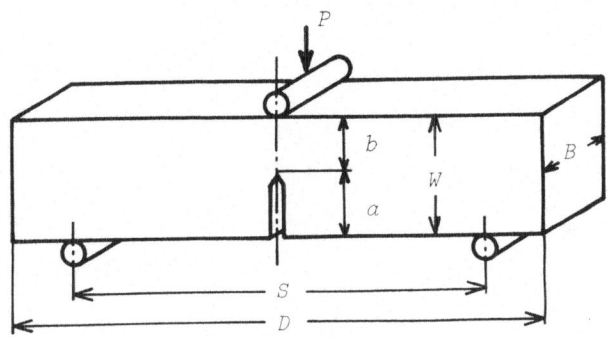

a=10, b=10, W=20, B=10,

S=80, D=110 (unit:mm)

Fig.3 The geometry of three-point bend bar.

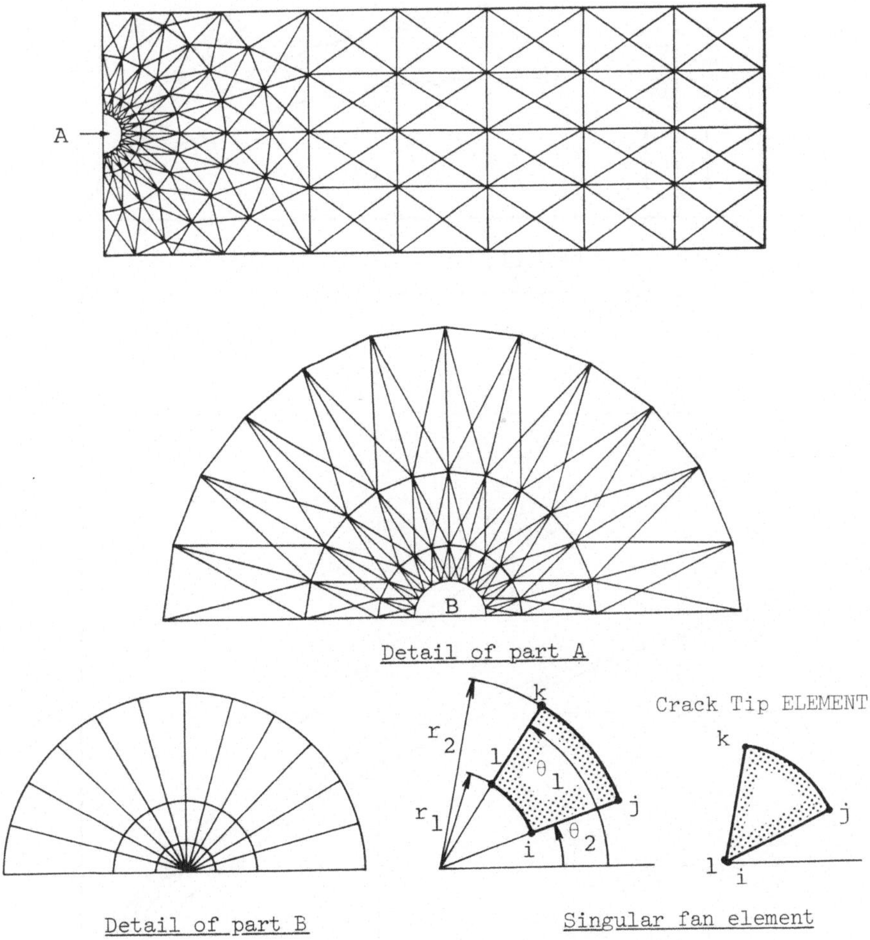

A
B
Detail of part A
Crack Tip ELEMENT
k
l
j
i
r2
r1
θ1
θ2
Detail of part B
Singular fan element

Fig.4 Nodal breakdown of the specimen.

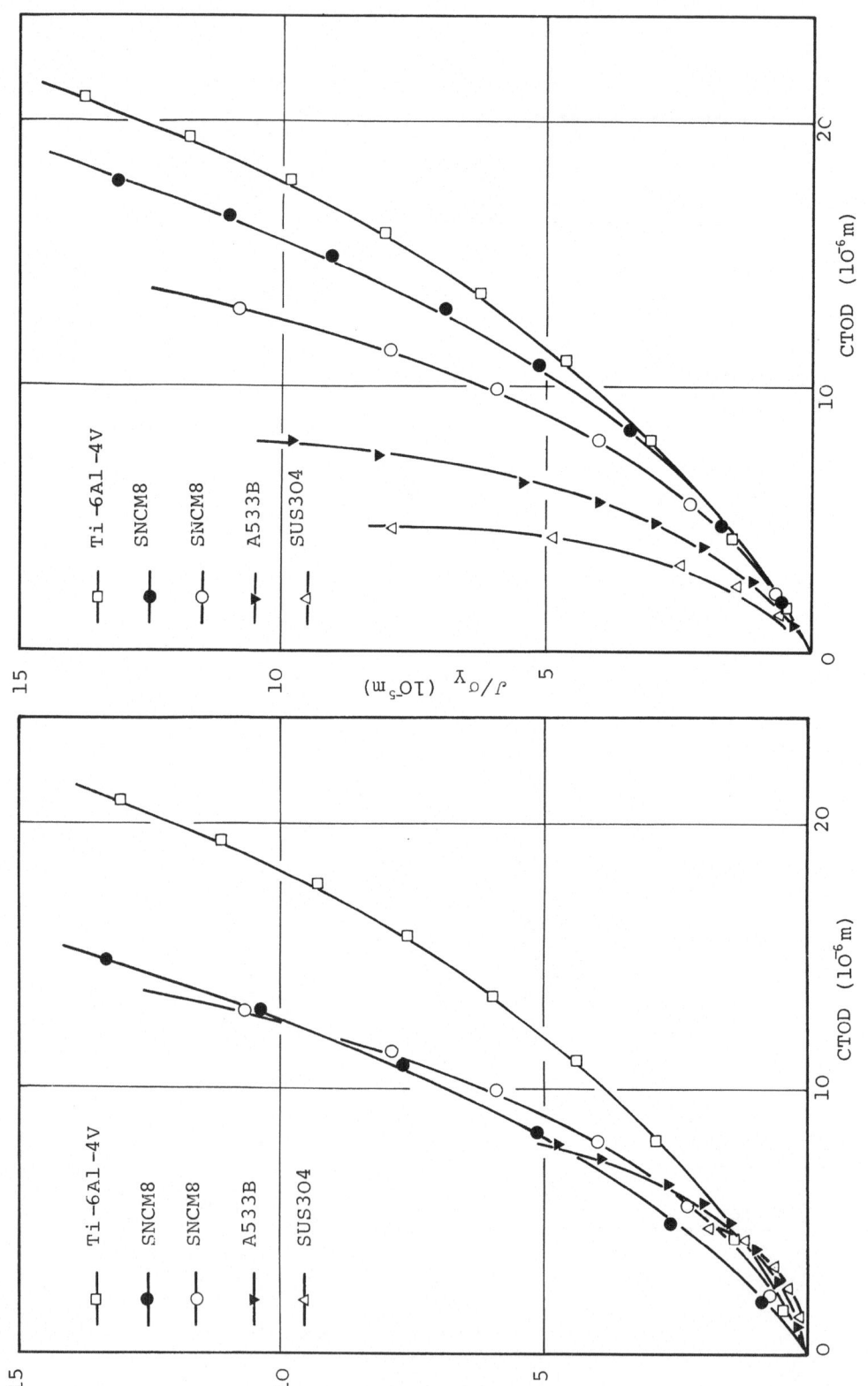

Fig.6 The relation between J/σ_Y and CTOD.

Fig.5 The relation between J and CTOD.

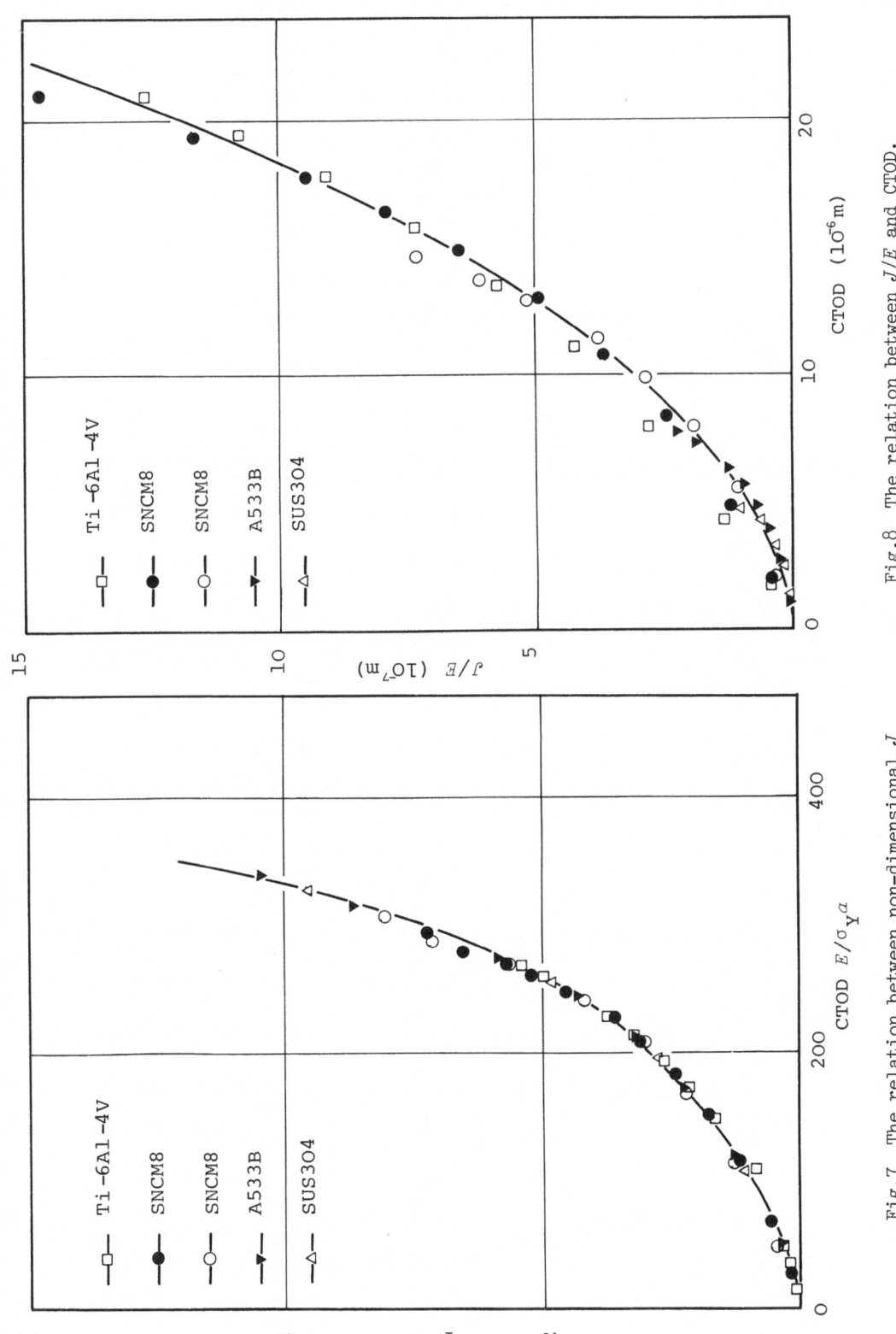

Fig.8 The relation between J/E and CTOD.

Fig.7 The relation between non-dimensional J and non-dimensional CTOD.

III. J-INTEGRAL – EXPERIMENTAL TECHNIQUES

THE J-INTEGRAL EVALUATION OF THE CRACK TIP PLASTIC BLUNTING AND THE ELASTIC-PLASTIC FRACTURE

H. Kobayashi, H. Nakamura, K. Hirano and H. Nakazawa
Associate Professor, Graduate Student, Assistant and Professor
Department of Physical Engineering, Tokyo Institute of Technology
Ohokayama 2-12-1, Meguro-ku, Tokyo, Japan

ABSTRACT

A fractographically-derived, subcritical, stretched zone width (SZW) as a measure of the crack tip plastic blunting has been examined quantitatively, and a relation between the SZW and the J-integral has been clarified for various materials, ranging from linear elastic fracture mechanics (LEFM) to elastic-plastic fracture mechanics (EPFM) regimes. Based upon the relation between SZW and J, a new elastic-plastic fracture toughness J_{Ic} test method has been proposed, and its validity and applicability have been confirmed.

A FRACTOGRAPHICALLY-DERIVED SUBCRITICAL STRETCHED ZONE WIDTH

Direct measurement of the crack tip opening displacement (CTOD) is generally difficult, and virtually impossible in a routine test. When a fatigue pre-cracked specimen is subjected to a monotonically increasing load, the crack tip blunts, and fracture initiates at the critical CTOD value ($CTOD_c$). As shown by Fig. 1, fractographic examination of this fracture surface reveals that a featureless transition region exists between the fatigue pre-crack and overload fracture regions. This region, referred to as a stretched zone, has been interpreted as the result of the crack tip plastic blunting, and the critical stretched zone width (SZW_c) obtained from a specimen broken by overload has been correlated with the $CTOD_c$ value as well as a plane-strain fracture toughness value (K_{Ic}) [1, 2].

As present authors have indicated in previous works [3], however, a subcritical stretched zone appears as the loading progresses, and its width (SZW) reaches a critical value (SZW_c) when fracture initiates. Moreover, SZW_c is not affected by any following stable crack growth. Figure 1 shows that it is possible to detect the subcritical stretched zone by subsequent fatigue cracking. A representative fractograph, showing the general features of the subcritical stretched zone between the fatigue pre-crack and subsequent fatigue marked regions, is shown in Fig. 2.

Figure 3 presents a schematic section profile of the subcritical stretched zone. Fractographic examinations have been conducted mainly with the help of a transmission electron microscope (TEM), using a two-stage plastic-carbon replica with chromium shadowing. A scanning electron microscope (SEM) has occasionally been used with it. Figure 3 shows that SZW measured by TEM gives a crack extension value (Δa), i.e.,

$$\Delta a = SZW . \tag{1}$$

However, it appears that SZW measured by SEM generally takes a maximum value (SZW_{max}) when the macroscopic fracture surface makes some oblique angle ($\beta = 35 \sim 45°$) with the electron beam, rather than being normal to the beam. Therefore, the crack tip blunting angle is assumed to be $2\beta = 70 \sim 90°$, as shown in Fig. 3. Then,

$$\Delta a = \cos\beta \; SZW_{max} \simeq (0.8 \sim 0.7) \; SZW_{max} , \tag{2}$$

$$CTOD = 2 \sin\beta \; SZW_{max} \simeq (1.1 \sim 1.4) \; SZW_{max} , \tag{3}$$

assuming that $\beta = 35 \sim 45°$.

A PROPOSED ELASTIC-PLASTIC FRACTURE TOUGHNESS J_{Ic} TEST METHOD

Recently, an elastic-plastic fracture toughness (J_{Ic}) criterion and its test method have been developed by Begley and Landes [4] and the ASTM task Group E24.01.09 [5]. In their test method, attention has been directed mainly to the processes of stable fracture. An assumed J versus Δa blunting line

$$\Delta a = CTOD/2 = J/2\sigma_{flow} \tag{4}$$

has been proposed, where σ_{flow} is the average of the yield stress (σ_{ys}, offset = 0.2 %) in tension and the tensile strength (σ_B). Furthermore, an experimentally determined J versus Δa resistance curve (R-curve) has been utilized, as shown in Fig. 4, where Δa is a total crack extension value after the onset of stable crack growth. The J_{Ic} value has been defined as a J value at the intersection of the R-curve and the blunting line.

The J_{Ic} value can also be obtained by examining the relation between SZW and J before and after the onset of stable crack growth. In previous works [3], the authors have proposed a new J_{Ic} test method based on this concept. The procedure of the proposed J_{Ic} test method is summarized as follows.

1) A bend type specimen is needed. This includes compact specimens or 3-point bend bars.

2) The specimen should be deeply notched with a/W = 0.6 and pre-cracked in fatigue (part I in Fig. 5(a)). The cyclic load during fatigue pre-cracking is set at 30 $\sim$ 40 % of the load, corresponding to the anticipated J value.

3) Statically load several specimens to different displacement values where the stable crack growth does not initiate (part II in Fig. 5(a)).

4) Partially unload each specimen and mark the front of the subcritical stretched zone formed at part II by subsequent fatigue cracking (part III in Fig. 5(a)). Then, pull the specimen apart by overload (part IV in Fig. 5(a)).

5) Measure SZW formed in part II. The fractographic examination should be performed at the midthickness of specimens where plane-strain conditions exist. Calculate a J value from the load versus load line displacement record (Fig. 5(b)) using the estimation procedure of Rice [6] or Merkle [7]. Plot SZW versus J and draw a best fit blunting line to the points before the onset of stable crack growth (Fig. 5(c)).

6) Pull a virgin specimen apart by overload and measure SZW_c.

7) Mark J_{Ic} as a J value at the intersection of the line $SZW = SZW_c$ and the extrapolation of the J versus SZW blunting line (Fig. 5(c)).

A RELATION BETWEEN THE SUBCRITICAL STRETCHED ZONE WIDTH AND THE J-INTEGRAL

More recently, we in Japan (Kobayashi, Iwadate, Miyamoto, Hijikata, Hashimoto, Ohtsuka, Kishi, Ohji, and Yokomaku) have carried out the J_{Ic} tests proposed by the authors. The objective of this chapter is to summarize Japanese J_{Ic} test results and to clarify the relation between the measures of the crack tip plastic blunting and the J-integral [8, 9].

Figure 6 presents all the results on a log-log plot of SZW obtained for the various materials as a function of J/E, where E is Young's modulus. Figure 6 shows that the change of E from 70.6 to 206 GPa has little influence on the relation between SZW and J/E, and there are no influences of σ_{flow} and the test temperature on the relation between SZW and J/E. Therefore, it is evident that the relation between CTOD and J does not obey Eq. (4). If we assume a relationship of the form

$$SZW = C_1 \, J/E \,, \tag{5}$$

the mean value becomes $\overline{C}_1 = 89$ and the deviation is $54.7 \leq C_1 \leq 143$ for 90 % confidence limits.

It seems rather strange that the SZW formed by crack tip plastic blunting does not depend on σ_{flow} but on E. However, a relation like this is well known in fatigue. In stage 2b, plane-strain, and LEFM fatigue crack growth, growth rates (da/dN) are in excellent agreement with striation spacings (S) and obey

$$da/dN = S = C_2 \, (1 - \nu^2) \, (\Delta K/E)^2 \,, \tag{6}$$

where the stress ratio (R) is about 0 and ΔK is the stress intensity factor range, ν is Poisson's ratio, and C_2 is about 9 on the assumption that $\nu = 1/3$ [10, 11]. Since the relation between J and the stress intensity factor (K_I) is given by

$$J = (1 - \nu^2) \, K_I^2/E \tag{7}$$

for the LEFM case, the following equation can also stand for the LEFM case.

$$S = C_2 \ J/E \tag{8}$$

For comparison, the striation spacings (S), when R is about 0 for various materials [11], have been plotted in Fig. 6 in relation to J/E, where all the data on S satisfy small scale yielding conditions. Figure 6 shows that the change of E has also little influence on the relation between S and J/E, and there is no influence of σ_{flow} on the relation between S and J/E. If we assume the relationship of Eq. (8), the mean value becomes $\overline{C}_2 = 9.4$ and the deviation is $6.0 \leq C_2 \leq 14.8$ for 90 % confidence limits.

That is to say, despite the difference between the monotonic load and the fatigue load, the widths of the subcritical stretched zone and of the striation formed by the same micromechanism, glide plane decohesion, are characterized by the same fracture mechanics parameter. However, as Fig. 6 illustrates, the widths of these two for the same value of J/E are as different as approximately one order of magnitude, $C_1/C_2 = 89/9.4 = 9.5$. The reason for the smaller width of the striation should be attributed to crack closure under the fatigue load [12].

CORRESPONDENCE OF THE SUBCRITICAL STRETCHED ZONE TO THE GIANT STRIATION

As stated above, the monotonic load (Fig. 7(a)) forms the subcritical stretched zone, and the fatigue load (Fig. 7(b)) forms the striation. On the other hand, it has been shown that variations in the fatigue load, such as a single peak overload (Fig. 7(c)), form giant striations. These can be considered the subcritical stretched zone [13]. Figure 8 presents the results on a log-log plot of the giant striation spacings (SZW) obtained for several materials as a function of $\Delta J_{max}/\Delta J_0$, where ΔJ_{max} and ΔJ_0 are the J-integral corresponding to the single peak overload and the fatigue load, respectively [14]. A curve in Fig. 8 represents the following equation

$$SZW = 9.4 \ \Delta J_0/E + 89 \ (\Delta J_{max} - \Delta J_0)/E \tag{9}$$

on the assumption that Eqs. (5) and (8) can stand. Equation (9) agrees approximately with the data. Then, for the monotonic load case (Fig. 7(a)), it can be supposed that the subcritical stretched zone formation begins by obeying Eq. (8) and shifts gradually to obeying Eq. (5) as the monotonic load exceeds the cyclic load during fatigue pre-cracking. In fact, an intersection of the J versus SZW blunting line with the J-axis corresponding to the subcritical stretched zone formation exists, as shown in Fig. 5(c), and the deviation from Eq. (5) becomes larger as the J value becomes smaller.

EVALUATION OF THE ELASTIC-PLASTIC FRACTURE TOUGHNESS J_{Ic} IN A533B-1

The relation between SZW and J in A533B-1 (quenching and tempering) is shown in Fig. 9 [15]. Open and solid symbols in Fig. 9 represent data before and after the onset of stable crack growth, respectively. The results indicate that all the data before SZW reaches SZW_c fall on a J versus SZW blunting line with little variation, in spite of the difference in specimen thickness (B = 19 and 1.5 mm) and materials (base metal, weldment and HAZ). The J versus SZW blunting line obtained here for A533B-1 (Q & T) agrees with Eq. (5). Also, there is very little variation in SZW_c, despite the difference in specimen thickness and materials. A horizontal line in Fig. 9 represents the mean value of these SZW_c's, and its value is 77 μm. The J_{Ic} value can be marked as the J value at the intersection of the line SZW = SZW_c and the blunting line. It is 0.172 MN/m.

The corresponding plane-strain fracture toughness ($K_{Ic}(J)$) can be calculated from J_{Ic} according to Eq. (7). It is 200 MPa√m. On the other hand, the valid K_{Ic} value of 160 MPa√m has been obtained according to the ASTM procedure E399 using a compact specimen with a full-thickness of 305 mm [16]. The $K_{Ic}(J)$ value obtained here agrees approximately with the K_{Ic} value, although a little difference exists between them due to possible variation in material properties. That is to say, on A533B-1, the proposed J_{Ic} test requires specimens with thickness of 1/15 or less of the ASTM procedure E399 valid K_{Ic} specimen, while the K_{Ic} test requires specimens with thickness nearly as large as pressure vessel. Furthermore, it should be noted that the $K_{Ic}(J)$ values of weldment and HAZ specimens are not low, compared with that of a base metal specimen.

Figure 10 shows the experimentally determined blunting line and R-curve in A533B-1 (normalizing and tempering). The blunting line of A533B-1 (N & T) agrees with that of A533B-1 (Q & T), shown in Fig. 9. However, the SZW_c value of 37 μm of A533B-1 (N & T) is very small compared with that of A533B-1 (Q & T). The following are the J_{Ic} and $K_{Ic}(J)$ values of A533B-1 (N & T): J_{Ic} = 0.0853 MN/m and $K_{Ic}(J)$ = 140 MPa√m. It should be noted that A533B-1, quenching and tempering, shows superior fracture toughness, compared with that of normalizing and tempering.

Figure 10 shows a large difference between the experimentally determined blunting line, Eq. (5), and the proposed one, Eq. (4). Furthermore, it is practically impossible to determine the R-curve according to the ASTM task Group procedure by using only those data points that fall within minimum crack extension (0.15 mm offset) and maximum crack extension (1.5 mm offset) lines. Also, the Δa values are taken as maximum values at the midthickness of specimens, although the ASTM task Group procedure recomends that the average Δa value should be used. Figure 10 clearly shows that the ASTM task Group procedure overestimates the J_{Ic} value.

Validity and applicability of the proposed J_{Ic} test method have also been confirmed in other materials: 4340, 10B35, and Ti-6Al-4V out of high strength materials, 304 and 5083-O out of low strength materials [8].

APPLICATION TO THE ASSESSMENT OF THE INTEGRITY OF THE NUCLEAR PRESSURE VESSEL

Recently, the ASME Boiler and Pressure Vessel Code, Section III, Appendix G and Section XI, Appendix A have been proposed [16]. In these appendices, LEFM is used to assess the effects of flaws on the integrity of nuclear power plant components. The K_{Ic} value in Section XI was determined by drawing a lower-bound curve to all available data and the following equation was used to represent K_{Ic}:

$$K_{Ic} = 36.5 + 3.084 \exp[0.036 (T - RT_{NDT} + 55.6)] \tag{10}$$

where K_{Ic} is in MPa√m and T and RT_{NDT} are in K. The K_{Ic} curve is terminated at 220 MPa√m (200 ksi√in). The implication is made that K_{Ic} remains constant at this value up to the pressure vessel operating temperatures. Under this upper shelf condition, LEFM is not applicable: the validity criterion of the ASTM procedure E399 restricts the specimen size that can be used to define K_{Ic}.

Figure 11 shows the results of the J_{Ic} test in A533B-1 (N & T) over a temperature range from room temperature to 673 K [17]. Figure 11 shows that upper shelf elastic-plastic fracture toughness exists clearly in this temperature region and its value is $K_{Ic}(J) = 140 \sim 120$ MPa√m. So that, the elastic-plastic fracture toughness $K_{Ic}(J)$ value obtained in A533B-1 (Q & T) at room temperature can be assumed as the value of the upper shelf. This assumed upper shelf fracture toughness value is very small compared with the terminated value of $K_{Ic} = 220$ MPa√m in the K_{Ic} curve of Section XI. That is to say, the K_{Ic} curve of Section XI does not necessarily provide a conservative estimation.

REFERENCES

(1) A. J. Brothers, M. Hill, M. T. Parker, W. A. Spitzig, W. Wiebe, and U. E. Wolff, "Correlation of Fracture Toughness, K_{Ic}, with Fractographically Derived Plastic Stretched Zone Width", ASTM STP 493, (1971), pp. 3-19.

(2) H. Kobayashi, H. Nakazawa, and A. Nakajima, "Correlation of Plane-Strain and Plane-Stress Fracture Toughness with Fractographically Derived Stretched Zone Width in High-Strength Steel", Strength and Structure of Solid Materials, Proc. Joint Japan-U.S.A. Seminar, Noordhoff, (1976), pp. 115-128.

(3) H. Kobayashi, K. Hirano, H. Nakamura, and H. Nakazawa, "A Fractographic Study on Evaluation of Fracture Toughness", Proc. 4th Intl. Conf. on Fracture, Waterloo Univ., Vol. 3, (1977), pp. 583-592.

(4) J. A. Begley, and J. D. Landes, "The J-Integral as a Fracture Criterion", ASTM STP 514, (1972), pp. 1-23.

(5) G. A. Clarke, W. R. Andrews, J. A. Begley, J. K. Donald, G. T. Embley, J. D. Landes, D. E. McCabe, and J. H. Underwood, "A Procedure for the De-

termination of Ductile Fracture Toughness Values Using J-Integral Techniques", J. of Testing and Evaluation, 7-1, (1979), pp. 49-56.

(6) J. R. Rice, P. C. Paris, and J. G. Merkle, "Some Further Results of J-Integral Analysis and Estimates", ASTM STP 536, (1973), pp. 231-245.

(7) J. G. Merkle, and H. T. Corten, "A J-Integral Analysis for the Compact Specimen, Considering Axial Force as well as Bending Effects", Trans. ASME, Ser. J, 96, (1974), pp. 286-292.

(8) H. Kobayashi, H. Nakamura, and H. Nakazawa, "The J-Integral Evaluation of Stretched Zone Width and Its Application to Elastic-Plastic Fracture Toughness Tests", Recent Researches on Mechanical Behavior of Solids, Univ. of Tokyo Press, (1979), pp. 341-357.

(9) H. Kobayashi, H. Nakamura, and H. Nakazawa, "A Relation between the Crack Tip Plastic Blunting and the J-Integral", Proc. 3rd Intl. Conf. on Mechanical Behaviour of Materials, Pergamon Press, Vol. 3, (1979), pp. 529-538.

(10) R. C. Bates, W. G. Clark, Jr., and D. M. Moon, "Correlation of Fractographic Features with Fracture Mechanics Data", ASTM STP 453, (1969), pp. 192-214.

(11) H. Kobayashi, "The Influence of Microstructure of Material and Fracture Mechanisms on Fatigue Crack Growth Resistance", J. of the Japan Society of Mechanical Engineers, 80-703, (June, 1977), pp. 492-497, (in Japanese).

(12) W. Elber, "The Significance of Fatigue Crack Closure", ASTM STP 486, (1971), pp. 230-242.

(13) R. W. Hertzberg, "On the Relationship between Fatigue Striation Spacings and Stretch Zone Width", Intl. J. of Fracture, 15-2, (April, 1979), pp. R69-R72.

(14) H. Kobayashi, H. Nakamura, A. Hirano, and H. Nakazawa, "A Fractographic Approach to the Influence of a Single Peak Overload on Fatigue Crack Growth", Fatigue '81, Univ. of Warwick, (1981), to be presented.

(15) H. Kobayashi, H. Nakamura, and H. Nakazawa, "The Elastic-Plastic Fracture Toughness and the Fatigue Crack Growth Resistance in Base Metals and Weldment of A533B-1 and 304 Steels", Proc. 4th Intl. Conf. on Pressure Vessel Technology, IME, (1980), in press.

(16) ASME Boiler and Pressure Vessel Code, Section III, "Nuclear Power Plant Components", and Section XI, "Rules for Inservice Inspection of Nuclear Power Plant Components", Division 1, ASME, 1977 Edition.

(17) K. Hirano, H. Kobayashi, and H. Nakazawa, "Elastic-Plastic Fracture Mechanics Study of Thermal Shock Cracking", Proc. 3rd Intl. Conf. on Mechanical Behaviour of Materials, Pergamon Press, Vol. 3, (1979), pp. 457-467.

118

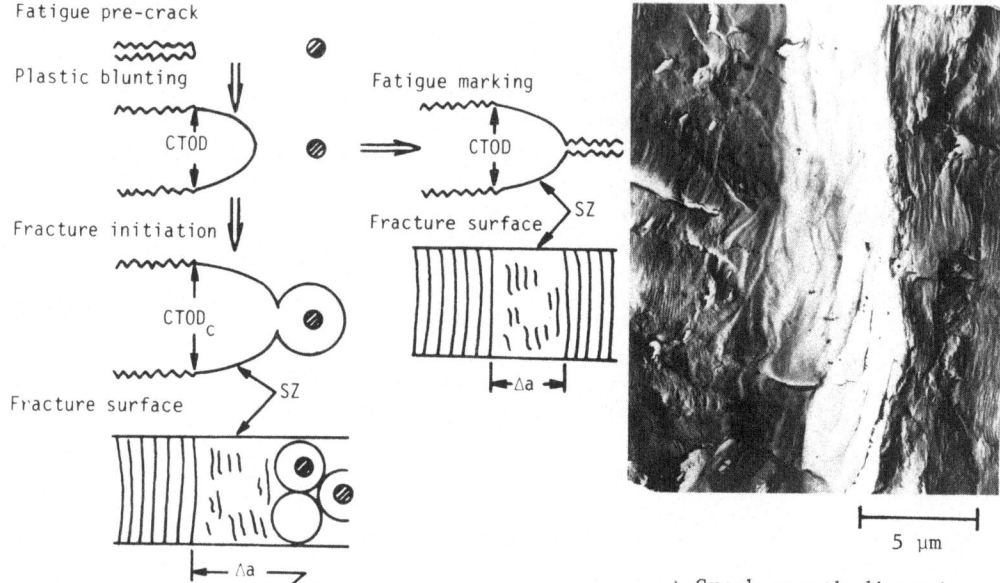

Fatigue pre-crack

Plastic blunting

CTOD

Fracture initiation

CTOD$_c$

Fracture surface

SZ

Δa

Fatigue marking

CTOD

Fracture surface

SZ

Δa

Fig. 1   Crack tip plastic blunting
and its detecting method.

→ Crack growth direction

Fig. 2   Fractograph showing
subcritical stretched
zone (Ti-6Al-4V).

5 μm

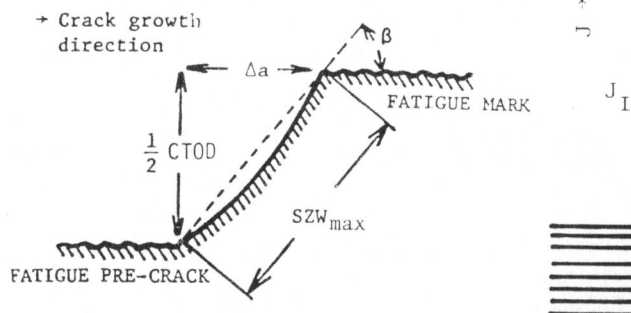

→ Crack growth
   direction

β

Δa

FATIGUE MARK

$\frac{1}{2}$ CTOD

SZW$_{max}$

FATIGUE PRE-CRACK

Fig. 3   Schematic section profile
of subcritical stretched
zone.

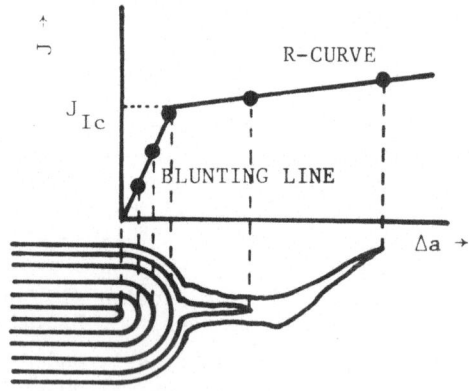

J

R-CURVE

J$_{Ic}$

BLUNTING LINE

Δa

Fig. 4   Blunting line and R-curve.

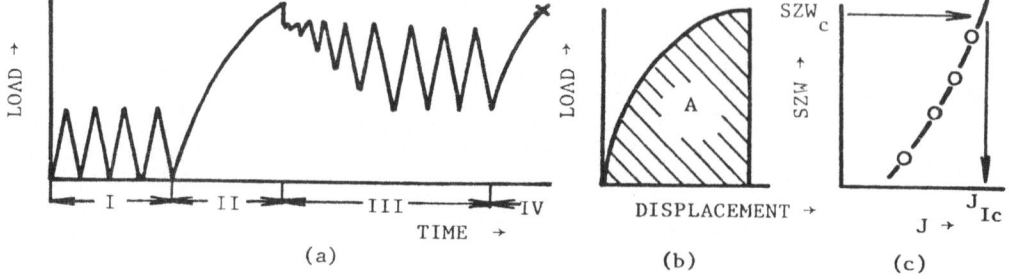

LOAD →

I   II   III   IV

TIME →

(a)

LOAD →

A

DISPLACEMENT →

(b)

SZW$_c$

SZW →

J →   J$_{Ic}$

(c)

Fig. 5   Procedure of proposed J$_{Ic}$ test method.

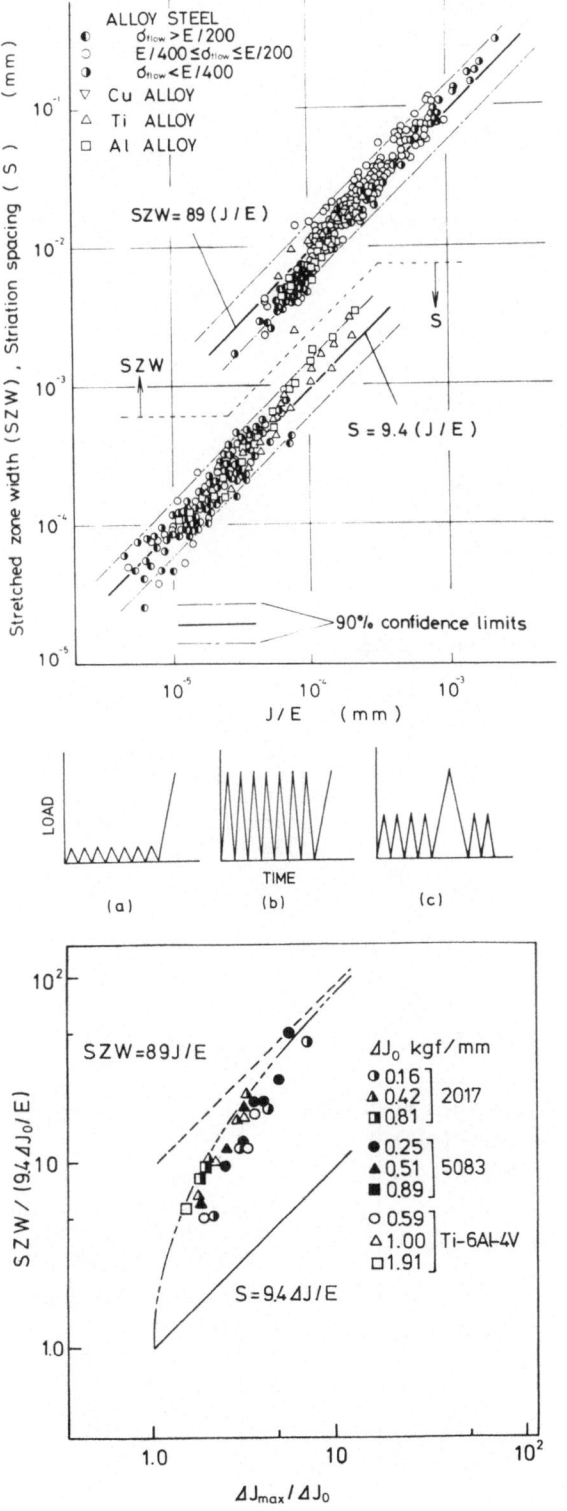

Fig. 6
Comparison between subcritical stretched zone widths, SZW, and striation spacings, S, as a function of J/E.

Fig. 7
Schematic illustration of cyclic loading histories for a given value of maximum load:
(a)   The monotonic load (>> the cyclic load during fatigue pre-cracking) forms the subcritical stretched zone.
(b)   The fatigue load forms the striation (S < SZW).
(c)   The single peak overload (> the fatigue load) forms the giant striation (S < the giant striation spacing < SZW).

Fig. 8
Influence of overload ratio, $\Delta J_{max}/\Delta J_0$, on the giant striation spacing, SZW.

120

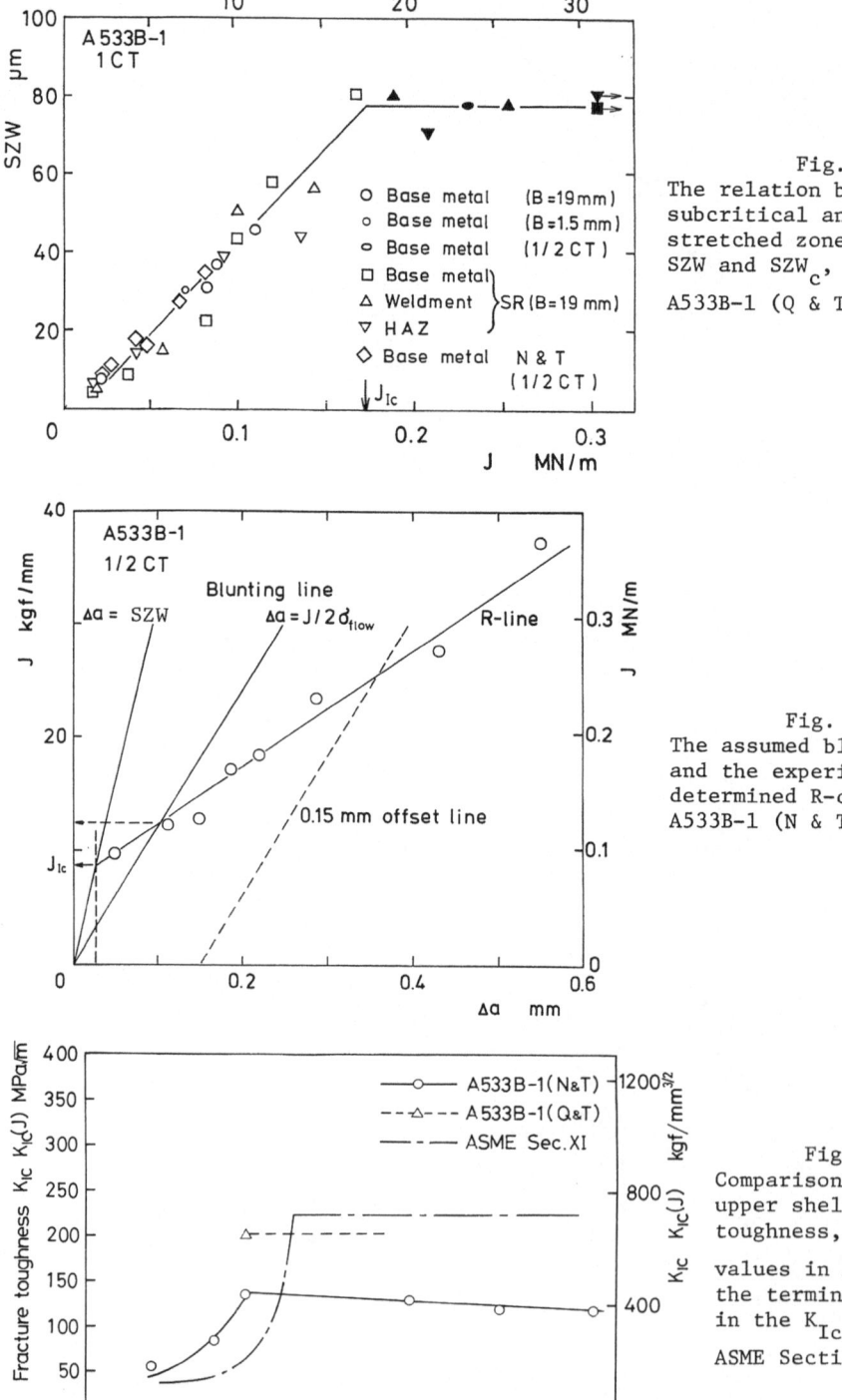

Fig. 9
The relation between the subcritical and critical stretched zone widths, SZW and SZW_c, in A533B-1 (Q & T).

Fig. 10
The assumed blunting line and the experimentally determined R-curve in A533B-1 (N & T).

Fig. 11
Comparison between the upper shelf fracture toughness, $K_{Ic}(J)$, values in A533B-1 and the terminated value in the K_{Ic} curve of ASME Section XI.

ELASTIC PLASTIC TOUGHNESS OF IRRADIATED STEELS
WITH THE SINGLE SPECIMEN COMPLIANCE TECHNIQUE

F. J. Loss
Naval Research Laboratory
Washington, D.C., 20375, USA

ABSTRACT

J-R curve trends of irradiated A533-B submerged arc weld metal containing a high copper impurity level have been characterized with the single specimen compliance technique. The R curve for these steels exhibits a power law behavior. This non-linearity precludes both an exact determination of J_{Ic} by the proposed ASTM multispecimen standard as well as the calculation of a single value of tearing modulus T in the region of J-controlled growth. A new experimental definition of J_{Ic} is proposed which accounts for the curvature in the R curve. The use of side grooves promotes a straight crack-front extension which results in T values which are less than those produced by smooth specimens. With side-grooved specimens irradiation has reduced T by a factor of 2.5 while a smaller reduction in J_{Ic} was observed under the same conditions. The phenomenon of an R curve interrupted by cleavage in the upper shelf region is also discussed.

INTRODUCTION

The beltline region of water reactor pressure vessels suffers a progressive degradation in fracture toughness with increasing neutron fluence. In addition, certain steels and weld deposits used in the beltline which contain a high copper impurity level also exhibit a high sensitivity to irradiation embrittlement. As a result, the initiation toughness K_{Ic} of these steels is projected to be below that which may be required to provide an acceptable margin for certain postulated accident conditions. Consequently, research studies are exploring the concept of tearing instability as defined by Paris and others [1] as a means to asssure structural reliability for cases in which crack initiation must be assumed. This concept hinges on the phenomenon of a J-R curve which exhibits an apparent increase in fracture toughness with crack extension within the upper shelf temperature regime.

Since space is limited for surveillance specimens in commercial power reactors, and in test reactors as well, it is necessary to explore the effects of irradiation on the basis of small specimens. In the upper shelf temperature region these specimens will behave in an elastic-plastic manner and care must be taken to assure that the trends described here are not masked by specimen size effects which may preclude application of the results to larger structures. The J integral offers a means to circumvent differences in gross plastic deformation between specimen and structural prototype through a geometry-independent characterization of the strain field near the crack tip. Therefore, our work has extended the J-integral application to irradiated structural steels in order to provide an experimental basis to assess the adequacy of the tearing instability approach for nuclear structures.

Because of the difficulty in obtaining and testing of irradiated steels, it is necessary to develop a single specimen method for R-curve measurement. In this regard investigations with the single specimen compliance (SSC) technique [2] are highlighted. This study describes the application of the SSC procedure to two A533-B submerged arc (S/A) weld deposits containing a high (0.35%) copper impurity level. These alloys are representative of those which are expected to exhibit a high embrittlement in service. The mechanical properties of these steels, including the Charpy V-notch (C_v) trends, are given in Refs. 3 and 4.

SINGLE SPECIMEN COMPLIANCE PROCEDURE

The SSC technique permits the development of the complete J-R curve from a single specimen. This method incorporates small unloadings (~10%) at various load-line displacements. These unload and reload segments are conducted under elastic conditions and thus can be used to infer crack length changes from the corresponding changes in specimen compliance (EB δ/P), where E is Young's modulus, B is the specimen thickness, and P and δ are the load and load-line deflection, respectively. A 25.4-mm-thick compact specimen (1T-CT) was selected for these studies because it can be irradiated easily and yet its relatively large size provides sufficient mechanical constraint to infer high levels of plane strain initiation toughness when the specimen itself exhibits elastic-plastic behavior. For example, the ASTM proposed criterion of b>25 J_{Ic}/σ_f will permit measurement of J_{Ic} close to 500 kJ/m^2 for irradiated material where b and σ_f are the unbroken ligament and flow stress, respectively.

The SSC method as developed by NRL is described in Ref. 3-4. A CT specimen conforming to ASTM E-399 geometry is employed. However, the notch region has been modified to permit the mounting of razor knife edges for the measurement of load-line displacement with a clip gage. In addition, a crack length-to-width ratio (a/W) of 0.6 was used. The magnitude of J is computed from the relationship:

$$J = \frac{1+\alpha}{1+\alpha^2} \frac{2A}{Bb_o} \tag{1}$$

where A is the specimen energy absorbed based on total deflection, b_o is the original unbroken ligament and the term $(1+\alpha)/(1+\alpha^2)$ is a modified Merkle-Corten correction

[5] to account for the tension component of the loading. In addition, the method incorporates a correction for specimen rotation which is required when the compliance slopes must be determined to a high precision. Another correction is required to account for crack extension as described by Ernst [6]. However, this correction is difficult to apply and the change in J is not large for the steels discussed here. Consequently this correction has been omitted.

Application of this method with 1T-CT specimens has produced a typical accuracy of ±0.2 mm in predicted crack extension vs that measured optically, provided the crack front is straight and the steel is relatively free from metallurgical homogeneity. The scatter in predicted crack extension increments on an individual R curve is only ±0.08 mm. Thus, the SSC technique presents a highly reliable and accurate means for R-curve determination.

J_{Ic} MEASUREMENT PROCEDURE

Since a standard procedure for the measurement of J_{Ic} with a single specimen does not exist, an attempt was made to apply the methods described by the ASTM multispecimen procedure for J_{Ic}. A typical R curve developed with the SSC procedure is illustrated in Fig. 1. Note that the R curve is nonlinear. Evidence of this curvature has also been shown in other investigations [2,7,8]. On the basis of the inherent accuracy of the SSC technique and the supporting evidence of the other investigations, it is clear that the R curve exhibits a power law behavior for the types of steels used in nuclear vessel construction. When defining the R curve between the dashed lines in Fig. 1 using only four specimens, as with the proposed ASTM multiple specimen technique, insufficient data are available with which to assess this curvature. One must then approximate the R curve by a least squares fit of the data for purposes of J_{Ic} determination. However, application of this procedure will result in an average value of tearing modulus T, defined as $(dJ/da)(E/\sigma_f^2)$, which may not be appropriate for structural application.

By representing the R curve as a power function, it is impractical to define J_{Ic} by the intersection of this curve with the blunting line ($\Delta a = J/2 \sigma_f$) as required by the proposed ASTM J_{Ic} standard. In other words, a J_{Ic} defined this way could be near zero as illustrated in Fig. 1. Consequently, NRL investigators [9] have evolved an alternative experimental definition for J_{Ic}. The logic behind the new procedure is that J_{Ic} should be determined at a small (non-zero) crack extension to preclude the difficult task of defining the point of crack initiation. This approach is analogous to the definition of yield stress vis a vis the proportional limit. The crack extension at the J_{Ic} point has been chosen (a) to permit a small amount of real crack extension in terms of crack opening displacement (COD), and (b) to require a minimum (0.15-mm) crack extension in order to account for the experimental scatter in the R-curve data. The permissible crack extension is defined in terms of COD so as to be physically reasonable for different

124

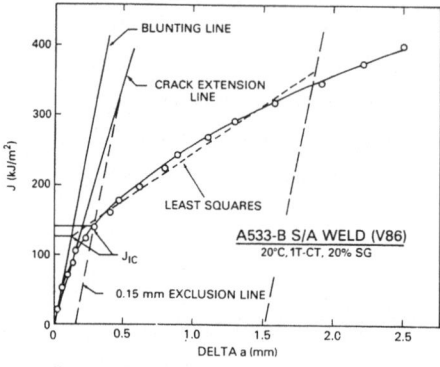

Figure 1. Nonlinear R curve described with SSC approach illustrating alternative definition of J_{Ic}

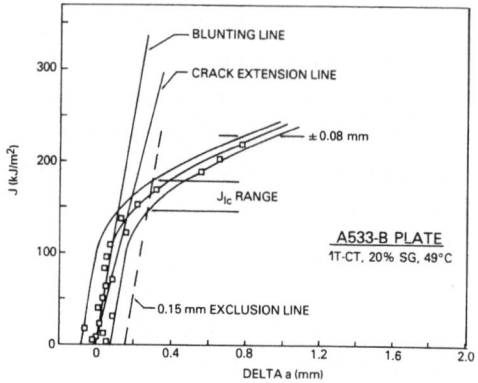

Figure 2. J_{Ic} definition illustrating the significance of data scatter

levels of toughness. For this purpose Tada [10] has suggested a criterion of $3J/4\sigma_f$. The minimum crack extension is defined by the 0.15-mm exclusion line and is consistent with the approach employed in the proposed ASTM multispecimen procedure for J_{Ic}.

As shown in Fig. 1, J_{Ic} is taken as that value which corresponds to the intersection of a smoothly drawn R curve through the data with either the crack extension or 0.15 mm exclusion lines, whichever produces the higher value. The data have also been represented with a power law of the form $J = C\Delta a^n$ where C and n are chosen to optimize the curve fit. For brittle (cleavage) failures at crack extensions less than the exclusion line, J_{Ic} is taken as the J level at failure. For the low toughness steels described here, J_{Ic} will be governed by the exclusion line; it can be shown from Fig. 1 that the latter will define J_{Ic} for J levels below 0.6 σ_f or approximately 330 kJ/m^2 for reactor vessel steels. As compared with the ASTM proposed multispecimen procedure, the alternative method proposed here can yield a comparable value of J_{Ic} (see Fig. 1) so that the two methods are not incompatible. Of greater importance is the fact that a power law description of the R curve permits assessment of the variation of T with crack extension.

Figure 2 illustrates the R curve for an A533-B steel specimen which exhibited a small crack extension beyond the 0.15-mm exclusion line before failure in the cleavage mode. This expanded plot highlights the ±0.08-mm scatter which can be produced with 1T-CT specimens and demonstrates the need for incorporation of a finite crack extension (e.g., 0.15-mm exclusion line) in the J_{Ic} measurement point. (This figure is typical of the largest scatter which has been observed with the current experimental technique.) The data scatter results in a J_{Ic} variation of approximately ±10 percent for the example shown.

EFFECT OF SIDE GROOVES

The crack-front extension typically exhibits a tunneled shape (Fig. 3) which results from the lack of mechanical constraint at the free surfaces. The latter permits plastic

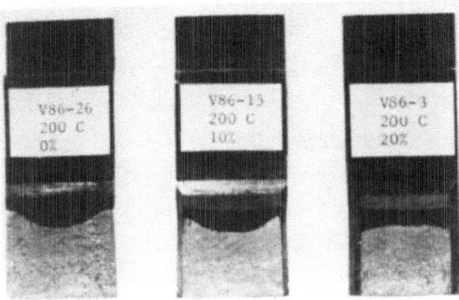

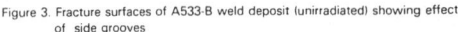

Figure 3. Fracture surfaces of A533-B weld deposit (unirradiated) showing effect of side grooves

Figure 4. Fracture surfaces of A533-B weld deposit (irradiated) showing effect of side grooves

deformation at these locations at the expense of crack extension. There is disagreement as to whether the tunneling is representative of the crack-front behavior of large cracks in a structure. Neverthless, this phenomenon does result in a three-dimensional effect which may not be properly accounted for by the two-dimensonal form of the J integral. It has been shown that side grooves in the specimens will eliminate the tunneling [4]. In Fig. 3, for example, 20 percent side grooves have completely eliminated the tunneling and have even resulted in a slight overcorrection (reverse tunneling). On the other hand, 10 percent side grooves have had little effect in straightening the crack front. In contrast to this behavior, 10 percent side grooves in irradiated weld deposit (Fig. 4) have resulted in a straight crack front. Apparently, a lesser degree of side grooving is required with the irradiated material because of its higher flow stress.

With the SSC technique a severe tunneling of the crack front will result in an underprediction of the crack extension determined optically [4]. This inconsistency between the crack length predictions vs optical measurements has been confirmed experimentally using specimens having a machined-tunneled shape (Fig. 5). This observed difference is reasonable in that the theoretical compliance relationship [11] used with the SSC technique to predict crack extension was derived on the basis of a straight crack-front extension. Once a straight crack extension is assured through the use of side-grooved specimens, the SSC method can predict the optically-measured crack front to a high precision as stated earlier. With side-grooved specimens the net thickness B_N is used in Eq. (1) to compute J. This is an approximation since an exact relationship for J does not exist for this case.

Figure 6 compares the J-R curves produced with the specimens illustrated in Fig. 3. Those specimens having 0 and 10 percent side grooves, which resulted in a tunneled crack, also produced identical R curves. However, the specimen having a straight crack front (20 percent side grooves) resulted in a distinctly lower R curve. A portion of this difference in the R curves is due to the crack-length prediction error associated with a tunneled crack front. However, this factor is insufficient to fully account for the

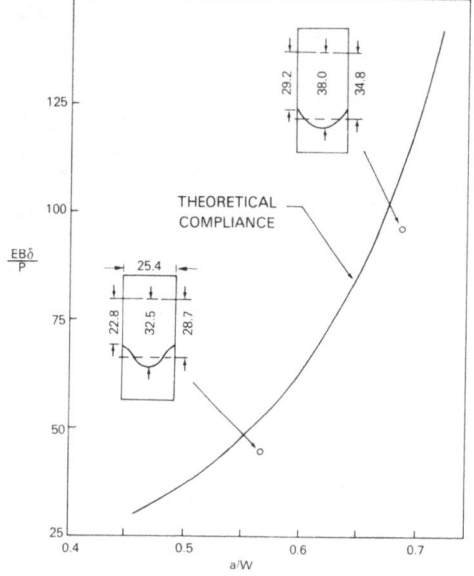

Figure 5. Illustration of departure from theoretical compliance when crack front is tunneled (dimensions in mm)

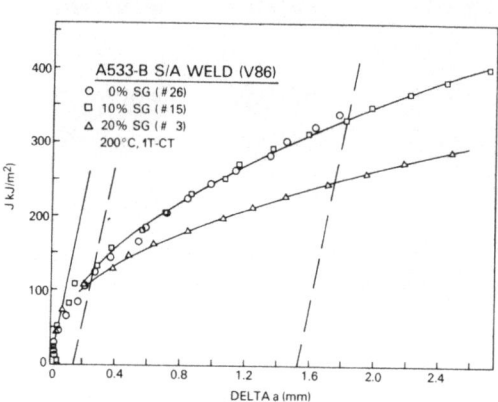

Figure 6. Comparisons of R curves as a function of side grooves for unirradiated specimens

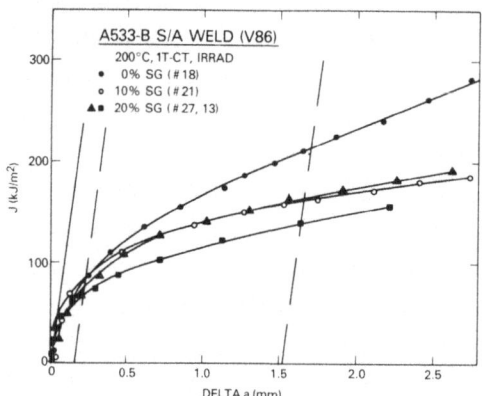

Figure 7. Comparisons of R curves as a function of side grooves for irradiated specimens

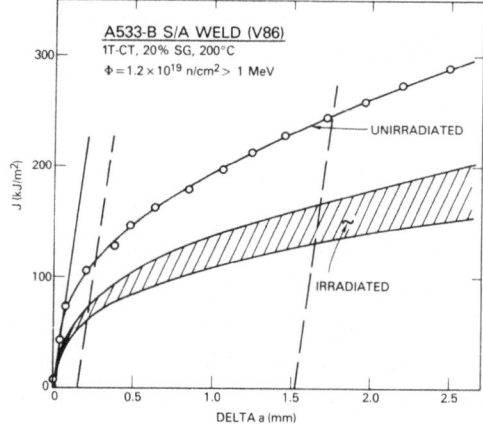

Figure 8. Comparison of R-curve trends in unirradiated and irradiated conditions

observed reduction in R-curve level. These results suggest that the R curve produced with the straight crack front may represent the trend expected with thicker sizes such as those found in nuclear pressure vessels. Consequently, the program emphasis was placed on specimens having 20 percent side grooves.

Figure 7 illustrates a similar comparison on the effect of side grooves for the irradiated specimens shown in Fig. 4. The R curves produced with a straight crack front, that is, with the specimens having 10 and 20 percent side grooves, also are below the R curve for the specimen having no side grooves.

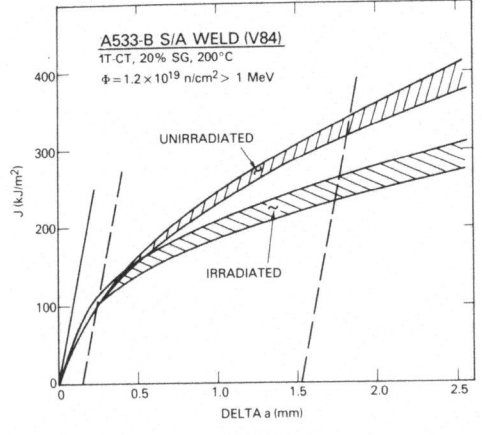

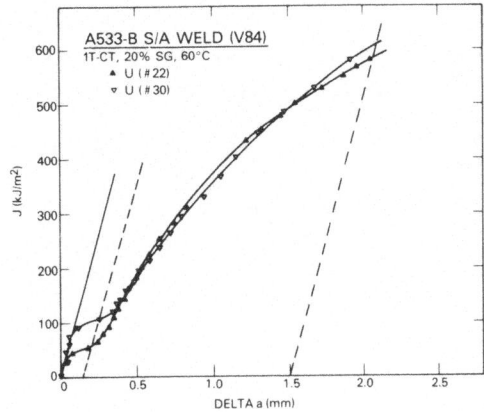

Figure 9. Comparison of R-curve trends in unirradiated and irradiated conditions

Figure 10. Sigmoidal-shaped R curves believed to result from material inhomogeneity

R-CURVE COMPARISONS

Figure 8 illustrates the effect of irradiation on the R curves for one A533-B submerged arc weld deposit (Code V86). A theoretical analysis by Shih [12] suggests that the region of J dominance will be maintained for these steels provided the crack extension is less than ~ 6 percent of the unbroken ligament. The latter condition is approximated by the dashed line at 1.5-mm crack extension. Because of the nonlinear R-curve behavior, the tearing modulus exhibits a continuous variation in this J-dominated region. With the unirradiated test, for example, T varies by a factor of 2.6, i.e., from 128 to 49 in the region between the dashed lines; for the irradiated tests decribed by the shaded band, T varies by a factor of 6, i.e., from 89 to 15.

For comparison purposes, an average value of T for each R curve was computed from a linear regression fit to the data between the dashed lines. On this basis, T was reduced with irradiation by a factor of 2.5. On the other hand, J_{Ic} was reduced somewhat less by a factor of 1.7. A tabulation of these data is given in Ref. 4.

The results of a similar study of R-curve behavior, using another submerged arc weld (Code V84) having a higher C_v upper shelf energy, are illustrated in Fig. 9. In sharp contrast to the reduction in J_{Ic} described in Fig. 8, these data show that J_{Ic} is unaffected by irradiation. On the other hand, both welds exhibit a reduction in the average value of T by a factor of 2.5.

The apparent insensitivity in J_{Ic} with irradiaton for Weld V84 may be linked to the inhomogeneity of this steel. Figure 10 illustrates the variability in the R curves at a small crack extension. The latter does not affect the value of T but does result in a large change in J_{Ic}. Because of the demonstrated accuracy of the SSC technique it is possible to attribute this S-shaped R curve to metallurgical factors.

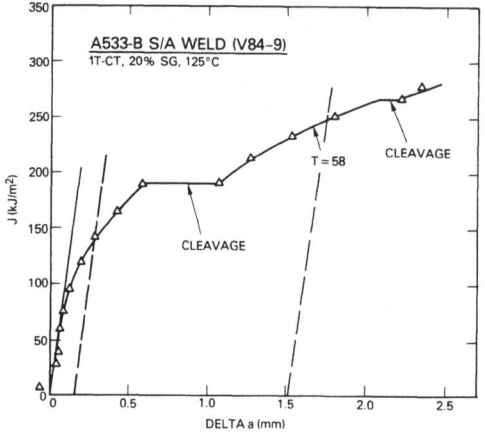

Figure 11. R curve illustrating cleavage instability followed by arrest and stable crack extension

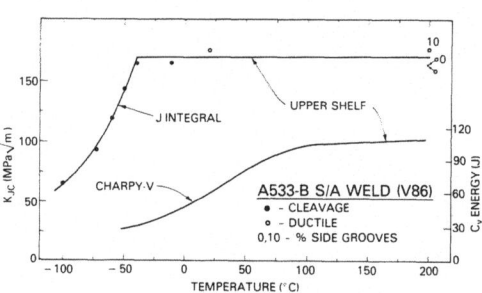

Figure 12. Comparison of J integral and C_v energy trends. Specimens had 20% side grooves unless otherwise indicated

Cleavage instability

Another type of R-curve variability deals with brittle (cleavage) fracture which is preceded by a small amount of ductile crack extension as illustrated in Fig. 11. The trend shown here provided sufficient stable crack extension to define J_{Ic}. As a consequence of this behavior it is possible to define an upper shelf behavior in terms of initiation toughness K_{Jc} under quasi-static loading (Fig. 12) even though this is followed by a brittle fracture. (In Fig. 12, K_{Jc} has been computed from the relationship $K_{Jc} = [E\ J_{Ic}/(1 - \nu^2)]^{1/2}$ where ν is Poisson's ratio.) Traditionally, the term "upper-shelf toughness" has been associated with the C_v test and defines a region of ductile fracture in which the specimen energy absorption is relatively independent of temperature. As illustrated in Fig. 12, the two upper shelf regions defined by K_{Jc} and C_v energy are displaced in temperature. This is a consequence of the strain-rate dependence of the brittle-ductile transition region for these low alloy steels.

An interruption of the R curve by cleavage has certain implications with respect to a structural application of the tearing instability concept. Specifically, it may be unconservative to apply this concept in that portion of the upper shelf region in which failure in a cleavage failure mode is possible. On the basis of the author's experience, the cleavage phenomenon will be avoided entirely within the upper shelf region as defined by a dynamic test, such as the Dynamic Tear test, and which subjects the material to a high mechanical constraint.

SUMMARY AND CONCLUSIONS

The concept of tearing instability is being investigated as a means to demonstrate the margin of safety against fracture for nuclear structures employing materials of low upper shelf toughness. With these materials structural integrity cannot always be predicated on an adequate margin against crack initiation. The present study has characterized the elastic plastic toughness of two reactor vessel steels with irradiation in terms of the J-R curve and the tearing modulus T. Knowledge of the material behavior in terms of these quantities is required to establish the adequacy of the tearing instability concept for this application.

Investigations with the single specimen compliance (SSC) technique for R-curve determination have shown this method to be a reliable and accurate means to characterize irradiated steels. Using this technique, studies with A533-B submerged arc weld deposit in the upper shelf temperature regime have shown that the R curve exhibits a power law behavior for reactor vessel steels. Realization of this fact has resulted in an alternative definition of J_{Ic} to that given by the ASTM proposed standard for J_{Ic}. While other studies have confirmed the nonlinear R-curve behavior, its significance with respect to tearing modulus has been highlighted by the present study. It is concluded that T is a function of crack extension and varies by a factor of two to six within the crack-extension region of J dominance. Recognition of the power law behavior may necessitate revisions in conceptual procedures for application of the tearing instability approach.

The characterization of conservative R-curve trends appears to necessitate a straight crack-front extension. The use of side grooves has been investigated to achieve this objective. It is shown that the R curves developed with a straight crack extension exhibit a lower slope, dJ/da, than that produced by specimens having a tunneled crack. While a 20 percent total side groove depth is sufficient to promote a straight crack extension for certain steels, the optimum depth of side grooves for other alloys may depend upon flow stress, relative crack depth, as well as product form.

With both of the steels investigated, irradiation to a fluence of 1.2×10^{19} $n/cm^2 > 1$ MeV has reduced average value of T by a factor of 2.5. On the other hand, the decrease in J_{Ic} was less, with one steel showing a factor of 1.7 reduction and the other steel exhibiting no effect of irradiation on J_{Ic}. Examples have been provided which suggest that the inhomogeneity in the steel can account for variability in the crack initiation toughness.

The experimental procedure for the measurement J_{Ic} developed here also provides an indicator of upper shelf toughness for specimens tested in a quasi-static mode. Within this upper shelf regime certain specimens will, nevertheless, exhibit a brittle fracture following a small amount of stable crack extension. This behavior is not unexpected since the upper shelf as defined by quasi-static tests is normally within the brittle-ductile transition region as shown by dynamic tests. Consequently, this phenomenon must be weighed carefully in structural application of the tearing instability concept.

ACKNOWLEDGMENT

The sponsorship of this research by the U. S. Nuclear Regulatory Commission, Division of Reactor Safety Research, is gratefully acknowledged. The author is also indebted to his colleagues, B. H. Menke and R. A. Gray, Jr., for the conduct of various phases of the experimental program.

REFERENCES

[1]   P. C. Paris, H. Tada, A. Zahoor, and H. Ernst, "The Theory of Instability of the Tearing Mode of Elastic-Plastic Crack Growth," ASTM STP 668, (Mar. 1979), pp. 5-36.

[2]   G. A. Clarke, W. R. Andrews, P. C. Paris, and D. W. Schmidt, "Single Specimen Tests for J_{Ic} Determination," ASTM STP 590, (1976), pp. 27–42.

[3]   F. J. Loss, Editor, "Structural Integrity of Water Reactor Pressure Boundary Components, Quarterly Progress Report for the Period April-June 1979," NUREG/CR-0943, NRL Memorandum Report 4064, (Sep. 28, 1979).

[4]   F. J. Loss, Editor, "Structural Integrity of Water Reactor Pressure Boundary Components, Annual Report, Fiscal Year 1979," NUREG/CR-1128, NRL Memorandum Report 4122, (Dec. 31, 1979).

[5]   G. A. Clarke and J. D. Landes, "Evaluation of J for the Compact Specimen," Scientific Paper 78-ID3-JINTF-P1, Westinghouse R&D Center, Pittsburgh, PA, (Jun. 12, 1978).

[6]   Hugo Ernst, P. C. Paris, Mark Rossow, J. W. Hutchinson, "Analysis of Load-Displacement Relationship to Determine J-R Curve and Tearing Instability Material Properties," ASTM STP 677, (1979), pp. 581-599.

[7]   W. R. Andrews and C. F. Shih, "Thickness and Side-Groove Effects on J- and δ-Resistance Curves for A533-B Steel at $93^{\circ}C$," ASTM STP 668, (1979), pp. 426-450.

[8]   A. D. Wilson, "Characterization of Plate Steel Quality Using Various Toughness Measurement Techniques," ASTM STP 668, (1979), pp. 469-492.

[9]   F. J. Loss, B. H. Menke, R. A. Gray, Jr., and J. R. Hawthorne, "J-R Curve Characterization of Irradiated Nuclear Pressure Vessel Steels," Proceedings of CSNI Specialists Meeting on Plastic Tearing Instability, NUREG/CP-0010, Nuclear Regulatory Commission, Washington, D.C., (Jan. 1980).

[10]   Tada, H., personal communication.

[11]   A. Saxena and S. J. Hudak, "Review and Extension of Compliance Information for Common Crack Growth Specimens," Intn. Journal of Fracture, Vol. 14 (1978), pp. 453-468.

[12]   C. F. Shih, "An Engineering Approach for Examining Crack Growth and Stability in Flawed Structures," Proceedings of CSNI Specialists Meeting on Plastic Tearing Instability, NUREG/CP-0010, Nuclear Regulatory Commission, Washington, D.C., (Jan. 1980).

A SIMPLE APPROXIMATION FOR THE ESTIMATION OF
J-INTEGRAL VALUE OF A SEMI-ELLIPTICAL SURFACE CRACK

T. ENDO*
M. SATO**
T. FUNADA**
M. TOMIMATSU**
Takasago Technical Institute, Mitsubishi Heavy Industries, LTD.
Arai Shinhama 2-1-1, Takasago, Hyogo, Japan

ABSTRACT

An attempt to estimate the J-integral of a semi-elliptical surface crack is discussed. Findings show that the J value obtained by combining the three-dimensional, elastic analysis with the two-dimensional, elastic-plastic analysis agreed with the J_{IC} value of the material, and the proposed method of J estimation is useful in practical application.

INTRODUCTION

In order to prevent unstable fractures in structures, an analysis considering cracks must be conducted. The linear elastic fracture mechanics is well established for linear elastic material behavior, and this concept is often used in practical applications. However, the application of this concept is limited to the case of the small-scale yielding condition. Several elastic-plastic fracture mechanics concepts, which can be applied to the large-scale yielding problem, have been proposed. They are the crack opening displacement concept, the J-integral concept, and so on. Among them, the J-integral proposed by Rice [1] seems to be the most available criterion because of its clear definition. Although the J-integral was originally defined in two-dimensional bodies, it is hoped to extend this concept to the region of the three-dimensional bodies. However, the three-dimensional, elastic-plastic analysis is very costly and laborious.

In this paper, a simple approximation procedure of the J value for a semi-elliptical surface crack without the three-dimensional, elastic-plastic analysis is discussed.

---

* Manager of Strength Research Laboratory
** Research engineer

PROPOSED J APPROXIMATION METHOD

Generally speaking, cracks which are contained in structures are mainly not of the through-thickness type, but of the elliptical or semi-elliptical type, which is necessary in the three-dimensional analysis. To date, various extensions of J-integral to the three-dimensional crack problem have been proposed [2 ~ 4], and they are mainly divided into two categories:

1. The area integral, which is calculated in the region of a surface surrounding all of a crack front [2]. This integral presents the average value of the crack tip singularity.

2. The line integral, defined in the plane normal to the crack front. The values of the integral vary along the crack front [3 ~ 4].

For the three-dimensional crack problem, the above integrals need the three-dimensional analysis. However, as with a semi-elliptical surface crack (which is one of the most popular types of crack), the three-dimensional, elastic-plastic analysis is not practical, because of the cost and the labor involved.

With this in mind, a simple method to approximate the J value of a semi-elliptical surface crack is proposed that combines the three-dimensional, elastic analysis and the two-dimensional, elastic-plastic analysis. The procedure follows.

A. Elastic analysis of a semi-elliptical surface crack

First, the authors conducted the three-dimensional elastic stress analysis using the finite element method. An estimated J value in the elastic condition was obtained from the stress intensity factor K_I, using the relation shown in Eqs. (1).

$$J = \frac{(1-\nu^2)K_I^2}{E} \tag{1}$$

Where E is Young's modulus, ν is Poison's ratio. K_I can be determined by the direct or the energy method.

B. Elastic-plastic analysis of single-edge crack

A single-edge cracked plate is substituted for a semi-elliptical surface crack at the deepest point, and the two-dimensional, elastic-plastic analysis is conducted for the model. J is calculated by using Eqs. (2) [1].

$$J = \int (wdy - T \cdot \frac{\partial u}{\partial x} \, dS) \tag{2}$$

C. J value of a semi-elliptical surface crack

The J value of a semi-elliptical surface crack in elastic-plastic condition is approximated as follows.

$$J_{3DP} = R \times J_{2DP} \tag{3}$$

Where

J$_{3DP}$: J value of a semi-elliptical surface crack in elastic-
plastic condition;

J$_{2DP}$: J value of a single-edge cracked plate in elastic-plastic
condition;

J$_{3DE}$: J value obtained by procedure A;

J$_{2DE}$: J value of a single edge cracked plate in elastic condition;

R   : The ratio of J$_{3DE}$ and J$_{2DE}$ in the same loading condition

EXAMPLE OF J-INTEGRAL APPROXIMATION

In this section, an estimation of J value was carried out in accordance
with the procedure mentioned previously. The model considered is a semi-
elliptical surface crack plate under uniform tensile stress (σ_∞), shown in
Fig. 1, where plate thickness t, crack depth a, and surface length 2c are set
at 50 mm, 12.5 mm, and 75 mm, respectively (aspect ratio of crack a/2c=0.167).

First, the three-dimensional, elastic stress analysis was carried out
using the finite element method (the Program MARC-CDC). The mesh division of
the model is shown in Fig. 2. The figure shows only a quarter of the model,
due to the symmetries of the X and Z axis. It contains 68 twenty-node isoparametric brick elements (Type 21 in MARC library) and 458 nodes. σ_Z stress
distribution in the plane of the crack versus the distance from the crack front
is shown in Fig. 3. In the vicinity of the crack front, the slope of stress is
nearly equal to −0.5 on a log scale, and it represents the singularity in the
stress distribution. J$_{3DE}$ value is obtained in accordance with the procedure
described in the previous section using K$_I$ (stress intensity factor), which is
determined from the nodal point displacement. The variation of J$_{3DE}$ along the
crack front is shown in Fig. 4. J$_{3DE}$ has the maximum value at θ=0, that is,
the deepest point of the crack. J$_{3DE}$ at this position is given by

$$J_{3DE} = 1.68 \times 10^{-4}\sigma_\infty{}^2 \tag{4}$$

where J$_{3DE}$ is in KJ/m^2 and σ_∞ is in MN/m^2.

Next, a single-edge cracked plate, which is substituted for a semi-elliptical surface crack, is considered. The plate width and crack depth are 50 mm
and 12.5 mm, respectively. The two-dimensional, elastic-plastic analysis was
carried out. The mesh division of a half of the model is shown in Fig. 5. It
contains 29 plane strain-eight node isoparametric elements (Type 27 in MARC
library) and 119 nodes. The dimension of elements adjacent to the crack tip
is 0.5 mm. The mechanical properties follow. Two cases of yield strength: σ_Y
are taken 483 MN/m^2 and 500 MN/m^2. Young's modulus E=20600 MN/m^2, Poisson's

ratio $\nu=0.3$, and the linear strain-harding slope of 0.01E was taken.

J_{2DP}, the J value of the single edge cracked plate, was calculated using Eqs. (2) along the path shown in Fig. 5. J_{2DE} (J value in elastic condition) was also determined in much the same way as J_{2DP} and given by

$$J_{2DE} = 3.71 \times 10^{-4}\sigma_\infty^2 \tag{5}$$

The ratio, R determined by using Eqs. (4) and Eqs. (5), was obtained as R=0.45. Then, the approximate J value of the semi-elliptical surface crack in the elastic-plastic condition is the product of R and J_{2DP} and shown in Fig. 6.

FRACTURE TESTS AND DISCUSSION

In order to examine the usefulness of the proposed J approximation, fracture tests were conducted using specimens machined from 240 mm plate of A533 Gr.B steel. The tensile properties and chemical composition of the steel are shown in Table 1. J_{IC} (critical J value at crack extension) was already measured in accordance with the experimental procedure in the Landes-Begley's Report [5]. J_{IC} values at room temperature (15°C) and at -20°C are 260 KJ/m^2, 101 KJ/m^2, respectively.

Test specimens similar to those shown in Fig. 1 were used. Prior to the test, the specimens were precracked by cyclic loading. The fracture tests were conducted at room temperature and -20°C, where yield strength σ_Y are 483 MN/m^2 and 500 MN/m^2, respectively. Two specimens were used at each temperature. In the room temperature test, one of the specimens (Specimen A) was loaded up to $\sigma_\infty=420$ MN/m^2, and the other (Specimen B) was up to $\sigma_\infty=460$ MN/m^2. At -20°C, two specimens (Specimen C, Specimen D) were loaded up to $\sigma_\infty=431$ MN/m^2, $\sigma_\infty=441$ MN/m^2, respectively. Then, all specimens were unloaded, heat tinted at 300°C, and broken apart at -196°C.

The fracture surface was observed by a microscope. The fibrous crack extension was not observed in Specimen A and Specimen C at all. On the other hand, crack extension slightly initiated at the interior, was seen in Specimen B and Specimen D, close to the deepest point of the crack. The amount of crack extension Δa was about 0.05 mm. These results show that the crack began to extend when $\sigma_\infty \fallingdotseq 460$ MN/m^2 at room temperature and $\sigma_\infty \fallingdotseq 440$ MN/m^2 at -20°C. Assuming that J value of the semi-elliptical surface crack reached J_{IC} under the above uniform tensile stress, the test data were shown in Fig. 6 with open circles.

The estimated stresses, which can be obtained using J versus σ_∞ curves (Fig. 6) with $J=J_{IC}$, are 460 $\sim$ 470 MN/m^2 (at room temperature) and 440 $\sim$ 450 MN/m^2 (at -20°C). Comparing the measured stress with the estimated one, the proposed J approximation seems to be applicable. However, the number of test specimens are still very limited. The authors believe that further investigations on the approximation of J, like this study, will be needed.

CONCLUDING REMARK

The authors discussed a simple approximation of the J-integral of a semi-elliptical surface crack problem. Then, the value of the estimated J at the initiation of the crack extension was compared with the J_{IC} value. The value estimated met with J_{IC} value. The proposed method seems applicable in this limited study.

REFERENCES

[1] J. R. Rice, "A Path Independent Integral and Approximate Analysis of Strain Concentration by Notches and Cracks", Trans. of ASME Ser. E Vol. 35 (1968), PP. 379-386

[2] K. Oji and S. Kubo, Trans. of JSME No. 750-1 (1975), PP. 199-202 (in Japanese)

[3] K. Kishimoto, S. Aoki and M. Sakata, Trans. of JSME No. 750-9 (1975), PP. 207-210 (in Japanese)

[4] W. S. Blackburn and A. D. Jackson, "An Integral Associated with the State of a Crack Tip in a Non-Elastic Material", Int. Journ. of Fracture, 13 (1977), PP. 183-200

[5] J. D. Landes and J. A. Begley, ASTM STP560, (1974), PP. 170-186

Table 1   Material Properties

(1) Mechanical Properties (at room temperature)

Yield Strength (MN/m^2)	Tensile Strength (MN/m^2)	Elongation (%)	Reduction of Area (%)
483	607	25.0	47.0

(2) Chemical Composition

(wt. %)

C	Si	Mn	P	S	Ni	Mo
0.19	0.23	1.40	0.008	0.007	0.63	0.55

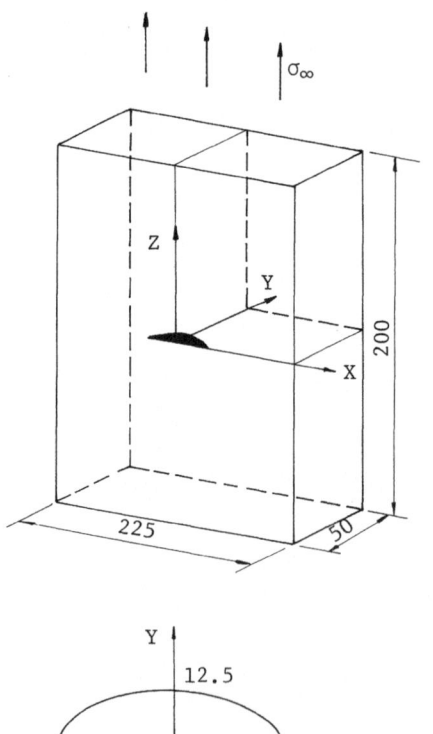

Fig. 1   The Model of Analysis

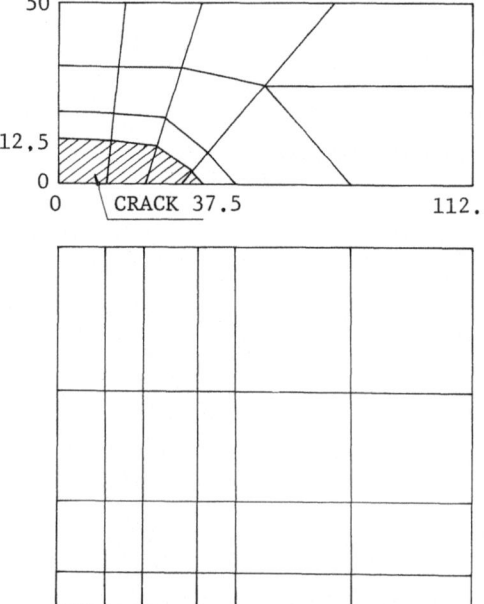

Fig. 2   Mesh Division of a Semi-elliptical Surface Crack

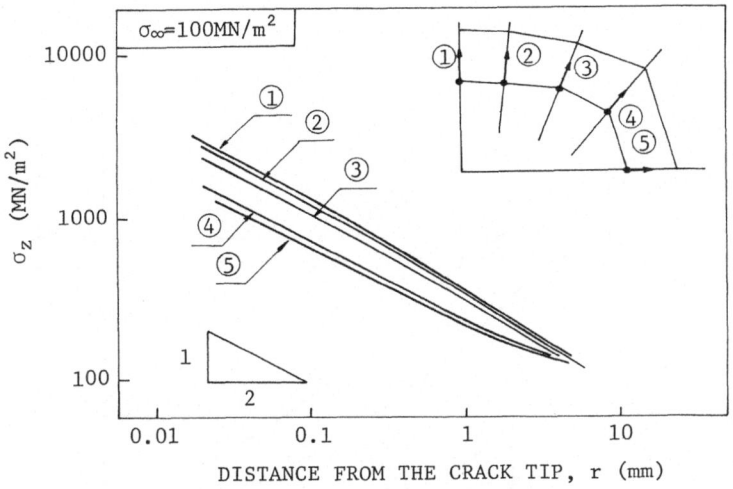

Fig. 3 σ_Z Stress versus the Distance from the Crack Front

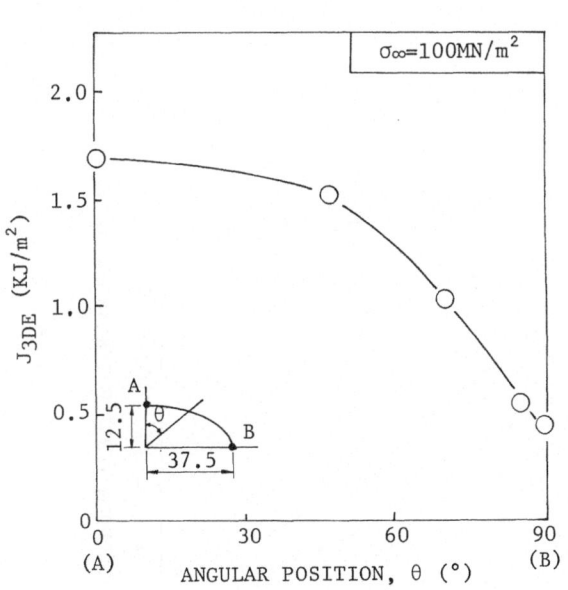

Fig. 4   The Variation of J_{3DE} along the Crack Front

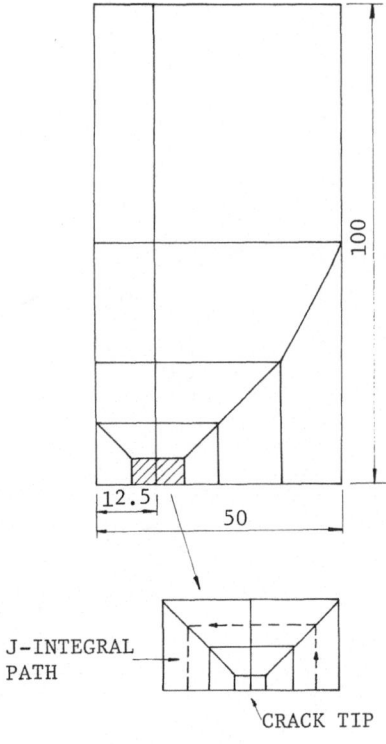

Fig. 5   Mesh Division of Single Edge Cracked Plate

138

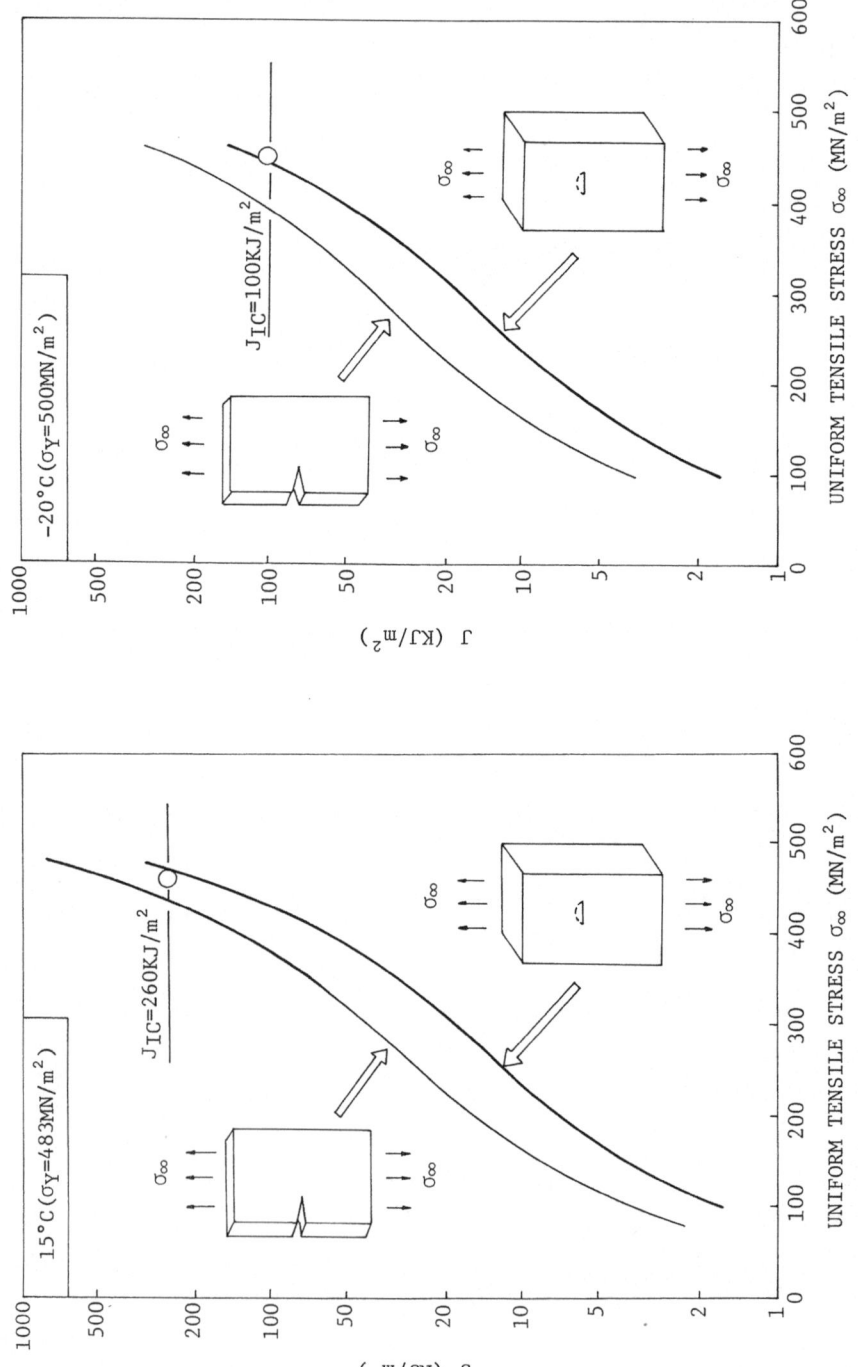

Fig. 6 Estimated J value of the semi-elliptical surface crack

AN EVALUATION OF PROCEDURES FOR J_{IC} DETERMINATION

N. Ohtsuka, M. Nakano, H. Ueyama
Civil & Applied Mechanics Research Department
Chiyoda Chemical Engineering & Construction Co., Ltd.
3-1-4 Ikegamishincho, Kawasaki-ku, Kawasaki 210, Japan

ABSTRACT

Several procedures to determine J_{IC}, the critical J-integral at the incipient crack growth, were investigated on a precracked, four-point bend specimen of A387QT 2-1/4Cr-1Mo steel and an SB42 low carbon steel. The R-curve method recommended by the ASTM Committee E24 was thought to overestimate often the J_{IC} value, because an apparently stable crack was observed at the midsection of the specimen corresponding to the point of J_{IC} determined by the method. Considering that a plane strain condition was sufficiently satisfied at the specimen midsection, the stretched zone width (SZW) method and the modified R-curve method, based on the measurement of the stretched zone width and crack growth at the midsection, were studied to give a more correct J_{IC} value. The compliance method and the acoustic emission (AE) method were also examined.

INTRODUCTION

The J-integral [1] is one of the most useful parameters of ductile fracture toughness due to its theoretical background and applicability to industrial materials. The two following values of J are proposed as the critical ones for evaluation of the toughness of the material.

(1) J_{IC} : the value of the J-integral at the onset of crack extension [2],

(2) J_C : the value of the J-integral at the onset of instability or at the maximum load [3].

Researches frequently discuss which value is reasonable for estimating the resistance of the material against fracture. In the usual design code for structures which have no severe defect, the allowable stress of the material is principally based on the yielding strength and the ultimate tensile strength. Consequently, the two critical values of the J-integral should be taken into consideration in the evaluation of cracks. Although the value of J_C may be determined more easily than J_{IC}, the J_C value is affected by test conditions such as the geometries of the test specimen. The application of the J_C value to the design should be conservative until the mechanism of crack propagation and the effect of the test conditions are clarified. On the other hand, the J_{IC} value may be considered to be a geometry-independent fracture toughness. The value of J_{IC} determined by a different J_{IC} test method, however, does not always coincide with others of the experimenter. In this study, the J_{IC} values determined by the R-curve method, the stretched zone width (SZW) method, the compliance method, and the acoustic emission (AE) method were experimentally compared and discussed on two kinds of material.

EXPERIMENT

The 2-1/4Cr-1Mo steel, ASTM A387 Gr.22, Cl.2 quenched (920°C x 1.5Hr, water cooled) and tempered 650°C x 2Hr, air cooled) (which is equivalent to the steel A542), and the low carbon steel for boiler and pressure vessel, JIS SB42 annealed (625°C x 4Hr, furnace cooled) [4], were examined. The chemical composition and mechanical properties of the materials are given in Table 1 and Table 2. The geometries of tensile smooth and bend (four-point bend) specimens are shown in Fig. 1. The bend specimen was precracked in fatigue in accordance with the ASTM E399-74 with a crack length to span ratio, a/w, equal to 0.6.

All the tests were conducted using the electric hydraulic fatigue testing machine at room temperature. In the beginning, the tensile tests by pin-loading were conducted on the two steels using the smooth specimen shown in Fig. 1 (a) and (b). The Displacement rate was 1.08 mm/min. for A387QT steel and 0.19 mm/min. for SB42 steel.

The bend specimens shown in Fig. 1 (c) and (d) were used for the four different J_{IC} tests that follow:

The R-curve method: The experimental procedure basically conformed to the recommended procedure for J_{IC} determination proposed by the ASTM Committee E24 [2]. Each specimen was loaded to different values at a displacement rate of 0.75 mm/min. Displacement, δ, between the loading fixtures and edge opening displacement, V_S, were measured by means of clip gauges. The value of J was calculated from the area, A, under the moment, M, vs. the rotational angle, θ, curve by [5]

$$J = 2A/(Bb) \qquad (1)$$

where B and b are the thickness of specimen and the remaining ligament length of the specimen. The following procedure was used to estimate the rotational angle, θ, from the concept of rotation [6]:

$$\theta = V_S \{a + 0.45(w-a)\} \qquad (2)$$

After unloading, the specimen was pulled apart at liquid nitrogen temperature, and the crack extension, Δa, was measured as an entire width including a stretched zone width, SZW, and an actual slow stable crack (tear dimple zone) width, DZW, by means of the scanning electron microscope (SEM). The crack extension, Δa, was calculated by using the average of nine evenly spaced measurement points from one side of the specimen to the other across the crack front. The J_{IC} value was calculated from the intersection of a linear fit to the J vs. Δa R-line and the crack blunting line [7]

$$J = 2\sigma_F \Delta a \tag{3}$$

where σ_F is the average of the yield strength and the ultimate tensile strength.

The SZW method: The method is principally based on the SZW method proposed in Ref. [8]. Examining the fracture surface of the specimen in SEM, used for the R-curve method, the average stretched zone width of the several measurements at the specimen midsection were determined as the value of SZW. The value of J_{IC} was determined as the intersection of the SZW vs. J curve line and the blunting line [8]

$$SZW = \alpha \, J/E \tag{4}$$

where E is the Young's modulus and α is a constant (89 is proposed in Ref. [8]).

The compliance method: A single specimen was partially unloaded after deformed different amount [9]. The M vs. θ (by eq. (2)) hysteresis data recorded on the Honeywell M96 magnetic tape recorder were reproduced after the test, and the change of the rotational angle, $\Delta\theta$, was obtained by an analogue processor based on the equation

$$\Delta\theta = \beta(\theta - \lambda_o M) \tag{5}$$

where λ_o is the compliance (θ/M) of the specimen before crack extension and β is an amplification factor ($\fallingdotseq 60$ in this case). Then, the increment of the compliance, $\Delta\lambda$, due to crack extension, measured on the unloading M vs. $\Delta\theta$ curve, was used to determine the crack extension, Δa, by the equation [9]

$$\Delta a = (w-a)(\Delta\lambda/\lambda_o)g(a/w)/2 \tag{6}$$

where $g(a/w)$ is the coefficient given as 0.872 for $a/w = 0.6$. The R-curve was drawn using the crack extension given by the equation (6).

The AE method: During the tensile and bend test, acoustic emission (AE) signals were monitored to detect the moment of crack growth in a nondestructive way. The Nortec AEMS-4 system was used as shown in Fig. 2. A couple of VZ-1/2 X1/4T AE transducers (PZT-5) were mounted on a specimen across the monitoring area with clamps. The assembly for testing a bend specimen is shown in Photo 1. The output from the transducer was amplified as much as 95dB and filtered by a 100 KHz high pass filter. Rate and energy pulses were obtained for each AE burst input above a fixed threshold level of 1.2 to 1.8 volts. A coincidence gate discriminated and identified AE signals from the monitoring area. The amplified AE signals were recorded on the magnetic tape recorder, the tape speed and the band width of which were 120 ips (3 m/sec) and 0.4 to 1500 KHz.

RESULTS AND DISCUSSION

Fracture surface

Photos 2 and 3 show typical macroscopic and microscopic fracture surfaces of the test specimen; the typical distribution of the stretched zone and the tear dimple zone through the thickness of the specimen are shown in Fig. 3. Fig. 3 indicates that both SZW and DZW are maximum at the specimen midsection, and that the variation of DZW with thickness is larger than that of SZW.

The R-curve method

From the intersection of the solid line and the dash line in Fig. 4, the J_{IC} value of A387QT steel was obtained as 250 N/mm and listed as $J_{IC}(A)$ in Table 3, whereas that of SB42 steel was indeterminable (the two lines didn't intersect each other). It should be noted that the crack extension in the figure is a through-the thickness average value. The requirement of the ASTM E24 procedure on 0.15mm offset line is ignored because the requirement gave too large a J_{IC} value for A387QT steel, shown in $J_{IC}(B)$ in Table 3. Observing the crack surface of the specimen at the value of $J_{IC}(A)$ and $J_{IC}(B)$, the crack extension was apparent at the midsection. Fig. 3 shows that the stable crack initiates earlier and propagates faster there. Considering that the stress and strain vary continuously along the specimen thickness and are symmetrical with respect to the midsection, a sufficiently small element at the midsection is assumed to nearly satisfy the plane strain condition. In the same way that the K_{IC} value is defined as the plane strain fracture toughness in ASTM E399-74, the value of J_{IC} should be defined as the smallest critical value of J_C at the plane strain condition, independent of the specimen thickness. From this point of view, the crack extension, Δa, at the specimen midsection was replotted to obtain the modified R-curve in Fig. 5. The $J_{IC}(R)$ values of A387QT and SB42 steel, determined from the intersections of the modified R-curve line and the experimentally obtained blunting line (equation (4) as mentioned later), were 174 N/mm and 94 N/mm which differed fairly from the $J_{IC}(A)$ values.

The SZW method

The SZW at the specimen midsection is plotted against the J value in Fig. 6. The SZW after crack initiation was considered nearly constant, regardless of the value of J; whereas the blunting line was represented not by equation (3) but by equation (4) whose proportional constant α was 85 for the material tested. The values of J_{IC} were determined as 168 N/mm for A387QT steel and 89 N/mm for SB42 steel by Fig. 6, and called $J_{IC}(S)$ in Table 3. Comparing the $J_{IC}(S)$ values with the $J_{IC}(R)$ values by the modified R-curve method, the two values almost coincide with each other.

The compliance method

Fig. 7 is a typical M vs. θ record in the compliance test of A387QT steel.

The circles in Fig. 8 show the R-curve obtained by the compliance method. For comparison, a specimen midsection value and a through-the-thickness average value of DZW, measured by SEM observation, are also plotted as solid and open squares in the figure. The crack extension measured by the compliance method fairly agrees with the through-the-thickness average DZW by SEM observation but the measurement accuracy was inferior. $J_{IC}(C)$, the J_{IC} value of A387QT steel determined by Fig. 8, is 180 N/mm as shown in Table 3.

The AE method

The AE rate count from smooth specimen during tensile test is shown in Fig. 9. Many AE signals were observed from the yielding of SB42 steel but almost no AE from A387QT steel at the monitoring threshold level, which indicates that between the two materials a remarkable difference exists in the acoustic activity by plastic deformation. Fig. 10 and 11 show the AE rate count, the cumulative AE energy, and load vs. cross-head displacement curves during the J_{IC} tests of precracked specimens. Double circles plotted in these figures are the corresponding J_{IC} points determined by the SZW method. Comparing these results, relatively few apparent AE signals were observed around the initiation of the slow stable crack on A387QT steel; whereas many AE signals were detected around the general yielding of SB42 steel. No remarkable increase of AE was observed at the crack initiation, shown in Fig. 11. AE signals, caused by the yielding of low carbon steel, are generally said to be characterized as low-amplitude, quasi-continuous emission, while high-amplitude burst emission characterizes crack growth [10]. Therefore, the high-amplitude AE signals were selected by reproducing the recorded AE signals with four different threshold levels, and the threshold levels which corresponded to the AE amplitude were replotted against the cross-head displacement in Fig. 12 and 13. These figures show that relatively high amplitude AE signals were detected at or near the initiation of the slow stable crack and could be distinguished from AE signals caused by yielding.

SUMMARY

Several techniques for the determination of fracture initiation were investigated. The results are summarized in Table 3. The R-curve method, which is recommended by the ASTM Committee E24 and is based on the measurement of the through-the-thickness average crack extension, gave fairly large values of J_{IC} on the materials investigated than those of $J_{IC}(S)$ and $J_{IC}(R)$, obtained by the SZW method and the modified R-curve method. Fig. 5 indicates that the stable crack has already propagated more than 0.5mm and 1.7mm at the specimen midsection at the point of $J_{IC}(A)$ and $J_{IC}(B)$ obtained by the R-curve method. Consequently, the method is considered to overestimate often the value of J_{IC},

144

especially when the requirement of 0.15mm offset line and equation (3) on blunting line are applied. Fig. 6 further indicates that the experimental blunting line fits not to equation (3) but to equation (4). The SZW method and the modified R-curve method are considered reasonable to determine the geometry-independent J_{IC} value because the SZW and the crack extension are measured at the specimen midsection where the plane strain condition is considered to be nearly satisfied. That is, if a specimen is loaded by forces which are distributed uniformly over the thickness and are symmetrical with respect to the crack plane and the crack extension is measured within a sufficiently smaller distance from the specimen midsection than the crack length and the radius of the crack front curvature, the measured value is considered to be in the sufficient plane strain condition. It is suggested that the two methods enable the determination of the value of J_{IC} using thinner specimens than that specified in the ASTM method. The minimum required thickness should be examined based on the condition to prevent the specimen from buckling or unsymmetrical deformation to the midsection. Either the compliance method or the AE method has the possibility of reducing the numbers of the test specimen. The compliance method, however, is less accurate than the other methods, so that plural specimens are essential to determine the value of J_{IC}. If troublesome noises are rejected, the AE method might be applicable to such materials as A387QT steel, which has a relatively low acoustic activity caused by plastic deformation but emanates high AE energy from crack growth. In the case of such materials as SB42 steel with high acoustic activity by plastic deformation, the AE signals by crack extension should be distinguished from those by yielding, using particular techniques such as the AE amplitude plot technique which was applied in this paper.

REFERENCES

[1] J. R. Rice, Trans. ASME, J. of Applied Mechanics, 35, (1968), PP. 379-386.

[2] G. A. Clarke, et al., J. of Testing and Evaluation, 7-1, (Jan., 1979), PP. 49-56.

[3] The JI Committee, Study on J Integral Fracture Criterion, Iron and Steel Department, the Japan Welding Institute, (Feb., 1979), 108 PP. in Japanese.

[4] N. Ohtsuka, et al., ASME Publication, 78-PVP-18, (June, 1978), 9 PP.

[5] J. R. Rice, et al., ASTM STP 536, (1973), PP. 231-245.

[6] British Standards Institution, DD19, (1972), PP. 4-21.

[7] L. D. Landes and J. A. Begley, ASTM STP 560, (1974), PP. 170-186.

[8] H. Kobayashi, et al, ICM3, 3, (Aug., 1979), PP. 529-538.

[9] G. A. Clarke, et al., ASTM STP 590, (1976), PP. 27-42.

[10] J. C. Spanner, Acoustic Emission Techniques and Application, Intex Pulishing Co., Evanston, (1974), PP. 37-95.

Table 1   The chemical composition of materials (%)

STEEL	C	Si	Mn	P	S	Cr	Mo
A387QT	0.12	0.28	0.53	0.010	0.008	2.23	0.90
SB42	0.12	0.31	0.71	0.013	0.027	—	0.06

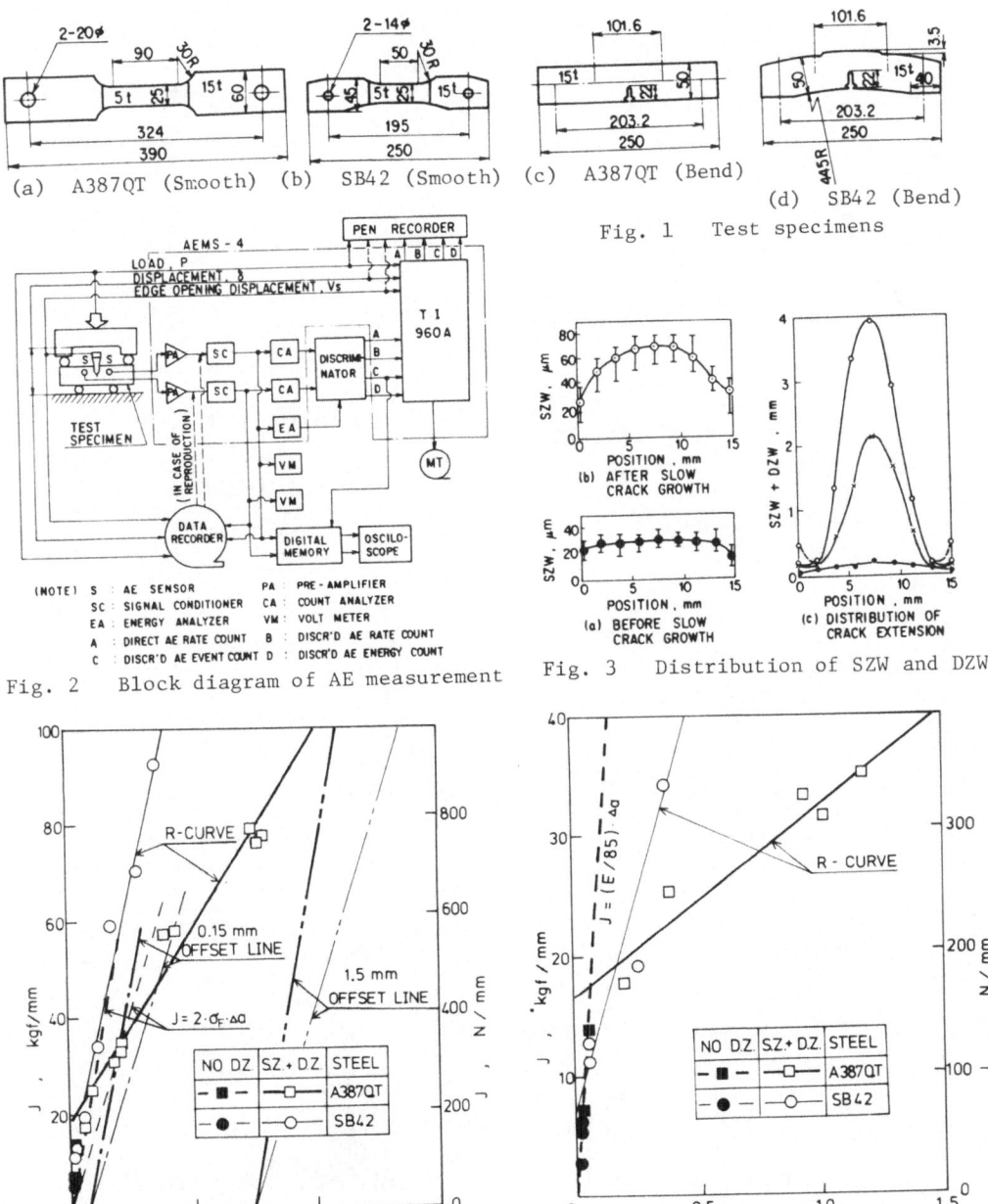

(a) A387QT (Smooth)   (b) SB42 (Smooth)   (c) A387QT (Bend)

(d) SB42 (Bend)

Fig. 1   Test specimens

Fig. 2   Block diagram of AE measurement

Fig. 3   Distribution of SZW and DZW

Fig. 4   R-curve of A387QT and SB42 steel

Fig. 5   Modified R-curve

146

Table 2　The mechanical properties of materials

STEEL	Y.S.　MPa	U.T.S.　MPa
A387QT	624	738
SB42	348	459

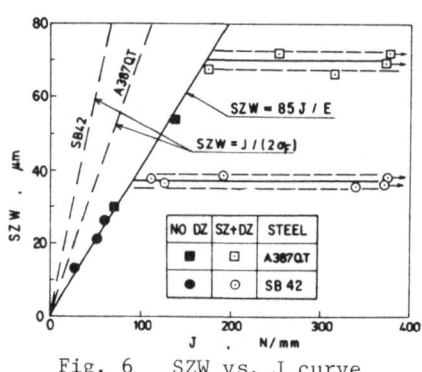

Fig. 6　SZW vs. J curve

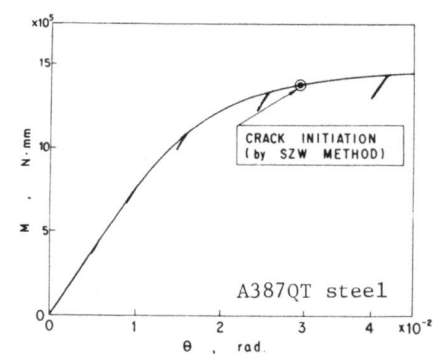

Fig. 7　M vs. θ curve of compliance test

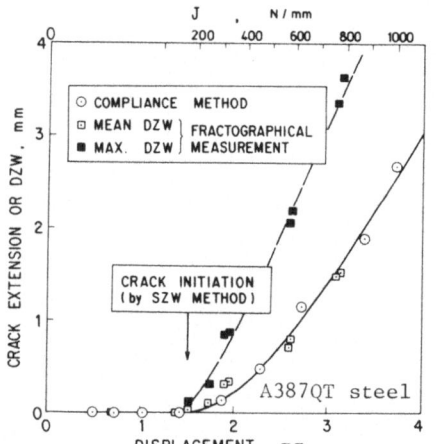

Fig. 8　R-curve obtained by compliance test for A387QT steel

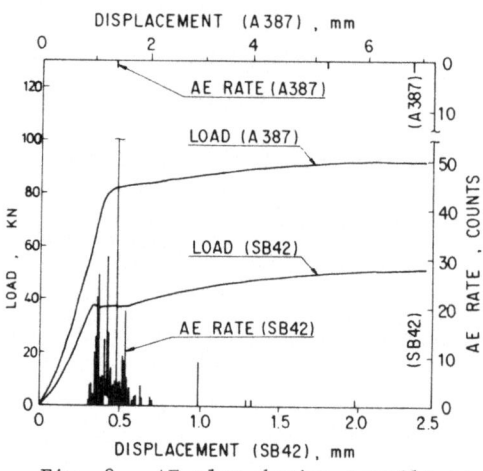

Fig. 9　AE plot during tensile test

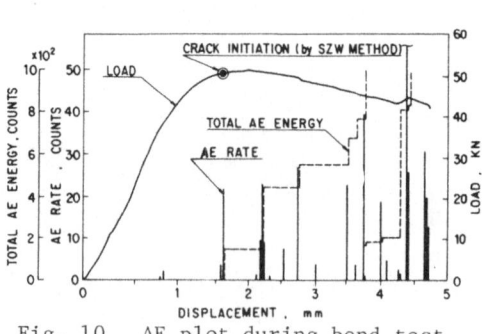

Fig. 10　AE plot during bend test of A387QT steel

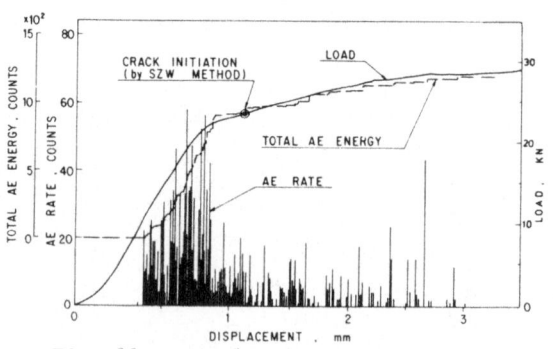

Fig. 11　AE plot during bend test of SB42 steel

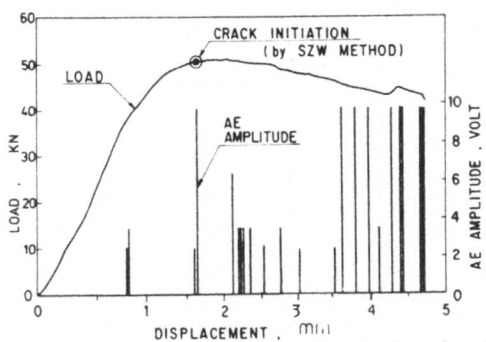

Fig. 12   AE amplitude plot during bend test of A387QT steel

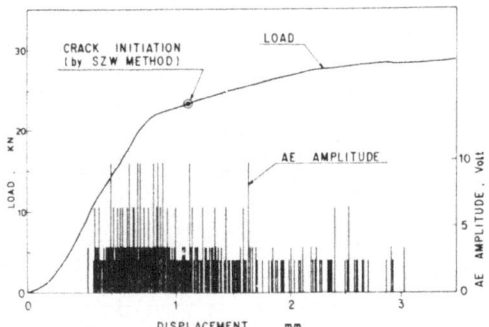

Fig. 13   AE amplitude plot during bend test of SB42 steel

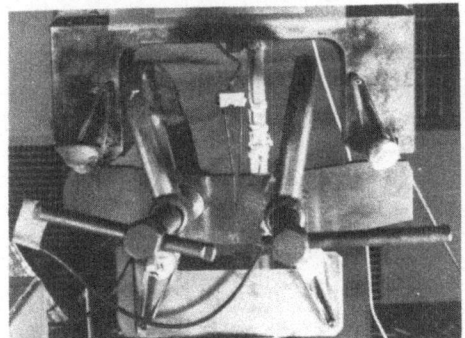

Photo 1   The assembly for test bend specimen

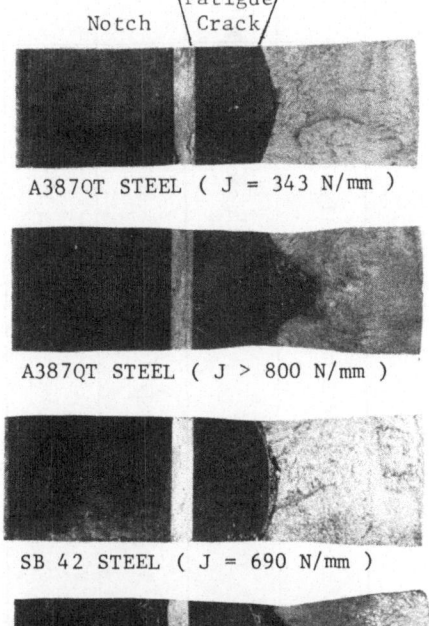

A387QT STEEL (J = 343 N/mm)

A387QT STEEL (J > 800 N/mm)

SB 42 STEEL (J = 690 N/mm)

SB42 STEEL (J > 1000 N/mm)

Photo 2   Typical macroscopic fracture surface

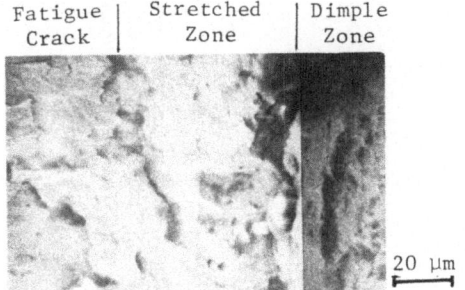

A387QT STEEL (J = 564 N/mm)

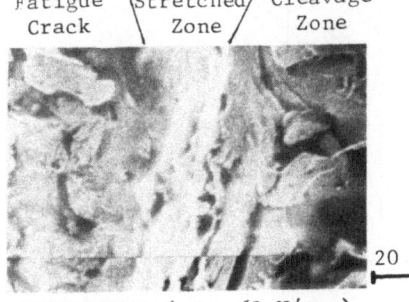

SB42 STEEL (J = 61 N/mm)

Photo 3   Typical microscopic fracture surface

Table 3 Comparison of J_{IC} values determined by different J_{IC} method (unit in N/mm)

Method	R-curve (ASTM E24)		Modified R-curve	SZW	Compliance	AE
Value of Δa	Average*		Midsection**	Midsection**	Average*	Local
Blunting Line	Eq. (3)	Eq. (3)	Eq. (4)	Eq. (4)	—	—
Notation	J_{IC} (A)***	J_{IC} (B)	J_{IC} (R)***	J_{IC} (S)***	J_{IC} (C)***	J_{IC} (E)
A387QT Steel — J_{IC} (± Scatter)	250 (±50)	436 (±30)	174 (±20)	168 (±20)	180 (±50)	179
A387QT Steel — Ratio to J_{IC} (S)	1.49	2.60	1.04	1.0	1.08	1.07
SB42 Steel — J_{IC} (± Scatter)	Indeterminable	>1000	94 (±20)	89 (±10)	—	90
SB42 Steel — Ratio to J_{IC} (S)	Indeterminable	> 10	1.05	1.0	—	1.01

(Note)
* through-the-thickness average value of a specimen
** value at the specimen midsection
*** not complied with a requirement of 0.15mm offset line

EVALUATING METHODS OF J-INTEGRAL
AND SCATTERINGS OF FRACTURE TOUGHNESS

Mitsuru Arii
Applied Metallurgy & Chemistry Department
Nuclear Energy Group
Toshiba Corporation
13-12, Mita 3-chome, Minato-ku, Tokyo, 108 Japan

Tadao Mori, Hideo Kashiwaya, Tetsu Yanuki, and Yoshiyasu Itoh
Metallurgical Engineering Group
Heavy Apparatus Engineering Laboratory
Toshiba Corporation
2-4, Suehiro-cho, Tsurumi-ku, Yokohama, 230 Japan

ABSTRACT

In recent work of fracture mechanics, the J-integral has been shown to be effective as a fracture criterion, and some experimental evaluating procedures of the J-integral have been proposed for fracture toughness testing. However, a number of important problems remained to be resolved for accuracy and the applicable range of the proposed evaluating procedures.

Therefore, convenient evaluation methods of the J-integral were newly derived in this report, and they were compared with other proposed methods for the accuracy of J-integral. As a result, the accuracy and the applicable range of proposed equations for elastic-plastic materials with deep or shallow crack were clarified by using the elastic-plastic finite element analysis. Then, using these convenient equations, fracture toughness tests were conducted, giving attention to scatterings of fracture toughness.

INTRODUCTION

A number of studies have been made of the unstable brittle fracture of cracked plate, and they may be clarified to assess the fracture strength at small scale yielding by using linear fracture mechanics.

In recent work, the major direction of fracture mechanics turns toward material-nonlinear problems. The J-integral, defined by Rice [1], attempts to extend linear fracture mechanics to large scale yielding fracture, and it has been used successfully as a J_{IC} criterion for various materials [10]. Rice has proposed the evaluating methods of J-integral from single load versus load point displacement relationships [2]. Up to the present, this single test evaluation of fracture toughness has been used to assess J_C or J_{IC}. However, Rice's convenient equations are not applicable to the shallowly cracked specimen, and their applicable ranges are not clear yet.

On the other hand, the J-integral is based on non-linear elastic materials and, therefore, its convenient evaluating equations are definite under this assumption. However, structural steels with crack reveal elastic-plastic behavior before general yield [8]. In this case, the meanings of convenient equations must be clarified and convenient equations must be simplified from the engineering point of view.

Therefore, the purpose of this report is to derive the new convenient evaluating equations of J-integral for elastic-plastic materials and to investigate their accuracy using finite element analysis. The effect of convenient equations of J-integral on the fracture toughness and its scatterings is experimentally investigated because data is not extensive.

CONVENIENT EVALUATING METHODS OF J-INTEGRAL

Definition of J-integral and derivation of its convenient equations

The J-integral proposed by Rice [1] is defined under the restriction of non-linear elastic materials for a two-dimensional deformation field and is evaluated (Fig. 1).

$$J = \int_\Gamma (U dy - \vec{T} \frac{\partial \vec{u}}{\partial x} ds) \qquad (1)$$

where U is the strain energy density function, Γ is an integral path and $\vec{T}$ and $\vec{u}$ are traction vector and displacement vector on Γ.

Hutchinson [3], Rice, and Rosengren indicated that the singularities of crack tip plastic stress and strain can be expressed as a function of the J-integral, and Rice has shown that the J-integral may be interpreted as the potential energy released rate for crack extension [1].

$$J = \frac{1}{t} \int_o^P \left(\frac{\partial u}{\partial c}\right)_p dp \quad \text{(constant load)}$$

$$J = \frac{1}{t} \int_o^u \left(-\frac{\partial p}{\partial c}\right)_u du \quad \text{(constant load point displacement)} \qquad (2)$$

Here, t is the crack front length; p is the load; u is the load point displacement, and c is the crack length. It is necessary to establish the relationship between the load; p and the load point displacement; u for the evaluating J-integral by use of eq. (2).

Rice [2] has assumed that the displacement between the load points u, may be regarded as the sum of the displacement without crack $U_{no\ crack}$ plus the displacement due to introducing the crack U_{crack} (or alternately as the elastic displacement plus the displacement due to plasticity). In this report, it is assumed that U_{crack} and $U_{no\ crack}$ can be separated into elastic components,

$U_{e.crack}$, $U_{e.no\ crack}$ and plastic components respectively $U_{p.crack}$, $U_{p.no\ crack}$ for elastic-plastic materials and the total displacement; U is

$$U = U_{e.crack} + U_{p.crack} + U_{e.no\ crack} + U_{p.no\ crack} \qquad (3)$$

Substitution of eq. (3) into eq. (2) leads to the equation

$$J = J_e + \frac{1}{t} \int_o^\Gamma (\frac{\partial U_{p.crack}}{\partial c})_p \, dp \qquad (4)$$

Here, J_e is the elastic component of the J-integral and is equal to the elastic strain energy released rate G.

Consider a plate with one side crack subjected to remotely applied in-plane bending of moment M (Fig. 2).

As was the case with eq. (3), the angle change between the points of moment application θ, can be separated into $\theta_{e.crack}$, $\theta_{e.no\ crack}$, $\theta_{p.crack}$, and $\theta_{p.no\ crack}$. It is assumed that $\theta_{p.crack}$ is

$$\theta_{p.crack} = g(\frac{M}{b^2 t}) \qquad (5)$$

The convenient equation of the J-integral for the specimen with a deep crack is obtained.

$$J_d = G + \frac{2}{bt} [\int_o^\theta M d\theta - \frac{1}{2} M\theta e] \qquad (6)$$

$$(\theta_e = \theta_{e.crack} + \theta_{e.no\ crack})$$

Also, if a remaining ligament is subject to bending, due to the applied load p, then eq. (6) yields

$$J_d = G + \frac{2}{bt} [\int_o^U p dU - \frac{1}{2} P.Ue] \qquad (7)$$

$$(U_e = U_{e.crack} + U_{e.no\ crack})$$

For low strain hardening materials with deep crack, eq. (6) and eq. (7) may be transformed by use of crack edge opening displacement δ_{edge}, as follows:

$$J_d = G + \frac{2}{bt} \frac{1}{c+\gamma b} [\int_o^{\delta edge} M d\delta edge - \frac{1}{2}M \, \delta e.edge] \qquad (8)$$

$$J_d = G + \frac{2}{bt} \frac{W}{c+\gamma b} [\int_o^{\delta edge} P d\delta edge - \frac{1}{2}P \, \delta edge] \qquad (9)$$

where $\gamma (\approx \frac{1}{3})$; rotational factor

$\delta_{e.edge}$; elastic component of crack edge opening displacement

On the other hand, for the specimen with a shallow crack, θ_{crack} at the general yielding (gross section yield [8]) is indicated as follows by considering the linear elasticity.

$$\theta_{crack} = (\frac{c}{w})^2 \, g'(\frac{M}{w^2 t}) \qquad (10)$$

Substitution of eq. (10) into eq. (4) leads to the equation

$$J_s = \frac{2}{ct} \int_o^M \theta_{crack} \, dM \qquad (11)$$

$$J_s = \frac{2}{ct} \int_o^P U\text{crack } dP \qquad (12)$$

For the specimen with a shallow crack at full yielding (net section yield [8]), we may rewrite eq. (11) and eq. (12) as

$$J_s = \frac{2}{ct} [\int_o^M \theta dM - \frac{1}{2} M \theta e.no \text{ crack}] \qquad (13)$$

$$J_s = \frac{2}{ct} [\int_o^P Udp - \frac{1}{2} P Ue.no \text{ crack}] \qquad (14)$$

In the same way, the convenient equations of the J-integral can be derived for the cracked plate in tension and the cracked round bar in tension. The results are shown in Table 1 with the assumptions to derive them.

Accuracy of convenient equations and derivation of more convenient
equations of the J-integral

The accuracy of the convenient equations is investigated, using a finite element elastic-plastic analysis based on the incremental strain theory. The stress-strain curve is used to simulate a mild steel with a yield stress; $\sigma_Y=231$ MN/m^2, strain hardening exponent; n=0.214, Young's modulus; E=205943 MN/m^2 and poisson ratio; ν=0.3. It is well known that the J-integral is independent on its integral path [5], and the independency of the J-integral on the various integral paths, except in the vicinity of a crack tip, is verified in this investigation. The accuracy of convenient equations is shown in Fig. 3. $J_d(J_s)/J$ is the ratio of the J-integral, which was assessed from the convenient equations and eq. (1) by using a finite element analysis.

As a matter of course, $J_d(J_s)/J$ assessed from J=G decreases at large scale yielding. In the deeply cracked three-point bend specimen, enough accuracy to evaluate J-integral is obtained by using the P-U method (eq. (6)) or P-δ_{edge} method (eq. (7)). In the internally deeply cracked plate in tension, the P-U method and P-δ_{center} method (Table 1) are applicable to evaluate J-integral at large scale yielding. In shallowly cracked plate, the P-U method is applicable.

On the other hand, in the deeply cracked bend specimen, Rice [2] and Kanazawa etc. [6] and Sumpter etc. [7] have proposed the convenient equation of the J-integral, respectively (see Table 2). The accuracy of J-integral for each convenient equations is shown in Fig. 4. Before full yield, the J-integral is evaluated of lower value by using Rice's equation [2]. It might be thought that Rice ignored the elastic-plastic behavior of cracked plate. On the contrary, the Kanazawa equation [6] assesses the J-integral at high value after full yield. The Sumpter equation [7] is not as applicable as other convenient equations; however, the Sumpter equation, which is not required in the area under P versus U (or δ_{edge}) curve, is useful in engineering sense.

Therefore, more convenient equations of J-integral are derived for the following stress and strain relation.

$$\begin{cases} \overline{\sigma} = E \, \varepsilon & \sigma \leq \sigma o \\ \overline{\sigma} = \sigma o \, (\overline{\varepsilon}/\varepsilon o)^n & \sigma > \sigma o \end{cases} \tag{15}$$

where $\sigma_o = E \, \varepsilon_o$ σ_o, ε_o: material constant

The assumption using analysis and its results are shown in Table 2.

For fully plasticity, Ohji etc. [4] have derived the convenient equation. The accuracy of author's convenient equations is shown in Fig. 4, and it can be evaluated with good results.

FRACTURE TOUGHNESS TESTING AND ITS SCATTERINGS

Specimen and experimental procedures

In the preceding discussion, it was shown that the convenient equations newly derived were sufficient to evaluate J-integral. Therefore, using those convenient equations, fracture toughness testing is conducted, and the scatterings of fracture toughness are investigated. From the safety point of view, in structures one of the important problems is to clarify the scatterings of fracture toughness for the assessment of fracture probability [9].

For the experimental work, low strength steel, SM41B of 19.4 mm thick was selected. The chemical compositions and mechanical properties at room temperature are listed in Table 3. The specimen configuration is shown in Fig. 5. All specimens were precracked by a fatigue testing machine and the fatigue crack extended 2∿3 mm from the mechanical notch tip. During the precracking process, the fatigue load level was sufficiently low to satisfy the ASTM condition (see Fig. 5).

The specimen was cooled with liquid nitrogen, dry ice, and alcohol; the temperature of specimen was measured with thermo-couples. All tests were conducted at the temperature range from -70°C to -196°C, and the tests regarding the scatterings were performed at -78°C, using 39 specimens.

Various evaluating methods of fracture toughness

For the deeply cracked three-point bend specimen, the following evaluating methods of fracture toughness were proposed.

(1) Critical stress intensity factor (ASTM E399-78)

$$Kc = \frac{PS}{BW^{3/2}} \, f(c/w) \tag{16}$$

$$f(c/w) = \frac{3(c/w)^{1/2} \, [1.99 - (c/w)(1-c/w)(2.15 - 3.939c/w + 2.7(c/w)^2]}{2(1+2c/w)(1-c/w)^{3/2}}$$

(2) Critical COD (FEM + Rotational Factor [12])

$$\delta c= \frac{\sigma_{YW}}{E}[0.003916+0.01392(\frac{\sigma_N}{\sigma_Y})+0.1224(\frac{\sigma_N}{\sigma_Y})^2-0.05658(\frac{\sigma_N}{\sigma_Y})^3+0.04321(\frac{\sigma_N}{\sigma_Y})^4]$$

$$(\frac{\sigma_N}{\sigma_Y} \leq 2.0)$$

$$\delta c= Vg/[1+(z+c)/\gamma(w-c)] \qquad (\frac{\sigma_N}{\sigma_Y} \geq 1.5) \qquad (17)$$

where $\sigma_N= 6PW/t(w-c)^2$

  z distance of clip gauge edge under the surface

  Vg clip gauge opening displacement

(3) Critical COD (Wells equation [11])

$$\delta c= \frac{0.45(w-c)}{0.45w+0.55c+z} [Vg - \frac{\gamma'\sigma_{YW}(1-\nu^2)}{E}] \quad (Vg \geq \frac{2\gamma'\sigma_{YW}(1-\nu^2)}{E})$$

$$\delta c= \frac{0.45(w-c)}{0.45w+0.55c+z} [\frac{Vg\ E}{4\gamma'\sigma_{YW}(1-\nu^2)}] \qquad (Vg < \frac{2\gamma'\sigma_{YW}(1-\nu^2)}{E}) \qquad (18)$$

$$\gamma' = 4.0484(\frac{c}{w})^4 -7.0261(\frac{c}{w})^3 +1.9282(\frac{c}{w})^2 +3.3956(\frac{c}{w})$$

(4) Critical COD (British standard [11])

$$\delta c = \frac{K_c^2}{2\sigma_Y E} + \frac{0.4(w-c)V_{gp}}{0.4W+0.6c+z} \qquad (19)$$

  where V_{gp} Plastic component of clip gauge opening displacement.

$$E'\{ \begin{array}{ll} = E/1-\nu^2 & \text{plane strain} \\ = E & \text{plane stress} \end{array}$$

(5) Critical J (Rice [2])

$$J_c \doteq \frac{2}{bt} \int_0^{U_{total}} PdU_{total} \qquad (20)$$

(6) Critical J (P-V_g method [12])

$$J_c = G_c + \frac{2}{bt} \frac{W}{c+(w-c)\gamma} \int_0^{V_{gp}} PdV_{gp} \qquad (21)$$

(7) Critical J (V_g method [12])

$$\frac{E'J_c}{c\sigma_Y^2} I_1(c/w) = [\frac{E'Vg}{W\sigma_Y} I_2(c/w)]^2 , (\frac{E'Vg}{W\sigma_Y} I_2(c/w)<2)$$

$$= 4[\frac{E'Vg}{W\sigma_Y} I_2(c/w)-1] , (\frac{E'Vg}{W\sigma_Y} I_2(c/w)\geq 2) \qquad (22)$$

  where $I_1(c/w)$, $I_2(c/w)$ are given, see [12]

(8) Critical J (P-δ_{edge} method)   see eq. (9)

(9) Critical J (P-U method)   see eq. (7)

(10) Critical J (P, δ_{edge} method)

$$J_c = \frac{K_c^2}{E'} + \frac{2}{1+n} \frac{P}{bt} \frac{W}{c+\gamma b} \qquad (\delta edge - \delta e.edge) \qquad (23)$$

Experimental results

The temperature dependency of fracture toughness, J_c and δ_c, are shown in Fig. 6 (a) and (b), respectively. J_c is evaluated by P-δ_{edge} method, and δ_c is evaluated by the Wells equation. The experiment for the scatterings of fracture toughness was performed at $-78°C$, so the transition phenomenon was comparatively remarkable. The frequency histograms of fracture toughness, J_c and δ_c at $-78°C$, are shown in Fig. 6 (a) and (b), respectively. The fracture toughness of this material at $-78°C$ has the large amount of scatter. The representative examples of the fractographs obtained in the experiment of the scatterings are shown in Fig. 7 (a) and (b). In spite of the same experimental conditions, one fatigue precrack (Fig. 7 (a)) propagates to cleavage fracture directly; the other fatigue precrack (Fig. 7 (b)) blunts the crack tip (i.e. the formation of the stretched zone) and propagates to cleavage fracture.

Assuming that the scatterings of fracture toughness show the Weibull distribution or the normal distribution [9], their mean value and standard deviation are shown in Table 4. The mean value obtained from the assumption of the Weibull distribution almost has a lower value than that of normal distribution; and for determining the effect of evaluating methods on fracture toughness, the critical COD has comparatively larger difference than critical J. However, it seems that the effect of evaluating methods on fracture toughness may be neglected in comparison with the scatterings of fracture toughness.

CONCLUSION

The major conclusions of this investigation are summarized as follows:

(1) The convenient equations of the J-integral were newly derived for the one side cracked plate subject to bending, the cracked plate in tension, and the cracked round bar in tension. Their accuracy in J evaluation was investigated by using the elastic-plastic finite element analysis based on the incremental strain theory.

(2) By using the convenient equation newly proposed in this report, fracture toughness was assessed for low strength steel, SM41B. The fracture toughness of this material at $-78°C$ has the large amount of scatter in comparison with fracture toughness caused by evaluating methods.

REFERENCES

(1) J.R. Rice "A Path Independent Integral and the Approximated Analysis of Strain Concentration by Notches and Cracks". Journal of Applied Mechanics. Vol.35, (1968), pp.379

(2) J.R. Rice, P.C. Paris "Some Further Results of J-Integral Analysis and Estimates", ASTM STP 536, (1973), pp.231

156

(3) J.W. Hutchinson "Singular Behavior at the End of a Tensile Crack in a Hardening Material", Journal of the Mechanical and Physics of Solids, Vol. 16, (1968), pp13

(4) K. Ohji, K. Ogura, and S. Kubo "The Convenient Equation of J-Integral at Fully Plastic and Its Application to Creep Fracture", Trans. of the Japan Society of Mechanical Engineers, Vol.44-382, (1978), pp.1831

(5) A.S. Kobayashi, S.T. Chiu, and R. Beeuwkes "A Numerical and Experimental Investigation on the Use of J-Integral", Engineering Fracture Mechanics, Vol.5, (1973), p283

(6) T. Kanazawa, S. Machida, S. Kaneda, and M. Onozuka "A Study on J-Integral as a Fracture Criterion", The Society of Naval Architects of Japan, Vol. 138, (1975), pp480

(7) J.D. Sumpter, and C.E. Turner "Method for Laboratory Determination of J_{IC}", ASTM STP601, (1976), p3

(8) M. Toyoda, Y. Itoh, and K. Satoh "Deformability in General Yielding Unstable Fracture of Weldments of Structural Steels", The Society of Naval Architects of Japan, Vol.145, (1979), pp212

(9) H. Okamura, and H. Itagaki, Statistical Treatment of Strength, Baihukan, (1979), 307pp. (in Japanese)

(10) H. Kobayashi, "The J-Integral Evaluation of Stretched Zone Width and Its Application to Elastic-Plastic Fracture Toughness Test", Trans. of the Japan Society of Mechanical Engineers, Vol.45-392, (1979), pp.336

(11) D.G.H. Latzko, Post-Yield Fracture Mechanics, Applied Science Publishers, (1979), 349 pp. (in U.S.A.)

(12) "The Plan of Joint Research for Fracture Toughness Criterion", Japan Welding Society, FTC Committee, FTC-53-1 (1978) 16pp

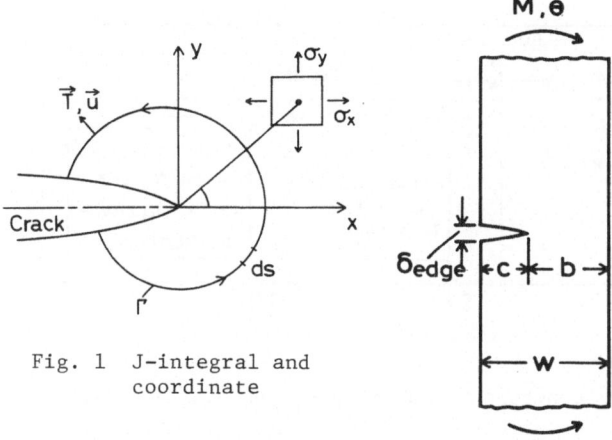

Fig. 1 J-integral and coordinate

Fig. 2 One side cracked plate subject to bending

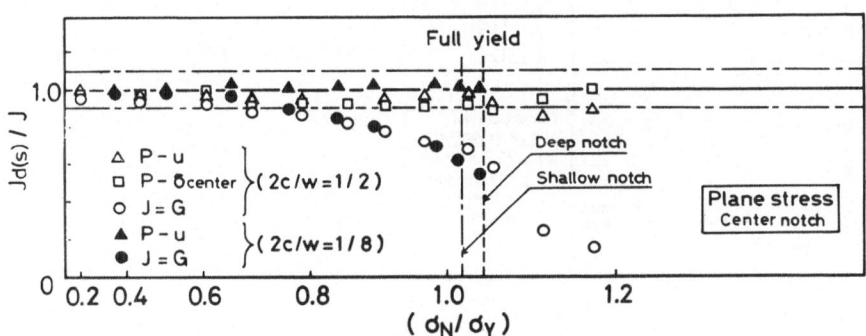

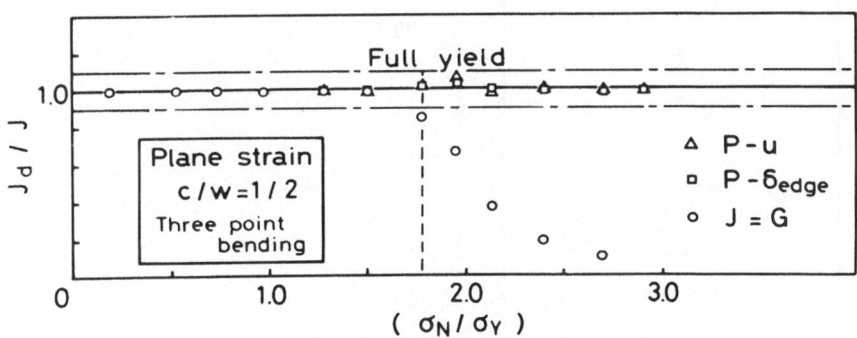

Fig. 3 Accuracy of convenient equations of the J-integral

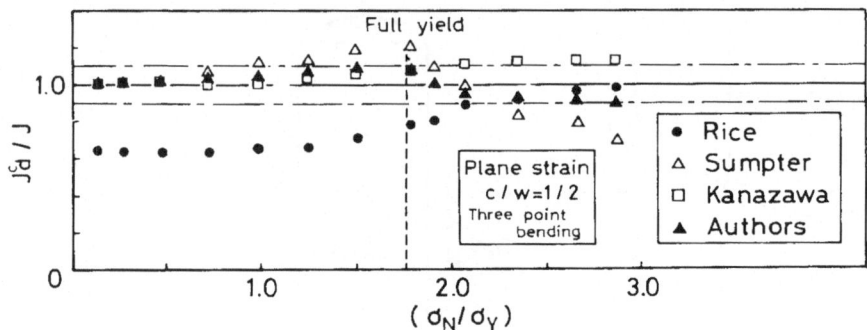

Fig. 4 Accuracy of convenient equations of the J-integral

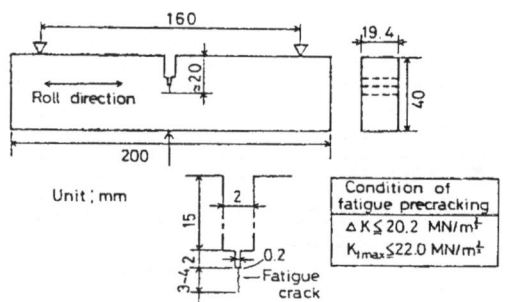

Fig. 5 Specimen configuration
and details of
precracking

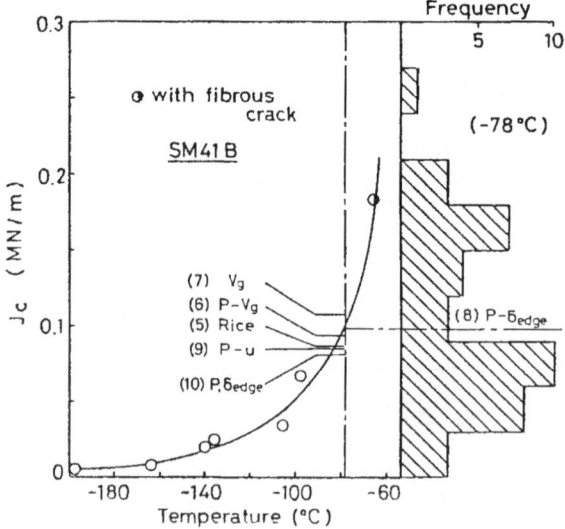

Fig. 6.a)

Temperature dependency
of J_c and its frequency
at -78°C

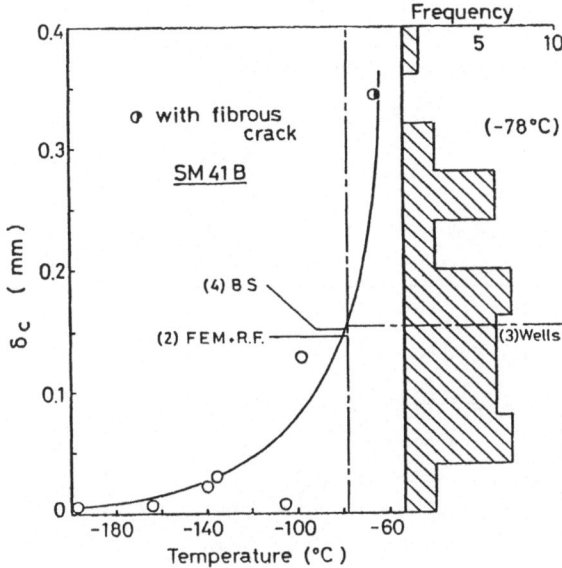

Fig.6 b)

Temperature dependency
of δc and its frequency
at −78°C

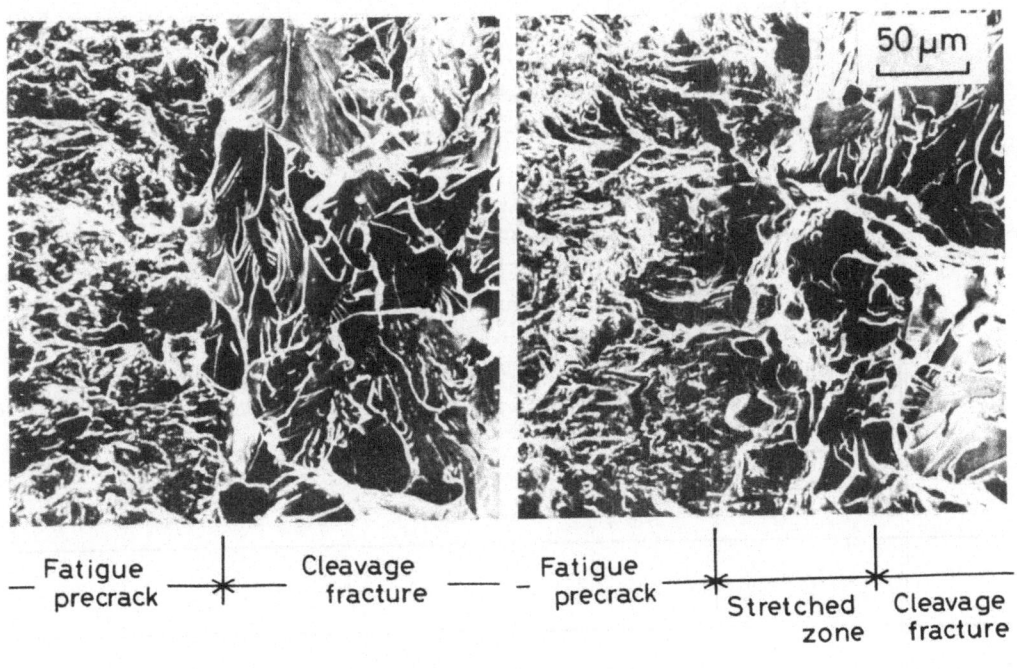

a)   b)

Fig. 7   Fractographs at −78°C

160

Table 1. Convenient evaluating equation of J-integral

		Assumption	J estimating formulas
Tensile notched specimen	Deep notch	$U_{\text{crack}} = b\, f(P/bt)$	$J_d = G + \frac{2}{bt}[\int_0^u Pdu - \frac{1}{2}Pu]$ (Rice) P-u $J_d = G + \frac{2}{bt}[\int_0^{\delta center} Pd\delta center - \frac{1}{2}P\delta center]$ P-δcenter
	Shallow notch	$U_{\text{crack}} = C\frac{C}{W} f'(P/wt)$	(General yielding) (Full yielding) $J_s = \frac{1}{ct}\int_0^P U_{\text{crack}}\,dP$ $J_s = \frac{1}{ct}[\int_0^P udP - \frac{1}{2}Pu_{e.\,no\,crack}]$ (Authors) P-u
Bending notched specimen	Deep notch	$\theta_{p.crack} = g(M/b^2t)$	$J_d = G + \frac{2}{bt}[\int_0^\theta Md\theta - \frac{1}{2}M\theta_e.]$ $(J_d = G + \frac{2}{bt}[\int_0^u Pdu - \frac{1}{2}Pu_e.])$ $(J_d = G + \frac{2}{bt}[\frac{W}{C+\delta b}\int_0^{\delta edge} Pd\delta edge - \frac{1}{2}P\delta e.\,edge])$ P-u (Authors) P-δedge
	Shallow notch	$\theta_{\text{crack}} = (\frac{C}{W})^2 g'(M/wt)$	(General yielding) (Full yielding) $J_s = \frac{2}{ct}\int_0^M \theta_{\text{crack}}\,dM$ $J_s = \frac{2}{ct}[\int_0^M \theta dM - \frac{1}{2}M\theta e.\,no\,crack]$ $(J_s = \frac{2}{ct}\int_0^P U_{\text{crack}}\,dP)$ $(J_s = \frac{2}{ct}[\int_0^u udP - \frac{1}{2}Pu_{e.\,no\,crack}])$ (Authors) P-u
Round bar with circumferential notch	Deep notch	$U_{p.crack} = r\,h(P/R^2)$	$J_d = G + \frac{1}{2\pi r^2}[3\int_0^u Pdu - Pu - \frac{1}{2}Pu_e]$ (Authors) P-u
	Shallow notch	$U_{\text{crack}} = C\frac{C}{2R} h'(P/R^2 r)$	(General yielding) (Full yielding) $J_s = \frac{1}{2\pi r C}\int_0^P U_{\text{crack}}\,dP$ $J_s = \frac{1}{2\pi r C}[\int_0^P udP - \frac{1}{2}Pu_{e.\,no\,crack}]$ (Authors) P-u

Table 2. More convenient equation of J-integral

	Assumption	J estimating formulas
Authors	$U_{p.crack} \propto P^{\frac{1}{n}}$ $U_{p.crack} \propto \varepsilon_0$	(Tensile notched specimen) $J_d^c = G + \frac{1-n}{1+n}\sigma_N(u - u_e.)$ $J_d^c = G + \frac{1-n}{1+n}\sigma_N(\delta center - \delta e.center)$ [P-u] [P-δcenter] (Bending notched specimen) $J_d^c = G + \frac{2}{1+n}\frac{D}{bt}(u - u_e.)$, $J_d^c = G + \frac{2}{1+n}\frac{D}{bt}\frac{W}{C+\delta b}(\delta edge - \delta e.edge)$ (Round notched bar) $J_d^c = G + \frac{2-n}{2+2n}\frac{D}{\pi r^2}(u - u_e.)$
	$U_{\text{crack}} \propto P^{\frac{1}{n}}$ $U_{\text{crack}} \propto \varepsilon_0$	(Tensile notched specimen) (Bending notched specimen) $J_s^c = \frac{n}{1+n}\frac{b}{C}\sigma_N U_{\text{crack}} F(C/W)$, $J_s^c = \frac{n}{1+n}\frac{D}{Ct}U_{\text{crack}} G'(C/W)$
Rice	$U_{\text{crack}} = g(P/b^2t)$	$J_d^c = \frac{2}{bt}\int_0^{U_{\text{crack}}} Pd u_{\text{crack}}$ (Bending notched specimen)
Kanazawa	$U_{\text{crack}} = sf(C/W)$ $g(P/b^2t)$ S ; Bending span	$J_d^c = \alpha\frac{Du_{\text{crack}}}{W} - (\frac{\alpha}{W} - \frac{2}{D})\int_0^{U_{\text{crack}}} Pd u_{\text{crack}}$ (Bending notched specimen) $\alpha = 2.2$ at $c/w = 1/2$
Sumpter		$J_d^c = G + \frac{2P_L}{bt}[\frac{W}{C+\delta b}](\delta edge - \delta e.edge)$ (Bending notched specimen) P_L ; Net yield load

Table 3 Chemical compositions and
Mechanical properties

Chemical compositions				Mechanical properties			
C X100	Si X100	Mn X100	S X100	Y. P. MN/m²	U.T.S. MN/m²	EL. %	RA. %
14	21	18	6	254	432	40	71

Table 4 Mean value and standard deviation
of fracture toughness

Evaluation method		Mean value	Standard deviation
J-integral (MN/m)	(2) Rice's eq.	0.084 (0.091)	0.048 (0.053)
	(3) P - Vg	0.091 (0.120)	0.061 (0.062)
	(4) Vg	0.106 (0.088)	0.051 (0.052)
	(5) P - δedge	0.096 (0.102)	0.057 (0.060)
	(6) P - u	0.083 (0.092)	0.035 (0.053)
	(7) P, δedge	0.079 (0.090)	0.044 (0.052)
COD (mm)	(8) FEM + R.F.	0.146 (0.148)	0.068 (0.073)
	(9) Wells eq.	0.154 (0.164)	0.094 (0.106)
	(10) BS eq.	0.152 (0.166)	0.101 (0.136)

( ): from the assumption of normal distribution

IV. STRESS, STRAIN AND COD AS FRACTURE CRITERIA

NOTCH ANALYSIS OF DUCTILE FRACTURE

Volker Weiss
Professor of Materials Science
Syracuse University, Syracuse, New York 13210, USA

Notch analysis of ductile fracture is based on the stress field analysis of notches by Neuber (1,2) and the slip-line field solution by Hill (3). The work at Syracuse (4-6) has shown that an estimate of a materials resistance to crack extension can be obtained by considering the elastic and plastic deformation processes in the crack tip region. Detailed studies of the strain field in the vicinity of the crack tip, e.g. by Liu and co-workers (7) and by Ogasawara, Adachi, Nagao and Weiss (8) show that the strain field ahead of the crack tip is reasonably well represented by:

$$\varepsilon_{r,\theta} = \varepsilon_{F,\alpha,\beta}^{\frac{2}{n+1}} \left(\frac{\rho^*}{\rho^*+2r}\right)^{\frac{1}{n+1}} f_{ij}(\theta) \tag{1}$$

where n is the strain hardening exponent, ρ^* the Neuber micro support effect constant and $\varepsilon_{F\alpha\beta}$ the maximum strain at the tip of an extending crack under the stress field, $\alpha = \dfrac{\sigma_2}{\sigma_1}$ and $\beta = \dfrac{\sigma_3}{\sigma_1}$, in that region. The experimental measurements indicate good agreement with $n \simeq 1$ i.e.

$$\varepsilon_{r,\theta} = \varepsilon_{F,\alpha,\beta} \left(\frac{\rho^*}{\rho^*+2r}\right)^{\frac{1}{2}} f_{ij}(\theta) \tag{2}$$

Thus the size of the plastic zone, r_p, in the direction of crack extension, is given by

$$r_p = \frac{\rho^*}{2} \left[\left(\frac{\varepsilon_{F\alpha\beta}}{\varepsilon_r}\right)^2 - 1\right] \tag{3}$$

where ε_r is the effective yield strain.

This leads to an estimate for the fracture toughness. The work to propagate the plastic zone a distance associated with the formation of a unit area of new crack surface is proportional to the square of the fracture strain at the crack tip, $\varepsilon_{F\alpha\beta}^{2}$. The model predicts that K_{IC}, the critical stress intensity factor at fracture, is proportional to the fracture strain at the tip of the crack. As this fracture strain is difficult to determine, Weiss and co-workers have proposed use of the fracture strain as obtained from an equi-biaxial bulge test, or a plane strain tensile test. For equi-biaxial tests on steels the data yield the relationship

$$K_{IC} = 162 \; \varepsilon_{F,\alpha=1,\beta=0} \; \text{Mnm}^{-3/2} \tag{4}$$

For aluminum alloys the data suggest

$$K_{IC} = 168 \; \varepsilon_{F,\alpha=1,\beta=0} \; \text{MNm}^{-3/2} \tag{5}$$

The fracture strain at the crack tip, $\varepsilon_{F\alpha\beta}$ critically determines the fracture toughness. For the coalescence mode of crack propagation this strain is made up of two components, the nucleation strain ε_{n} and the propagation strain ε_{p}. Conditions for void nucleation have been analyzed by Argon and co-workers (9) for void formation at inclusions. The work indicates that void nucleation occurs at a critical mean stress, σ_{m}, which is given by

$$\sigma_{m} = \sigma_{rr} - Y(\varepsilon,T) \tag{6}$$

where σ_{rr} is the inclusion – matrix separation stress and Y is the local flow stress of the matrix. McClintock (10) has analyzed the conditions for void coalescence and obtained

$$\varepsilon_{F} = \frac{(1-n) \; \ln V_{F}^{-1/3}}{\sinh[(1-n)(\sigma_{1}+\sigma_{2}/2\bar{\sigma}/\sqrt{3})]} \tag{7}$$

where the V_{F} is the volume fraction of voids, $(V_{F})^{-1/3}$ therefore the intervoid distance. Examination of both relationships show that the nucleation strain as well as the coalescence strain is critically dependent on the stress state. The temperature dependence of the void nucleation strain is a consequence of the temperature dependence of the flow stress. This causes a ductile to brittle transition in bcc materials without any change in the microscopic fracture mode, namely void coalescence. Because of the ease with which one can conduct bulge tests or plane strain tensile tests, the method of assessing the fracture toughness of a material, and its dependence on variables such as temperature, strain rate, and environment has been applied to other high strength steels and TRIP steels. Preliminary results and a tentative analysis of these studies are presented.

REFERENCES

(1) H. Neuber: Kerbspannungslehre. First Edithion 1937, Second Edition 1958,
     Berlin, Gottingen, Heidelberg.

(2) H. Neuber: Theory of Stress Concentration for Shear-Strained Prismatical
     Bodies With Arbitrary Non-Linear Stress-Strain Law. Trans. Am. Soc. Mech.
     Eng. J. Appl. Mech., 28 (1961) pp. 544-550.

(3) R. Hill: Mathematical Theory of Plasticity. Oxford (1950).

(4) V. Weiss: Notch Analysis of Fracture. In: Liebowitz, H.: Fracture,
     Vol. 3, New York and London, (1971) pp. 227-264.

(5) V. Weiss, Y. Kasai, and K. Sieradzki, "Microstructural Aspect of Fracture
     Toughness, ASTM STP 605, American Society for Testing and Materials, (1976)
     pp. 16-33.

(6) V. Weiss and H. Neuber, "Recent Advances in Notch Analysis of Fracture and
     Fatigue", Ingenieur-Archiv 45 (1976) S. 281-289, Springer-Verlag (1976).

(7) H. W. Liu, "An Analysis on Fatigue Crack Propagation, NASA CR-2032 (1972).

(8) M. Ogasawara, M. Adachi, M. Nagao, and V. Weiss, "Crack Initiation at
     Notches in Low Cycle Fatigue", Proc. III Intl. Conf. Fracture, Munich,
     Germany, V. 4, (April 1973) p. 511.

(9) A. S. Argon, J. Im, and R. Safoglu, Metallurgical Transactions, Vol. 6A,
     (1975) pp. 825-837, pp. 839-851.

(10) F. A. McClintock, "Plasticity Aspects of Fracture", In: Liebowitz, H.:
     Fracture, Vol. 3, New York and London, (1971) pp. 47-225.

A CRITERION FOR ELASTIC-PLASTIC FAILURE OF AN EDGE-CRACKED MEMBER
UNDER COMBINED LOADING OF MEMBRANE AND BENDING STRESSES

H. Okamura*, K. Kageyama**, T. Takano***
*Professor, **Graduate Student, ***Research Associate
of Mechanical Engineering
University of Tokyo, Bunkyo-ku, Tokyo 113, Japan

M. Kurotobi
Mukogawa Plant, Kubota Limited, Ohama-cho, Amagasaki 660, Japan

ABSTRACT

In order to establish a design criterion for the strength of an edge-cracked structural member under the combined loading of membrane stress and bending stress, the collapse load and the J-integral of the cracked member of pressure vessel steel have been experimentally studied in the present paper. It is concluded that the collapse load of the cracked member with a shallow crack, which is of interest in structural design and safety assessment, can be satisfactorily evaluated by existing analytical solutions for a deep crack, and that the strength of the member in the whole range from brittle to ductile failure can be estimated by a universal failure curve, based on the two-criteria approach from the fracture toughness and the collapse load.

INTRODUCTION

In some structural design codes, such as ASME Pressure Vessel Code Sec. III, the design of the energy-related, structural component is usually and essentially based on the collapse load in the un-cracked state of the member, but partly on the strength for brittle fracture in its cracked state. However, the strength might be important to evaluate for ductile failure of a cracked member in the design of most structural components. The criterion for brittle fracture has been well established, and we can predict the fracture strength of the cracked member with a heavy section by the fracture toughness K_{IC} in the small-scale yielding condition. On the other hand, one of the most convenient design criterion of ductile failure of a cracked member with a thin section will be given by the collapse load. Once both criteria are revealed, the strength of a cracked member with a medium section migth be estimated by a two-criteria approach, using the K_{IC} and the collapse load.

Some analytical solutions have been obtained for the collapse load of an edge-cracked member under the combined loading of the axial force and the bending moment. The present authors [1] have also investigated the load-carrying capacity of a cracked column under an eccentric compressive load experimentally and analytically. From the theoretical perspective, these analytical solutions cannot be applied to a shallow crack, and no other solution is currently known. In the structural design as well as in the assessment of safety, it might be important to reveal how the collapse load varies when a given length of crack is induced. From this point of view, the collapse load of the specimen of pressure vessel steel with a shallow edge crack is experimentally investigated under the combined loading of the axial force and the bending moment.

Rice [2] has proposed the J-integral as a fracture characterizing parameter in the case of both small- and large-scale plasticity. Many recent works demonstrate the validity of the fracture toughness J_{IC} as the criterion for the initiation of crack growth, independent of the type and the size of specimen. Once the J_{IC} criterion is established, the elastic-plastic behaviour of defects in a real structure can be predicted from the results of a small, easily performed, laboratory test. In order to apply the J-integral technique to a structural design, the method of the J-integral analysis for a cracked member under the combined loading conditions must be established. In the present paper, the two-criteria method proposed by Dowling and Townley [3] is adopted, and the effects of the combined loading is investigated experimentally, using the J_{IC} concept.

ANALYTICAL RESULTS ON THE COLLAPSE LOAD AND ITS LIMITATION

Theoretical solutions for a deep crack

Some results [1,4-6] have been obtained on the collapse limit of the member of rigid-perfectly plastic material with a deep crack under combined loading of the axial force and the bending moment. These will be briefly summarized here for the convenience of comparison with the experimental results.

For an edge-cracked specimen with width W, crack length a, and thickness b, the membrane stress P'_m and the bending stress P'_b (in the sense of stress intensity) are defined, by using the ligament width $W' = W-a$, the axial force F, and the bending moment M', referred to the center line of the ligament width:

$$P'_m = F/(bW') \quad , \quad P'_b = 6M'/(bW'^2) \tag{1}$$

(A) Extension of Green's solution: A so-called "perfect solution" for pure bending by Green [7] can be easily extended to the case in which the membrane stress is superimposed. The collapse stresses normalized by the tensile yield strength S_y are given by Eq. (2) in the range of small P_m'/S_y ratio.

$$P_b'/1.5S_y = 1.261(1+P_m'/S_y)^2 - 2(1+P_m'/S_y)(P_m'/S_y) \quad , \quad (-1 \leq P_m'/S_y \leq 0.551) \qquad (2)$$

(B) The Rice's solution: In the range in which the modified Green slip line cannot be applied (that is, $0.551 \leq P_m'/S_y \leq 1$), the slip line field which gives an upper bound has been proposed by Rice, and the minimum value in the upper bounds has been obtained numerically [5,6]. An approximate expression of the collapse stresses obtained in the present work is given by the following polynomial.

$$P_m'/S_y = 1.00 - 0.00864\xi + 0.976\xi^2 - 3.35\xi^3 + 3.81\xi^4 - 1.47\xi^5 \quad , \quad (0.551 \leq P_m'/S_y \leq 1) \qquad (3)$$

where $\xi = P_b'/1.5S_y$, and the error of the expression is lower than 2%. The Rice solution connects smoothly with the modified Green solution at $P_m'/S_y = 0.551$.

(C) The lower bound solution: A well-known, lower bound solution is given by

$$P_b'/1.5S_y = 1 - (P_m'/S_y)^2 \quad , \quad (-1 \leq P_m'/S_y \leq 1) \qquad (4)$$

As is indicated in Fig. 3, the lower bound (C) , shown by a broken line, gives very low value compared with the upper bounds (A) and (B). Considering the fact that the upper bound (A) is very close to the experimental results for pure bending [7], we can expect that the upper bound (A) is a good approximation of the collapse limit as far as a crack is deep.

Critical crack depth

The analytical solutions (A) and (B) are applicable to a crack deeper than a certain critical value. Taking account of the effect of axial force on the analysis for pure bending by Green [8] and Ewing [9], the critical crack depth ratio $(a/W)_c$ for the upper bound (A) is given as follows:

$$(a/W)_c = 0.297(1+P_m'/S_y)/(1+0.297P_m'/S_y) \quad , \quad (-1 \leq P_m'/S_y \leq 0.551) \qquad (5)$$

$(a/W)_c$ equals 0.297 for pure bending and increases as P_m'/S_y increases. On the other hand, the rigorous solution of the collapse load of an un-cracked member coincides with the lower bound (C). In the range from the un-cracked state to the critical crack depth, no analytical solution is obtained. In the structural design and for the assessment of safety, the strength of the member with a shallow crack is important to know. Therefore, the collapse load of the specimen with a shallow crack is experimentally investigated and compared with the analytical solutions for a deep crack mentioned above.

MATERIALS AND SPECIMEN CONFIGURATION

Materials

The materials tested are the two kinds of pressure vessel steel, JIS SQV2A and JIS SGV49N, equivalent to ASTM A533B Cl.1 and ASTM A516 respectively. Their mechanical properties and chemical compositions are shown in Tables 1 and 2.

The specimens were cut from the plate perpendicularly in a rolling direction.

Specimen configuration

In order to apply both the membrane stress and the bending stress simultaneously, the edge-cracked specimens were loaded eccentrically by pins. The specimen configuration and the eccentricity are shown in Fig. 1. The crack-like notch is induced by a grinding cutter, whose width and root diameter are approximately 0.2 mm. The combinations of the eccentricity e and the crack depth ratio a/w are shown in Table 3.

EXPERIMENTAL RESULTS ON COLLAPSE

During the tests, the axial displacement at the loading point u and the angular deformation between the end of specimen Θ were measured. The bending moment was calculated from the axial force and the lateral deflection.

Fig. 2 shows some examples of the load-displacement diagram. The collapse could be defined in many ways; but for convenience's sake, the criteria of twice displacement and twice compliance, both approximately indicating the onset of gross deformation, were applied in the study. Using the criterion of twice displacement, the collapse limit is determined by the load at the displacement of twice of that at the proportional limit on load-displacement diagram. The criterion of twice compliance defines the collapse limit by the intersection of the load-displacement curve and the straight line whose compliance is twice as large as that in the proportional range of the load-displacement diagram. The collapse load determined by these criteria and the maximum load are also shown in Fig. 2.

Fig. 3 shows the collapse points determined by the twice-displacement criterion and the maximum P_m' points on the trajectory of P_m' versus P_b' , together with the lower bound and the upper bound for a deep crack. After the specimen collapses near the upper bound curve, the ratio P_b'/P_m' decreases because of the lateral deflection. Then, P_m' reaches its maximum point. No remarkable difference exists between the two criteria of collapse, shown in Fig. 4. The results of compressive tests are also plotted in the figure.

These experimental results on two materials show that the upper bound analytical solutions for a deep crack can be used as a satisfactory design criterion for the collapse or the onset of the gross deformation of the member with a shallow crack with a/W of 0.1-0.3, and the cracked member can sustain the load far beyond the analytical upper bound.

COLLAPSE LIMIT DIAGRAM FOR STRUCTURAL DESIGN

In the design of structural components, the membrane stress P_m and the bending stress P_b in the un-cracked state are usually defined for the full thickness W.

$$P_m = F/bW \quad , \quad P_b = 6M/(bW^2) \qquad (6)$$

where M is the bending moment referred to the center line of full thickness.

Since it was experimentally verified that the upper bound solutions can be approximatelly applied to a shallow crack, the upper bound curves are shown in Fig. 5 as a function of the crack depth ratio a/W on P_m versus $P_m + P_b$ diagram, instead of P_m' versus P_b' diagram for the convenience of the structural design. The outer closed curve of parabolas is the collapse limit of an un-cracked member given by Eq. (4). The inner hexagon shows the design limit in the ASME Pressure Vessel Code Sec. III in the normal operating condition.

It is concluded from this diagram that the member can sustain the design load without appreciable plastic deformation, as long as the crack depth is shallower than a quarter of the thickness and if the member is designed according to the ASME design code.

THE TWO-CRITERIA UNIVERSAL FAILURE CURVE

The evaluation of the J-integral in a structural member is one of the most important problems in the application of the J_{IC} criterion to the structural design. In the present paper, the validity of the two-criteria approach proposed by Dowling and Townley [3] is investigated from the viewpoint of a J-integral evaluation for the cracked member under bending and tension.

Let P_m^c be the collapse stress and P_m^f be the fracture stress estimated by using the Linear Elastic Fracture Mechanics through K_{IC} under a given P_b/P_m ratio. The membrane stress P_m at fracture for the same P_b/P_m ratio may be postulated by a straightforward extension of the universal failure curve, based on the J_{IC} criterion in the Dugdale model. The universal failure curve will be given by the relation

$$P_m = P_m^c \frac{2}{\pi} \cos^{-1}[\exp\{-\frac{\pi^2}{8}(P_m^f/P_m^c)^2\}] \qquad (7)$$

Eq. (7) is derived from the theoretical result of an infinite inner-cracked plate subjected to uniform tensile stress at infinity, and the effects of the finite boundaries and the bending stress are not account for. The applicability of Eq. (7) is experimentally investigated in the case of the edge-cracked member under the combined bending and tensile loading.

The basic assumptions in the investigation follow: 1) The initiation of crack growth can be reasonably estimated by the J_{IC} criterion, independent of the specimen size and configuration.   2) The collapse is defined as the onset of gross deformation.   the criterion of twice compliance is adopted here. 3) The critical condition of fracture is determined by the initiation of crack growth or the collapse.   4) The specimen of similar shape and different size must behave similarly until the initiation of crack growth, in a sense of continuum mechanics.

Under these assumptions, the brittle-ductile transition can be determined from the experiment on the small specimen, in which the stable crack growth does not occur before the collapse.   The J-integral is determined experimentally from the elastic-plastic behaviour of specimens by a specimen compliance method generalized by Begley and Landes [10].   The normalized relation between J and P_m can be expressed as Eq. (8) for the given a/W and P_b/P_m ratios, by using a function f which can be determined experimentally or analytically,

$$\sqrt{\frac{E'J}{\pi a P_m^{c_2}}} = f(P_m/P_m^c) \tag{8}$$

Under the assumptions mentioned above, Eq. (8) is applicable to the specimen of similar shape with different size, and Eq.(9) is satisfied at fracture.

$$\sqrt{E'J} = C\sqrt{\pi a} \ P_m^f = K_{IC} \tag{9}$$

where C is a constant for a given combination of a/W and P_b/P_m.   Substituting Eq. (9) into Eq. (8), we obtain the relation between P_m^f/P_m^c and P_m/P_m^c at fracture for the given a/W and P_b/P_m ratios.

$$P_m^f/P_m^c = f(P_m/P_m^c)/C \tag{10}$$

Fig. 6 shows the predicted fracture curves obtained from the experimental results on A533B Cl.1 under several combinations of the bending and the tension. The range of e/W and a/W are 0.01-0.66 and 0.2-0.25 respectively.   The numerical analysis by the finite element method (FEM) is also carried out for the compact specimen (CT) with the mechanical properties of A533B Cl.1 and the crack depth ratio a/W of 0.52.   The theoretical curve given by Eq. (7) and the numerical result are also drawn in Fig. 6.

Within the range of experiments, the fracture stress under a combined loading of bending and tension can be estimated from the universal failure curve in Eq. (7) with a certain accuracy, as long as the J_{IC} criterion is valid.

CONCLUSIONS

The collapse limit and the load carrying capacity of the specimen with a relatively shallow edge-crack is experimentally investigated by using the two kinds of pressure vessel steel under the combined loading of the axial force and the bending moment. It is concluded from the experimental results that the collapse load of the member with a shallow crack can be satisfactorily evaluated by the upper bound analytical solutions for a deep crack. The collapse limit diagram as a structural design criterion is also shown in terms of the crack depth ratio.

In addition, the possibility of a straightforward application of the two-criteria approach is experimentally investigated in the combined loading condition within the framework of the J_{IC} concept, and it is estimated that the strength of the cracked member in the whole range from brittle to ductile failure can be approximately evaluated by a universal failure curve obtained from the fracture toughness and the collapse load.

ACKNOWLEDGEMENTS

The authors wish to express their gratitude to Prof. H. Shibata, University of Tokyo, and the other members of EDR Research Project. The present work has been done as part of that project.

REFERENCES
(1) H. Okamura, K. Watanabe and T. Takano, "Deformation and Strength of Cracked Member under Bending Moment and Axial Force", Eng. Fract. Mech., 7 - 3, (7-1975), pp. 531-539.

(2) J. R. Rice, "A Path Independent Integral and the Approximate Analysis of Strain Concentration by Notches and Cracks", Trans. ASME J. Appl. Mech., 35 - 2, (6-1968), pp. 379-386

(3) A. R. Dowling and C. H. A. Townley, "Effect of Defects on Structural Failure: A Two-Criteria Approach", Int. J. Pres. Ves. and Piping, 3, (1975), pp. 77-107.

(4) M. Shiratori and T. Miyoshi, "A J-Integral Estimation of CT Specimen Based on the Analysis of the Slip Line Field", Trans. JSME Ser. A, 45 - 389, (1-1979), pp.50-56, (in Japanese).

(5) M. Shiratori and T. Miyoshi, "Plastic Constraint of SEN Specimen and its J-Integral", Preprint of JSME, No. 780-9, (7-1978), pp. 63-68, (in Japanese).

(6) D. J. F. Ewing and C. E. Richards, "The Yield-Point Loads of Singly-Notched Pin-Loaded Tensile Strips", J. Mech. Phys. Solids, 22, (1-1974), pp. 27-36.

(7) A. P. Green, "The Plastic Yielding of Notched Bars due to Bending", Quart. J. Mech. Appl. Math., 6, (1953), pp. 223-239.

(8) A. P. Green, "The Plastic Yielding of Shallow Notched Bars due to Bending", J. Mech. Phys. Solids, 4 - 3, (1956), pp. 259-268.

(9) D. J. F. Ewing, "Calculations on the Bending of Rigid/Plastic Notched Bars" J. Mech. Phys. Solids, 16, (6-1968), pp. 205-213.

(10) J. A. Begley and J. D. Landes, "The J Integral as a Fracture Criterion", ASTM STP, 514, (1972), pp. 1-20.

Table 1 Mechanical properties

Material	Yield Strength of 0.2% Offset (MN/m^2)	Tensile Strength (MN/m^2)	Elongation (%)	Reduction of Area (%)
A516 (SGV49N)	325	477	32.7	67.1
A533B Cl.1 (SQV2A)	487	599	26.6	70.5

Table 2 Chemical compositions

Material		C	Si	Mn	P	S	Ni	Cr	Cu	Mo	V
A516 (SGV49N)		.16	.27	1.11	.016	.005	.20	.01	.01	.19	.05
A533B Cl.1 (SQV2A)	Top.	.17	.26	1.39	.007	.007	.68	.20	.04	.49	trace
	Bot.	.17	.27	1.38	.008	.006	.67	.20	.04	.49	trace

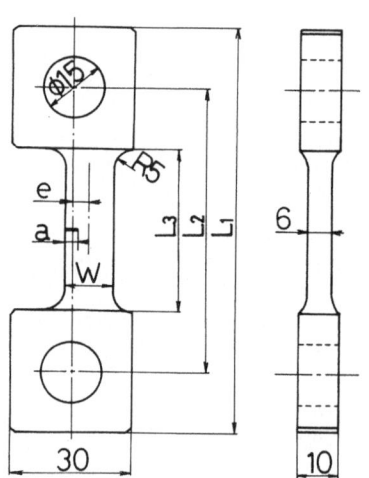

material	L₁	L₂	L₃	W
A516	100	70	40	12
A533B Cl.1	96	66	36	12

type	1	2	3	4	5
e	0	1	2	4	8

unit ; mm

Fig. 1 Specimen configuration

Table 3 Test condition

No.	A516 (SGV49N) e(mm)	a/W	A533B Cl.1 (SQV2A) e(mm)	a/W
1	-0.42	0.153	-0.02	0.147
2	0.02	0.249	0.14	0.201
3	0.02	0.215	0.10	0.246
4	0.01	0.279	0.21	0.301
5	0.92	0.238	0.79	0.158
6	0.90	0.198	0.85	0.316
7	0.96	0.303	0.82	0.194
8	0.87	0.193	0.84	0.320
9	0.90	0.154	1.84	0.210
10	1.86	0.148	1.93	0.252
11	1.90	0.219	1.95	0.286
12	1.87	0.319	1.86	0.152
13	1.90	0.262	3.88	0.164
14	1.89	0.231	3.93	0.203
15	1.90	0.196	3.83	0.252
16	1.91	0.149	3.88	0.300
17	1.86	0.290	7.83	0.151
18	3.95	0.182	7.76	0.203
19	3.93	0.226	7.93	0.243
20	3.89	0.250	7.90	0.304
21	3.87	0.295	-1.80	0.303
22	8.04	0.358		
23	7.96	0.240		
24	-1.94	0.150		
25	-2.13	0.287		
26	-4.05	0.308		
27	-4.08	0.160		
28	-3.98	0.400		
29	-4.05	0.507		
30	-8.10	0.296		
31	-8.07	0.500		
32	-8.06	0.148		

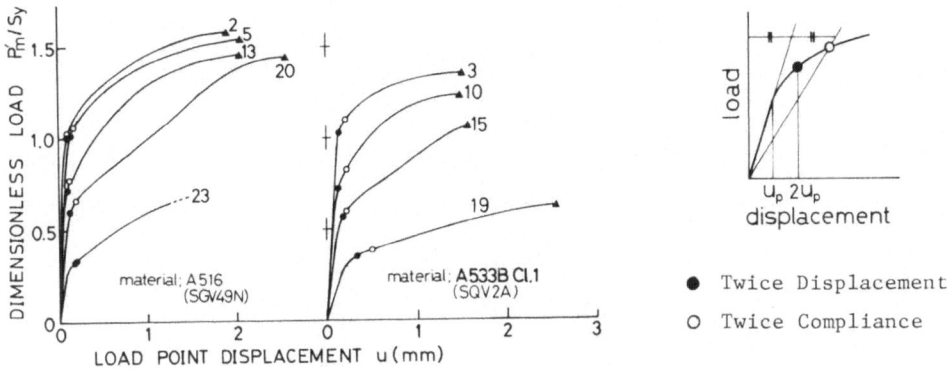

Fig. 2 Examples of load-displacement diagram

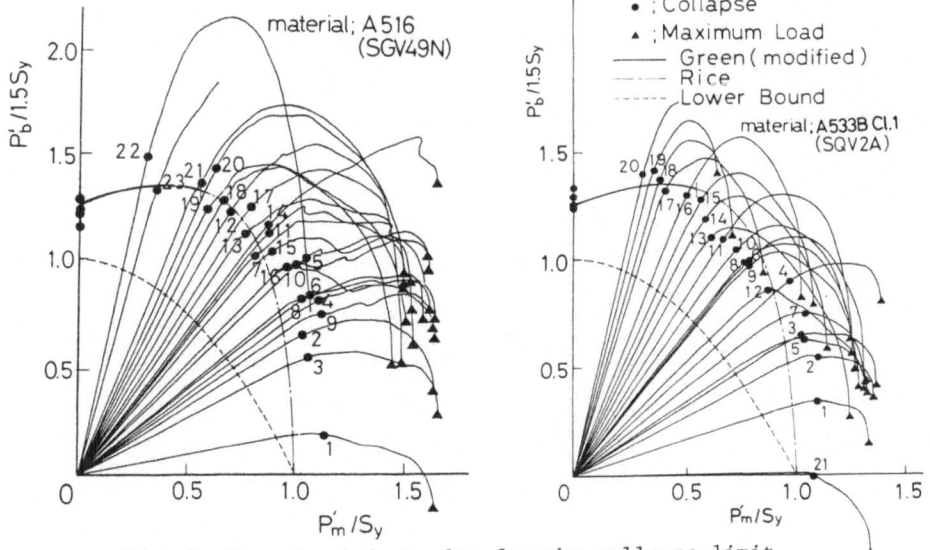

Fig. 3 Experimental results for the collapse limit
and the maximum membrane stress

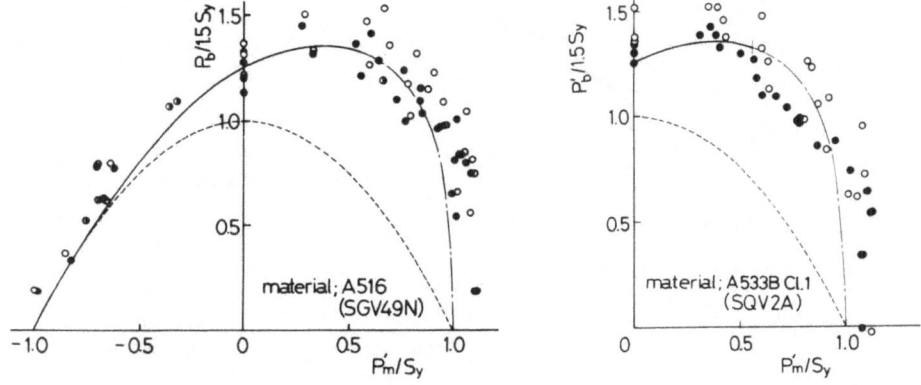

Fig. 4 Comparison of the two criteria for collapse limit

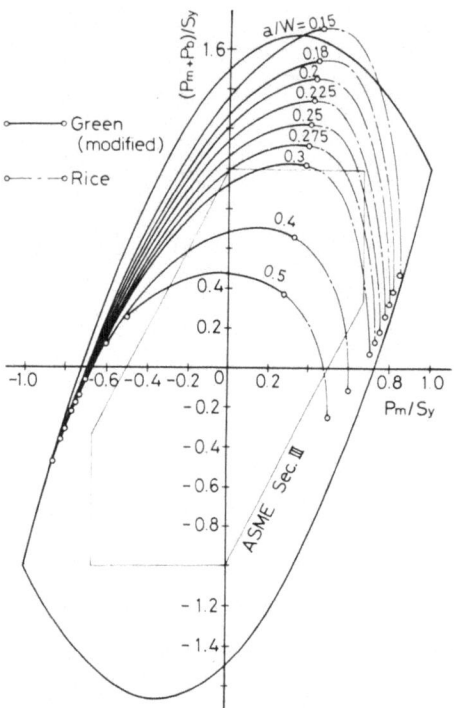

Fig. 5   Collapse limit diagram for structural design

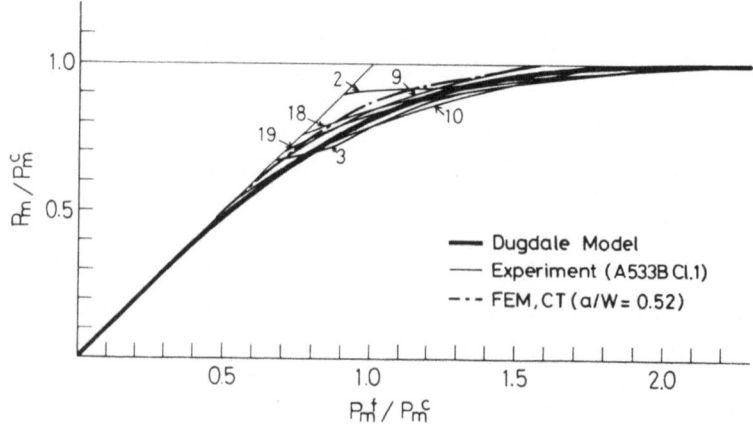

Fig. 6   Failure curves under several combined loading conditions

INFLUENCE OF MEAN STRESS ON DUCTILE FRACTURE PROPERTIES

T. Suzuki* and S. Yanagimoto**
*Researcher, **General Manager
Labo. II in Research Department,
Products Research & Development Laboratories
Nippon Steel Corporation
5-10-1, Fuchinobe, Sagamihara,
Kanagawa 229, Japan

ABSTRACT

The following facts were revealed by this investigation on ductile fracture behavior.

(1) Fracture strain depends on both specimen direction and stress condition in material.

(2) To estimate the ductile fracture behavior under multiaxial stresses, $\sigma_m/\bar{\sigma}$ can be used as the parameter of stress conditions with the linear relationship between $\sigma_m/\bar{\sigma}$ and $\bar{\varepsilon}_f$.

INTRODUCTION

It is well-known that the fracture mechanics presented by Irwin et al. [1] clearly explains the fracture stress of a brittle fracture, which occurs in structures made of high tensile steel or in structures used at low temperature. However, when structural members made of medium strength steel are used near room temperature, the fracture mode is quite different from that of brittle fracture; that is, the members are subjected to large plastic deformation even if they have notches, and they fracture by necking. The safety considerations for such steel structures make the clarification of fracture behavior after large plastic deformation necessary. Generally, when a steel structure fractures, structural parts relevant to the fracture are often strongly constrained. In ductile fracture as well as brittle fracture, the fracture behavior depends, to a large extent, on the stress conditions. Therefore, it is important to investigate the fracture behavior under multiaxial stresses and to determine the effect of stress conditions on fracture stress quantitively. From this viewpoint, ductile fractures are investigated from the actual aspects of fracture behavior under multiaxial stress. Specimens of various shapes are used.

TEST METHODS

To investigate the effect of triaxiality of stress on ductile fracture, various types of mechanical tests were carried out, such as the uniaxial tensile test, plane strain tensile test, torsion test, biaxial tensile test, and notched tensile test. The uniaxial tensile, the plane strain tensile, and notched tensile tests were performed by using the Instron type testing machine. Strain rates of these tests were 0.04 min^{-1}, 0.1 min^{-1}, and 0.1 min^{-1}, respectively. In the torsion test, rotational speed was 1 r.p.m., and in the biaxial tensile test, the strain rate was about 0.1 min^{-1}.

Specimen

The medium strength steel for tests has a chemical composition shown in Table 1. The specimen was heated at $950^{o}C$ for two hours and then cooled in the furnace for a uniformity of property. Test pieces cut out from any position of specimen show little variation in data. The microstructure and mechanical properties of the specimen after heat treatment are shown in Photo. 1 and Table 2.

Test items

1. Flow stress

The flow stress of the specimen was measured by the compression test to investigate its plastic deformation behavior. The size of the test piece was 20 mm in diameter and 30 mm in height. Strain was measured at each 10% reduction in height. As strain increased, the test piece took a barrel shape. Then, the strain was calculated in terms of the maximum diameter. Flow stress was calculated as the load over a maximum cross sectional area, and the measurement was done up to such a high strain as 1.4 of the value of the natural strain. The flow stress of the present specimen can be approximated by the power low as follows:

$$\bar{\sigma} = 80.5 \, (\bar{\varepsilon} + 0.002)^{0.167} \tag{1}$$

where $\bar{\sigma}$ and $\bar{\varepsilon}$ are equivalent stress and strain, respectively.

2. Uniaxial tensile test

Fracture characteristics of the specimen under uniaxial stress were measured with test pieces of such a shape as shown in Fig. 1. The fracture strain was calculated from the minimum sectional area of the necking part of the test piece, and the average fracture stress was determined by the fracture load over the minimum sectional area. The minimum longitudinal (profile) radius of curvature at the neck was measured under off-load conditions at various stages up to fracture.

3. Plane strain tensile test

Fracture characteristics of the specimen under plane strain conditions

were measured for test pieces, shown in Fig. 2. The fracture strain was calculated from the minimum value of the test piece thickness at three positions
where the plane strain condition was predominant. The curvature at the neck was
also measured in a method similar to the uniaxial tensile test.

    4. Torsion test

    The shape of the test piece is shown in Fig. 3. The amount of strain was
calculated from the twist angle of the marking line, inscribed in the axial
direction of test piece. The shearing stress was calculated from the following
equation and the torque **curve.**

$$\tau = \{3T + \theta \left(\frac{dT}{d\theta}\right)\}/2\pi r^3 \tag{2}$$

where   T : Torque

         θ : Twist angle per unit length

         r : Radius of test piece

         τ : Shearing stress on test piece surface

    5. Biaxial tensile test

    Fracture characteristics of the specimen under the biaxial tensile stress
condition were investigated by fracture tests of a thin wall cylinder subjected
to inner pressure and axial force. The shape of the test piece is shown in
Fig. 4. Various stress conditions were generated within the test piece by varying the ratio of inner pressure to axial force.

    6. Tensile test for notched test pieces

    To investigate fracture characteristics of the specimen under high const-
rained stress conditions, tensile tests were carried out on the notched test
pieces shown in Fig. 5. The notch tip radius was 2, 5, or 10 mm.

    Calculation of fracture stress and fracture strain

    1. Uniaxial tensile test

    In the fracture test of the uniaxial tensile test, the neck causes multi-
axial stress conditions in the test piece. The stress at the neck on the central axis was estimated by the following Bridgman equations [2];

$$(\sigma_{rr})_{r=0} = (\sigma_{\theta\theta})_{r=0} = \bar{\sigma} \ln \{1+(a/2R)\} \tag{3}$$

$$(\sigma_{zz})_{r=0} = \bar{\sigma} [1+ \ln \{1+(a/2R)\} \tag{4}$$

where a : Radius of the narrowest portion of the neck

         R : Minimum longitudinal radius of curvature at the neck

    $(\sigma_{rr})_{r=0}$: Radial stress

    $(\sigma_{\theta\theta})_{r=0}$: Circumferential stress

    $(\sigma_{zz})_{r=0}$: Longitudinal stress

The relation between flow stress and average fracture stress $(\sigma_{zz})_{av}$ is :

$$(\sigma_{zz})_{av} = \bar{\sigma} \{1+(2R/a)\} \ln\{1+(a/2R)\} \tag{5}$$

where $\bar{\sigma}$ is the equivalent stress.

2. Plane strain tensile test

The stress at the neck on the center of thickness was estimated by the following equations, which were obtained from Bridgman's idea applied to plane strain condition:

$$(\sigma_t)_{x=0} = (2/\sqrt{3})\, \bar{\sigma}\, \ln\{1+(a/2R)\} \tag{7}$$

$$(\sigma_\ell)_{x=0} = (2/\sqrt{3})\, \bar{\sigma}\, [1+\ln\{1+(a/2R)\}] \tag{8}$$

$$(\sigma_w)_{x=0} = (1/\sqrt{3})\, \bar{\sigma}\, [1+2\ln\{1+(a/2R)\}] \tag{9}$$

where   a : A half of the thickness at the thinnest portion

      R : Minimum longitudinal radius of curvature at the neck

      $(\sigma_t)_{x=0}$: Through-thickness stress

      $(\sigma_\ell)_{x=0}$: longitudinal stress

      $(\sigma_w)_{x=0}$: lateral stress

The relationship between the flow stress $\bar{\sigma}$ and the average fracture stress $(\sigma_\ell)_{av}$ is expressed as

$$(\sigma_\ell)_{aver} = (2/\sqrt{3})\, \bar{\sigma}\, \left[\sqrt{1+(2R/a)}\, \ln \frac{\sqrt{1+(2R/a)}+1}{\sqrt{1+(2R/a)}-1} - 1\right] \tag{10}$$

3. Notched tensile test

Since there is no analytical formula capable of estimating the stress conditions within the test piece which has a notch as shown in Fig. 6, the stress distribution in the test piece under general yielding were calculated using the finite element method. As Fig. 7 indicates, there is no large difference in the trend of stress distribution and absolute values between the calculated stress and those of Bridgman's equations corrected for plane strain conditions, equations (7) – (10). Therefore, the internal stress value in the specimen was estimated by equations (7) – (10). The minimum longitudinal radius of curvature at the neck (R) in the equations corresponds to the value at the fracture.

Results of experiments

[Anisotropy of ductile fracture characteristics in material]

The results of experiments with various test pieces are listed in Table 3. Anisotropy of fracture strain in the material is shown in Fig. 7, in which ε_1 ε_2, and ε_3 represent the strain to the rolling direction (L), transverse direction (C), and through-thickness direction (Z), respectively. Note that fracture strain depends on both the specimen direction and the stress condition in the material ; i.e., anisotropy in the fracture strain is strongly affected by stress conditions. The lowest fracture strains are found in CL (axial direction C, notch direction L) and ZL (axial direction Z, notch direction L) test pieces for plane strain tensile tests. In these test pieces, the fracture strain is only 0.4 –0.5, expressed in the equivalent strain in which the value is apro-

ximately a half of the minimum strain value of the uniaxial tensile test (direction Z). The directions of maximum shearing stress in the CL and ZL test piece coincide with the particular direction of the material, which makes an angle of 45° with C and Z directions, respectively. These results indicate that the direction of maximum shearing stress greatly influences the anisotropy of fracture strain.

The microstructure of the specimen used in these tests has a banded structure as shown in Photo. 1, in which pearlite colonies are stretched largely in direction L and also in direction C. Observation of the microstructure showed that microcracks occur in the inclusion and in the pearlite of the banded structure, and the growth rate of the microcrack after its formation depends on the direction of maximum shearing stress, as shown Photo. 2. Therefore, the anisotropy of the fracture can be explained by the difference of the growth rate of the microcrack. The difference of the growth rate can be explained as follows. Draw the line along maximum shearing direction. The segment between the neighboring two pearlite raws varies with maximum shearing direction. Shearing length in ferrite is defined by the segment length. The microscopic plastic constraint is thought to be less for longer shearing length. In other words, the microscopic structure and the distribution of microcracks exert effects on ductile fracture characteristics as well as the stress conditions in the material.

Discussion

[Ductile fracture characteristics and stress conditions]

Various studies have been presented concerning effects of the stress state on ductile fracture characteristics, insisting that

(1) Fracture occurs when the maximum tensile stress in the test piece exceeds a certain limit [3].

(2) Fracture occurs when the maximum shearing stress in the test piece exceeds a certain limit [4], [5].

(3) Fracture occurs when the amount of work by the maximum tensile stress in the test piece exceeds a certain limit [6].

(4) The component of hydrostatic pressure contributes to void growth as the major factor, and fracture is affected by the factor of $\sigma_m/\bar{\sigma}$ [7], [8].

These four theories were examined on the basis of test results, with each test piece of the direction L listed in Table 3. Findings show that the value of $\bar{\varepsilon}_f$ can be explained by the last theory. The results are shown in Fig. 8, in which σ_m, $\bar{\sigma}_f$, and $\bar{\varepsilon}_f$ represent the mean stress, equivalent stress, and equivalent plastic strain at fracture, respectively. They are defined as follows:

$$\bar{\sigma} = 1/\sqrt{2} \{ (\sigma_1 - \sigma_2)^2 + (\sigma_2 - \sigma_3)^2 + (\sigma_3 - \sigma_1)^2 \}^{\frac{1}{2}} \tag{11}$$

$$\sigma_m = (\sigma_1 + \sigma_2 + \sigma_3)/3 \tag{12}$$

$$\bar{\varepsilon} = \sqrt{2}/3 \{ (\varepsilon_1 - \varepsilon_2)^2 + (\varepsilon_2 - \varepsilon_3)^2 + (\varepsilon_3 - \varepsilon_1)^2 \}^{\frac{1}{2}} \tag{13}$$

As indicated in the graph, a linear relationship obviously exists between $\sigma_m/\bar{\sigma}$ and $\bar{\varepsilon}_f$, though there are some dispersions.

Conventional investigation of the stress effect exerted on ductile fracture has revealed that, as opposed to brittle fracture, the stress history in the deformation process which leads to fracture, as well as the stress at fracture, play important roles in actual fracture. In the present experiments, it was expected that the form of the integral formula would be more effective for describing the ductile fracture behavior as proved in Ohyane's equation [7]. However, due to some statistical variation, isolations of the effect on deformation history was impossible. In other words, the effect of the deformation history may be small in comparison with that of the hydrostatic pressure component as far as the present experiment is concerned. From the foregoing argument, it is concluded that the factor $\sigma_m/\bar{\sigma}$ serves as the useful parameter of the stress conditions in the first approximation to describe the ductile fracture behavior under multiaxial stresses.

Conclusions

The following facts were revealed by the present investigation on ductile fracture behavior and the effects of stress conditions on ductile fracture behavior for medium strength steel.

(1) Test pieces of a material differ largely in ductile fracture characteristics for their cutting directions. Stress directions in material largely affect the anisotropy in fracture strain. That is, when the maximum shearing stress has a special direction in the material, ductile fracture strain decreases severely. The extreme decrease is strongly related to the microstructure (banded structure and inclusion distribution) of the material.

(2) To estimate the ductile fracture behavior under multiaxial stresses, it is useful to take $\sigma_m/\bar{\sigma}$ as the parameter of stress conditions in the first approximation, which comes from the linear relationship between $\sigma_m/\bar{\sigma}$ and $\bar{\varepsilon}_f$. The gradient of the linear curve and the intersection of the curve on the abscissa are material constants that are directly affected by the feature of the microstructure of the material. The effect of the microstructure remains to be analyzed.

REFERENCES

(1) G. R. Irwin, Fracture I, Handbuch der Physik VI., Flügge Ed., (1958)558 pp

(2) P. W. Bridgman, "The Stress Distribution at the Neck of a Tension Specimen", Trans. of the American Society of Metal, 32 (1944), pp 553-574

(3) E. Orowan, Rep-Progr. Phys 12(1948) 185

(4) O. Mohr, Abhandlungen aus dem Gebiete der Technischen Mechanik, 2nd ed., Berlin (1914)

(5) J. Marin, Mechanical Behavior of Engineering Materials, Prentice Hall, (1963) 126

(6) M. G. Cockcroft, "Ductility and the Workability of Metals", J. of the Institute of Metals, 96, (1968), pp. 33-39

(7) M. Ohyane, "On the Criteria of Ductile Fracture", J. of the Japan Society of Mechanical Engineers, 75-639, (1972), pp. 596-601 (in Japanese)

(8) J. R. Rice and D. M. Tracey, "On the Ductile Enlargement of Voids in Triaxial Stress Fields", J. Mech, Phys. Solids, 17, (1969), pp. 201-217

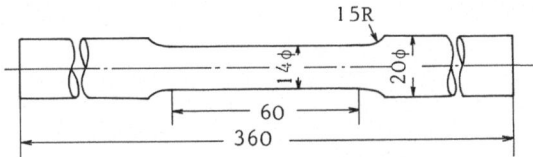

Fig. 1. Uniaxial tensile specimen

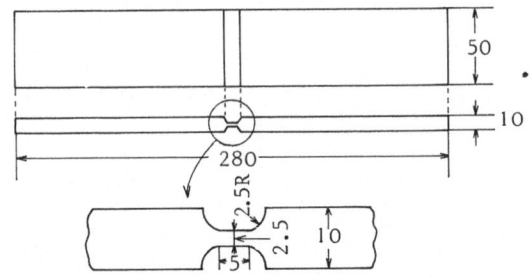

Fig. 2. Plane strain tensile specimen

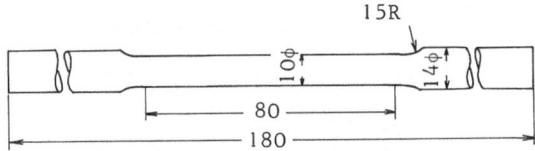

Fig. 3. Torsion specimen

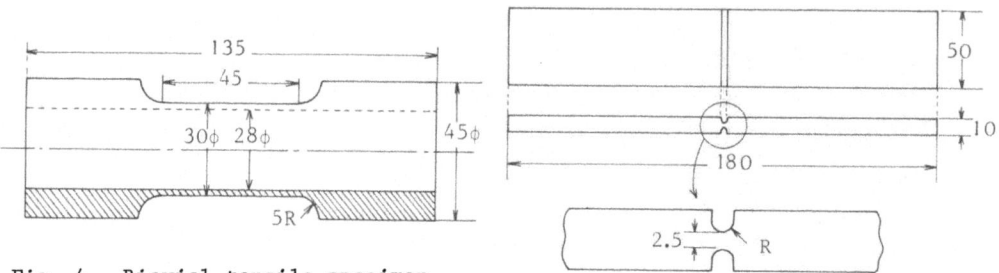

Fig. 4.　Biaxial tensile specimen

Fig. 5.　Notched tensile specimen

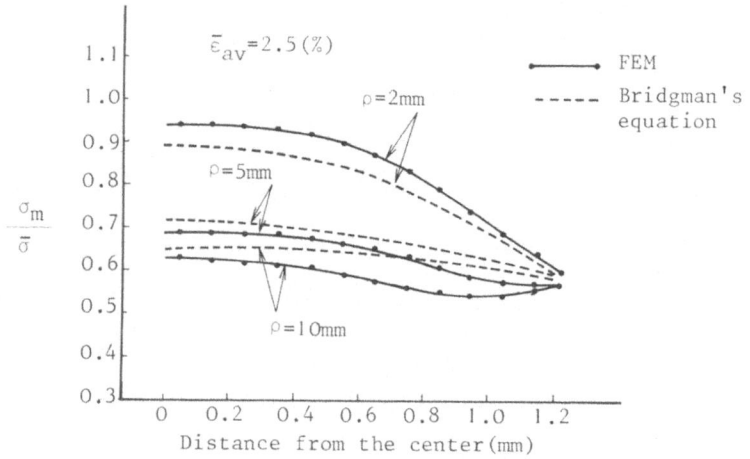

Fig. 6.　Stress distibution in the specimen

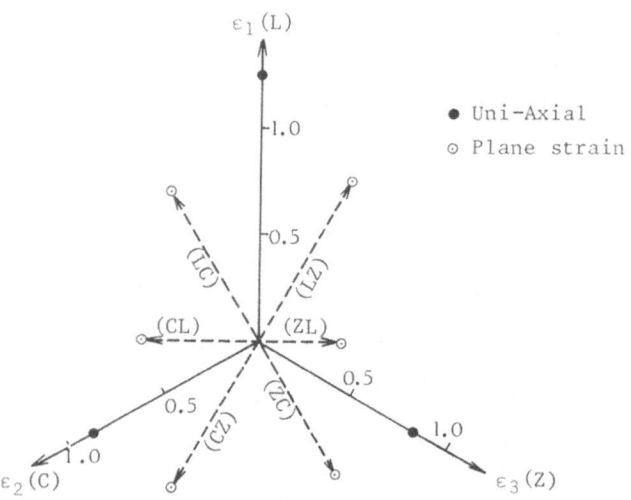

Fig. 7.　Fracture strain $(\bar{\varepsilon}_f)$ in different direction

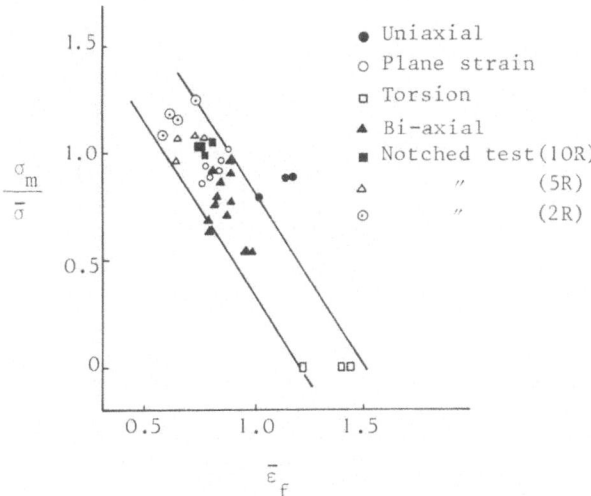

Fig. 8. The relation between $\sigma_m/\bar{\sigma}$ and $\bar{\varepsilon}_f$

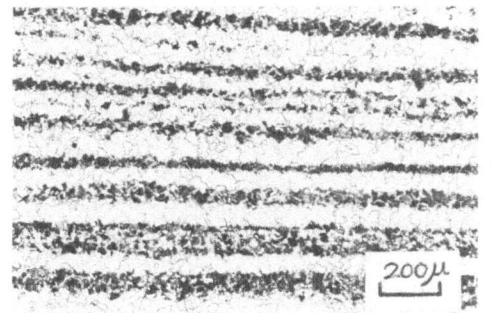

Photo. 1. Microstructure of specimen

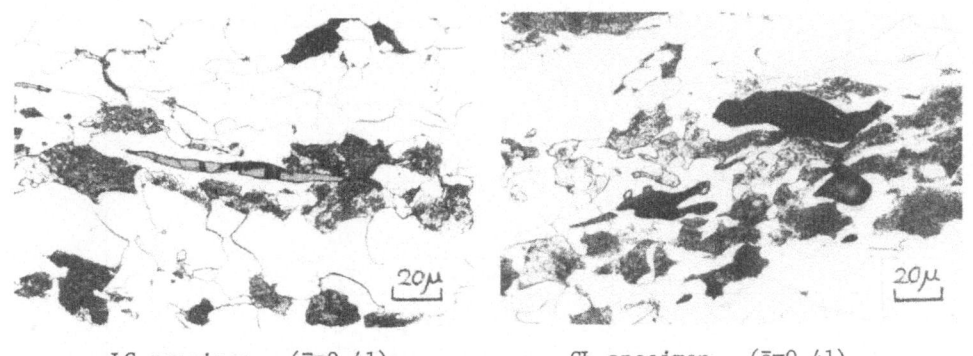

LC specimen $(\bar{\varepsilon}=0.41)$ CL specimen $(\bar{\varepsilon}=0.41)$

Photo. 2. The appearance of void growth
(Plane strain tensile specimen)

Table 1   Chemical Composition of Material (%)

	C	Si	Mn	P	S	V
HY50	0.16	0.46	1.35	0.016	0.010	0.04

Table 2   Mechanical Properties of Material

Direction	Y.P.(kg/mm^2)	T.S(kg/mm^2)	El(%)	ψ(%)
L	32.5	49.7	36.5	68.1
C	32.4	49.4	35.2	58.6
Z	31.3	49.2	33.8	56.7

Table 3   Test Results

Test Method	Direction	$\bar{\varepsilon}$	$\bar{\sigma}$	σmax	τmax	σm	σm/$\bar{\sigma}$	Remarks
Uniaxial	L	1.14	82.3	124.8	41.2	70.1	0.85	
tensile	C	0.88	78.0	105.6	39.0	53.8	0.69	
test	Z	0.84	77.5	100.5	38.8	48.8	0.63	
	LC	0.79	77.2	113.0	44.4	68.6	0.89	
	LZ	0.87	78.2	120.0	45.1	74.9	0.96	
Plane strain	CL	0.54	73.3	90.1	42.3	47.8	0.65	
tension	CZ	0.80	77.3	112.4	44.4	68.0	0.88	
test	ZL	0.40	70.5	76.0	38.0	38.0	0.54	
	ZC	0.72	76.0	104.2	43.9	60.3	0.79	
	L	1.37	85.5	49.4	49.4	0	0	
Torsion test	C	1.20	83.0	47.9	47.9	0	0	
	Z	1.06	80.9	46.7	46.7	0	0	
Tensile test	L	0.77	76.8	122.8	44.3	78.5	1.02	R=10mm
with notched	L	0.70	75.7	122.9	43.8	79.1	1.04	R=5mm
specimen	L	0.65	75.0	131.0	43.3	87.7	1.17	R=2mm

THICKNESS EFFECTS ON FRACTURE CRITERIA

H. W. Liu
Professor of Materials Science
Syracuse University, Syracuse, New York 13210, USA

ABSTRACT

Many engineering materials are tough and ductile, and the thicknesses of the structures are often thinner than that which necessitates a plane strain analysis. This paper summarizes briefly our studies on the thickness effects on the characteristics of crack tip stresses and crack tip deformation. Near tip stresses, near tip strains, the relative opening displacements between the upper and the lower crack surfaces, and those between the upper and the lower boundaries of the crack tip strip necking zone have been investigated both theoretically and experimentally. The results suggest that the thickness contractions of the fracture surfaces measured from broken specimens, crack opening displacements, and plane stress characterization can be used to correlate with fracture toughness. The choice of a specific fracture criterion depends upon the value of the necking parameter, $(K/\sigma_Y)^2/t$. t is plate thickness.

INTRODUCTION

When a small crack tip plastic zone is imbedded in a massive and thick plate, the constraint on thickness contraction induces a state of high tensile stress and low plastic strain, and the state of stresses and strains approaches that of plane strain. A large plastic zone in a thin plate causes crack tip necking, and a state of low tensile stress and high plastic strain prevails. The fracture process in a cracked plate is controlled by the stresses and strains at the crack tip. Since plate thickness greatly affects the crack tip stresses and strains, it is reasonable to expect that the choice of a fracture criterion depends on thickness.

The primary physical parameter, η, for the measure of the necking tendency is $(K/\sigma_Y)^2/t$ or $(JE/\sigma_Y^2)/t$, which is also a measure of the crack tip plastic zone size relative to plate thickness, t. The reciprocal of the necking parameter is the constraint parameter, a measure of the constraint on thickness contraction. These two parameters control the state of stresses and strains at a crack

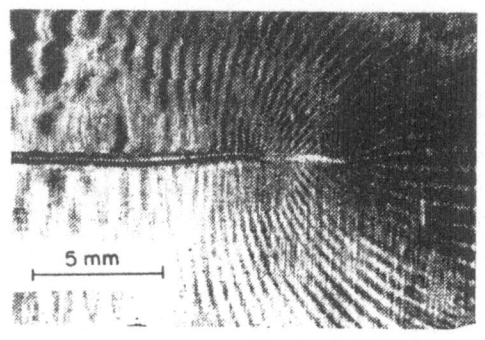

a

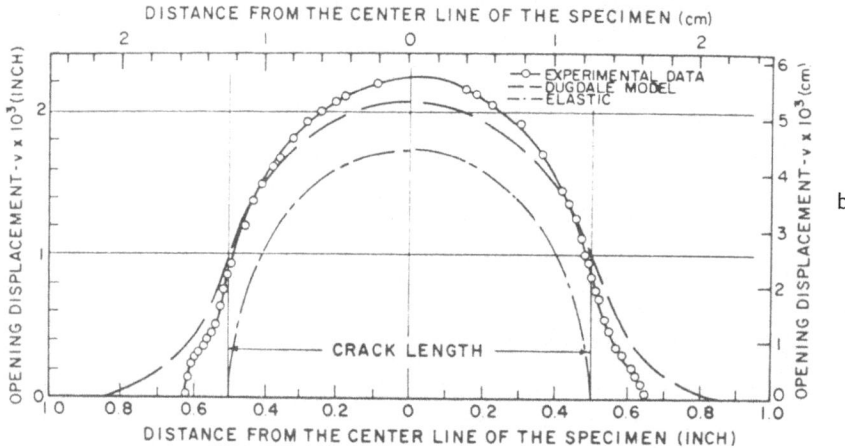

b

Fig. 1 (a) Moire pattern of a steel specimen. (b) Opening displacement along crack line. σ, 380 MN/m^2; σ_Y, 627 MN/m^2; E, 221 x 10^3 MN/m^2; thickness, 0.3 mm; width, 152.4 mm; slot length, 25.4 mm.

tip. It is well known and recommended by ASTM, that for thick plates with constraint parameters greater than 2.5, the plane strain correlation can be used. Many engineering materials are tough and ductile, and the thicknesses of the structures are thinner than that which necessitates plane strain analyses. Based on the results of the numerical calculations by FEM and the limited amount of experimental measurements, various regions of different fracture correlations between small laboratory specimens and large engineering structures are proposed. The choice is thickness dependent.

THICKNESS CONTRACTION AS A FRACTURE CRITERIA

In very thin plates, Schaeffer, Ke, and Liu [1] have observed crack tip strip necking as shown in Figure 1a. Figure 1b shows the measured relative opening displacements between crack surfaces and between the upper and the lower boundaries of the strip necking zone, including that at the crack tip. The measured values agree well with the values calculated according to the

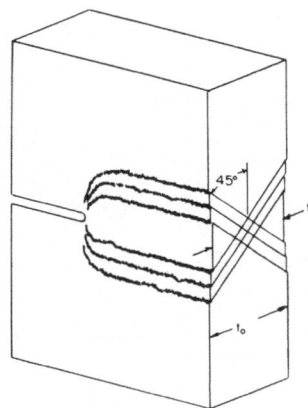

Fig. 2a   Formation of strip necking
zone by shear deformation.
Opening displacement in the
strip necking zone equal to
thickness contraction.

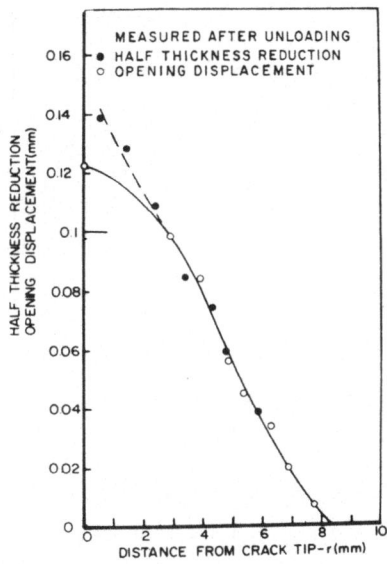

Fig. 2b   Measured opening displacements
and thickness contractions.

Dugdale strip yielding model. In the strip necking zone, the shear deformation takes place primarily on two sets of intersecting planes which are inclined 45°, both to the loading direction and to the plane of the specimen,
Figure 2a. According to this simple model of necking deformation, the relative opening displacements in the strip necking zone is equal to the thickness
contraction, $(t_o - t)$. Figure 2b shows that the measured opening displacements
in the strip necking zone are, indeed, equal to thickness contractions.

When stable crack growth takes place, the thickness in the region ahead of
the crack tip contracts. Once the crack tip passes a point along the fracture
path, the thickness at the point ceases to contract. In other words, the
thickness at a point measured from a broken specimen must be the thickness when
the crack tip reached the point.As explained by the model in Figure 2a, at the
moment when the crack tip reached the point, the crack tip opening displacement,
CTOD, was equal to thickness contraction, and CTOD was related to K or J by the
Dugdale model. Then,the value of K or J at the moment could be calculated directly from the thickness contractions measured from broken specimens.

$$J = \frac{K^2}{E} = \sigma_Y \text{ CTOD} = \sigma_Y \Delta t = \sigma_Y t_o \varepsilon_z \qquad (1)$$

where $\Delta t = (t_o - t)$, $\varepsilon_z = \Delta t/t_o$, and t_o is the original plate thickness. If
the relations of Equation (1) are applicable to small specimens in general
yielding, the fracture toughnesses of ductile materials can be measured from

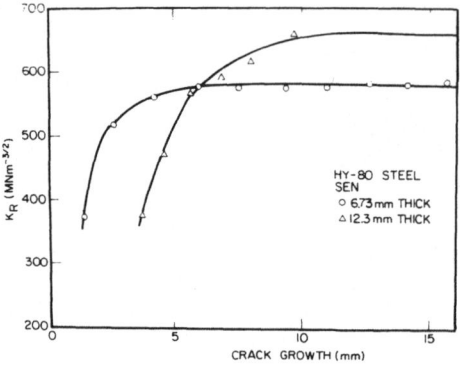

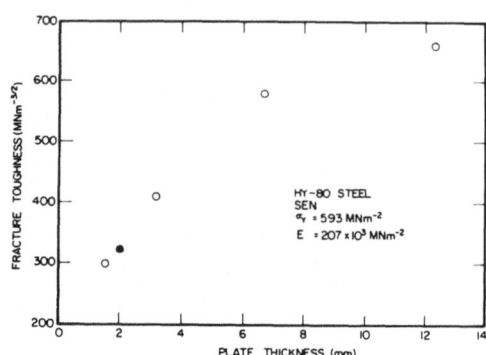

Fig. 3 R-curve for HY-80 steel

Fig. 4 Fracture toughness as a function of plate thickness.

thickness contractions of small broken specimens.

Ke and Liu [2] have measured the thickness contractions of the fractured single-edge-notched specimens made of HY-80 steel.   From   the   contractions measured in the broken specimens, the resistance curves of specimens can be constructed as shown in Figure 3 [3,4]. The plateaus of these two resistance curves give the fracture toughness values of 660 and 580 MNm$^{-3/2}$ for 12.3 and 6.73 mm thick specimens, respectively.

The fracture toughness of the steel is a function of plate thickness as shown in Figure 4. The solid point in the figure is obtained from the thickness contraction measurements of the two Clausing specimens. The geometry of the specimen is shown in Figure 5. The shear deformation of the Clausing specimen is restricted by the specimen geometry to two sets of intersecting planes, each inclined 45° both to the direction of loading and to the plane of the specimen. This is the same orientation as that in the strip necking zone of the cracked thin specimens. Therefore, the paths of the plastic deformation leading to fractures of these two types of specimens are the same. Consequently,

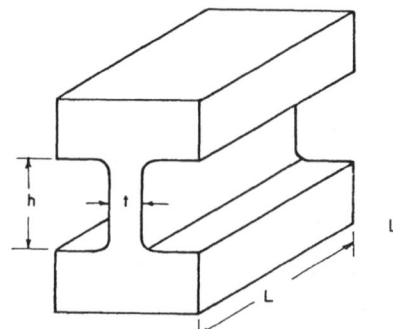

$L > h > t$

Fig. 5 Geometry of Clausing specimen.

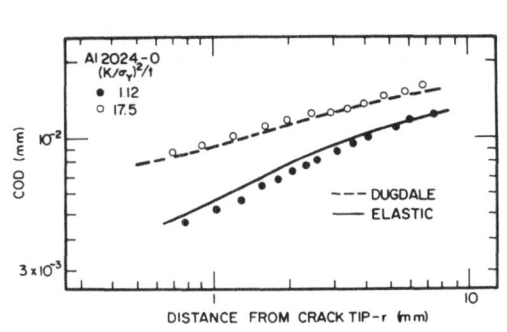

Fig. 6 Thickness effect on COD;
        Al 2024-0.

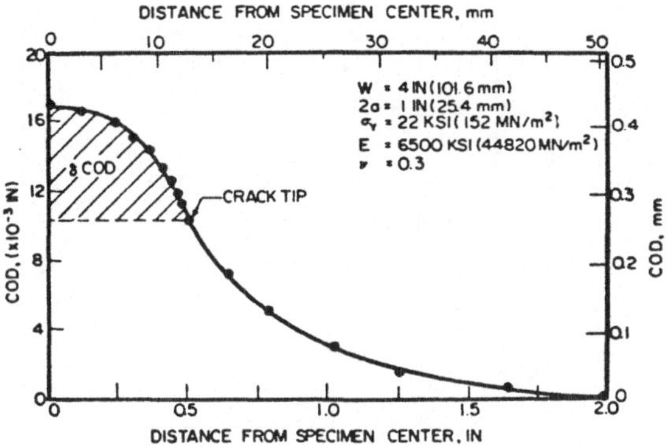

Fig. 7 COD of center cracked plate in
        general yielding, strip yield-
        ing simulation.

the fracture tougnesses obtained from the contraction measurements of these two
types of specimens must also be the same. Indeed, the data point of the Clausing specimens fits well with the data from the single-edge-notched specimens.

The strip necking zone size has to be much larger than the plate thickness, t, in order to have a strong agreement between the measured opening displacements in the strip necking zone and the calculated values according to the
Dugdale strip yielding model. The length of the strip necking zone of the
specimen in Figure la is nearly 13 times the specimen thickness, and η is 48.
The agreement is excellent. Therefore, the thickness contraction can be used
for fracture toughness measurements when η is equal to or greater than 48.

CRACK OPENING DISPLACEMENT

For thicker specimens at lower K levels, the value of η is lower, and the
strip necking zone is often too small to compare favorably with the Dugdale
strip yielding model. Crack tip necking begins to form at η equal to 18. For
the specimens with η greater than 18 but less than 48, it is not certain that
the thickness contraction can be used for fracture toughness measurements.

Crack opening displacements in specimens made of Al 2024-0, Al 2024-T3,
and Al 2024-T351 aluminum alloys with tensile yield strengths, 54, 310, and
386 MN/m^2, respectively, were measured by the moire method [5]. Figure 6
shows that the measured COD on two specimens made of Al 2024-0 aluminum alloy,
one thin and one thick. Both specimens were loaded to the same K-value, 4.53
MNm$^{-3/2}$. The values of η are 17.5 for the thinner specimen and 1.12 for the
thicker. The Dugdale model agrees very well with the thinner specimen, despite
the fact that strip necking zone does not exist in the specimen. The measured
CODs differ from those of plane stress elasto-plastic calculations. The same

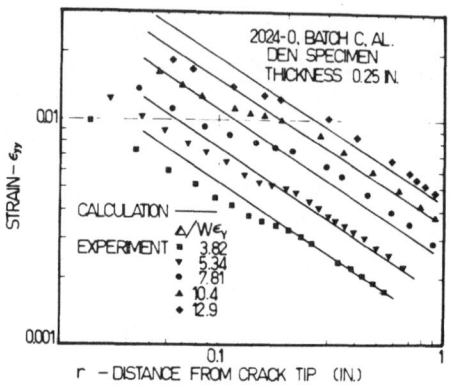

Fig. 8 Comparison of calculated and measured ε_{yy}.

conclusion can also be derived from the tests on the other two aluminum alloys [5]. If the conclusion is applicable to the case of general yielding, the COD can be used to measure fracture toughness.

The magnesium specimen tested by Kobayashi et al. [6] has shown that, beyond general yielding, the upper and lower crack surfaces move like two rigid surfaces except in the small area near the crack tip where the deformation of the crack surfaces is concentrated. The results of Kobayashi et al. indicate that the difference between the crack opening displacement at a distance r away from crack tip, COD(r), and that at crack tip, CTOD, remain unchanged, once general yielding is reached.

$$\delta COD(r) = COD(r) - CTOD \tag{2}$$

$\delta COD(r)$ has been calculated by the finite element method with Dugdale's strip yielding model. The result is shown in Figure 7. Once $\delta COD(r)$ is calculated and COD(r) is measured, the value of CTOD can be obtained from Equation (2), and the equivalent K or J is

$$J = \frac{K^2}{E} = \sigma_Y \text{ CTOD} = \sigma_Y [COD(r) - \delta COD(r)] \tag{3}$$

This is applicable, if Dugdale's strip yielding model is applicable, when $18 < \eta < 48$.

PLANE STRESS CHARACTERIZATION

Crack tip strains, ε_{yy}, have been measured on aluminum specimens in general yielding, and the measured values have been compared with the values calculated by the finite element method [7]. The region close to a crack tip is "stiffened" by the triaxial state of stress, and the stiffening effect reduces the plastic strain in the region. The linear size of the stiffened region is approximately equal to the plate thickness, as shown in Figure 8. Beyond this stiffened region, the measured strains agree very well with the plane stress

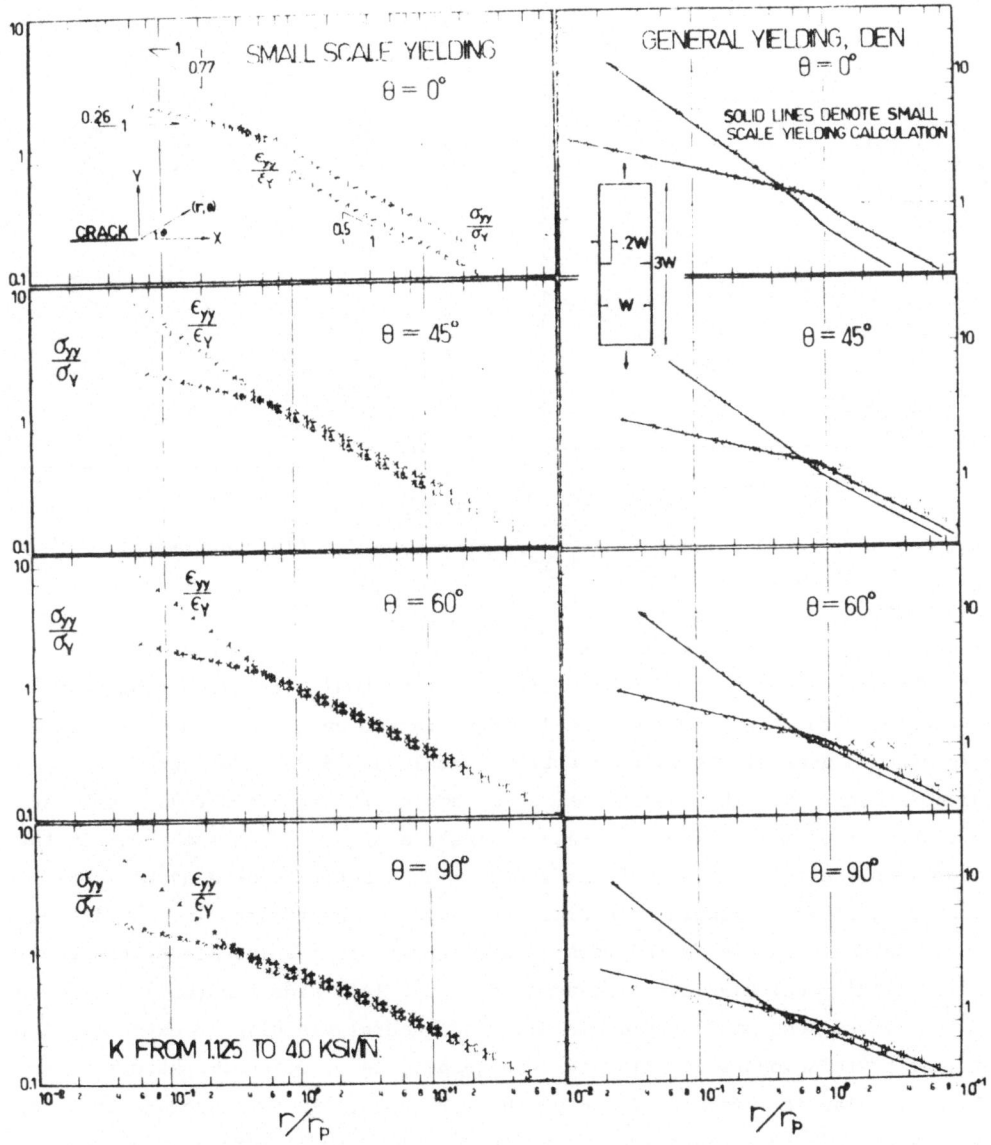

Fig. 9 Correlations of near tip stresses and strains in the
loading direction between the small scale yielding
and double-edge-notched specimen loaded into the
region of general yielding.

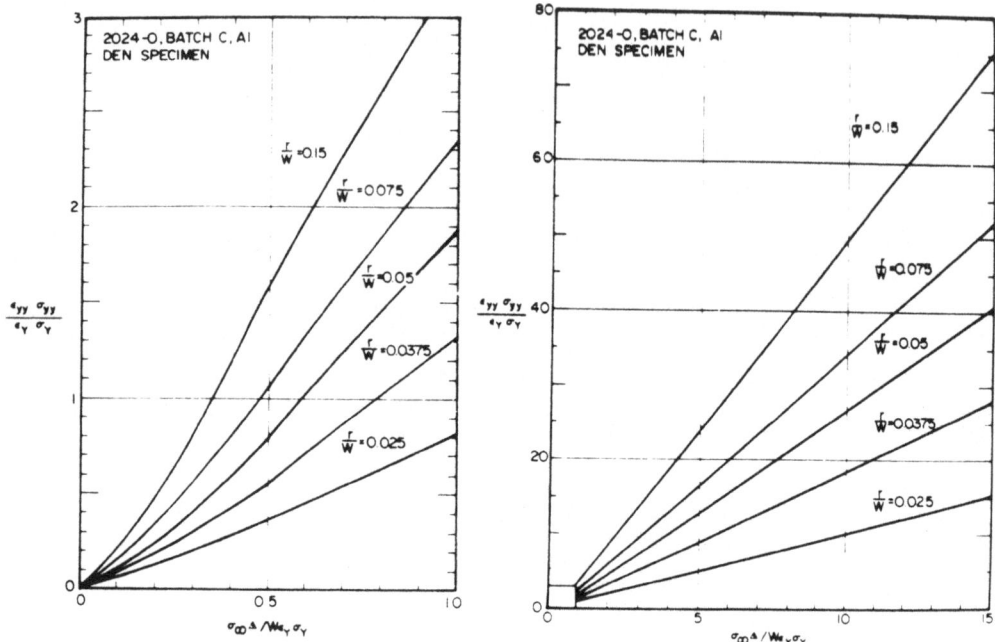

Fig. 10 Near field parameter, $\sigma_{yy}\varepsilon_{yy}/\sigma_Y\varepsilon_Y$ versus far field parameter, $\sigma_\infty\Delta/W\sigma_Y\varepsilon_Y$.

elastic-plastic finite element calculations.

The stresses and strains near crack tips in small scale yielding as well as that near the crack tips of double-edge-cracked and single-edge-cracked specimens in general yielding have been calculated [8,9]. The results are shown in Figure 9. The quantities σ_{yy}/σ_Y and $\varepsilon_{yy}/\varepsilon_Y$ at $\theta = 0^\circ$, 45°, 60°, and 90°, at various K-levels in the case of small scale yielding are shown in the figures in the left column. The corresponding quantities of a double-edge-notched specimen in general yielding are shown in the figures in the right column. The solid curves in the right side figures are the results of the small scale yielding calculations transposed from the left side figures. The correlation between the small scale yielding and general yielding is extremely good. Similar correlations also exist for the quantities of effective stresses and effective plastic strains. Based on these results, one can conclude that for a given state of crack tip stresses and strains in a small sample in general yielding, there exists an "identical" state of stresses and strains in a large sample in small scale yielding. The values of K or J of the small sample is equivalent to that of the large sample, which is known from the finite element calculations.

The product of the near tip stress, σ_{yy}, and the near tip strain, ε_{yy},

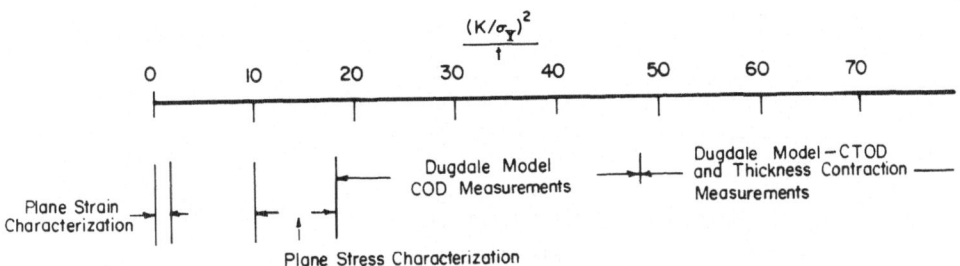

Fig. 11 Thickness effects on the choice of fracture criteria.

correlates well with the product of the applied global stress, σ_∞, and the imposed elongation, Δ, of the gage section as shown in Figure 10. The product of the near tip stress and near tip strain is linearly proportional to J. Therefore, from the data, we have

$$J = \frac{K^2}{E} = 34.5 \; (\frac{\sigma_\infty}{\sigma_Y})(\frac{\Delta}{W\varepsilon_Y}) \tag{4}$$

where σ_Y and ε_Y are yield stress and yield strain, and W is the characteristic size of the sample. In this case, W is the width. In order to have a meaningful plane stress correlation, the size of the plastic zone has to be larger than the plate thickness. Furthermore, the two dimensional plane stress characterization will not be valid, once necking takes place. Therefore, it can be used only when $10 < \eta < 18$.

In summary, as shown in Figure 11, when η is more than 48, thickness contraction can be used to measure K or J; when the necking parameter is more than 18 but less than 48, the COD can be used to measure equivalent K or J; the plane stress equivalence of the K or J of a small sample in general yielding and that of a large sample in small scale yielding can be used to obtain the fracture toughness of a ductile and tough material, when $10 < \eta < 18$. The boundaries between various regions are tentative. Additional experimental and theoretical works are needed to firmly establish the boundaries.

ACKNOWLEDGEMENT

This work was conducted at the George Sachs Fracture and Fatigue Research Laboratory, Syracuse University, Syracuse, New York. The author wishes to thank the Army Research Office for their financial support, Contract No. DAAG29-78-G-0069, and Mrs. H. Turner and Mr. R. Ziemer for the preparation of the manuscript.

REFERENCES

[1] B. Schaeffer, J. S. Ke, and H. W. Liu, "Deformation and the Strip Necking Zone in a Crack Steel Sheet", J. of Experimental Mechanics (April 1971).

[2] J. S. Ke and H. W. Liu, "Thickness Effects on Crack Tip Deformation at Fracture", J. of Engineering Fracture Mechanics, 8 (March 1976) pp. 425-436.

[3] H. W. Liu and A. S. Kuo, "Fracture Toughness of Thin and Tough Plates", Int. Journ. of Fracture, 14 (1978) pp. R109-R112.

[4] H. W. Liu and C. Y. Yang, "Strip Yielding Model and Fracture Toughnesses of Thin and Tough Plates", Proceedings of: International Conference - Fracture Mechanics in Engineering Application, Bangalore, India (March 26-30, 1979).

[5] A. S. Kuo and H. W. Liu, "An Experimental and FEM Study on Crack Opening Displacement", Proceedings of the Fourth Intl. Conference on Fracture, Waterloo, Canada, 3, ICF4 (June 19-24, 1977) pp. 311-321.

[6] A. S. Kobayashi, W. L. Engstrom, and B. R. Simon, "Crack Opening Displacements and Normal Strains in Centrally Notched Plates", Exp. Mech., 9, No. 4 (1969) pp. 163-170.

[7] W. L. Hu and H. W. Liu, "Crack Tip Strain - A Comparison of FEM Calculations and Measurements", Proceedings of the Ninth National Symposium of Fracture Mechanics, Pittsburgh, Pa., August 1975, ASTM STP 601, Crack and Fracture (1976).

[8] W. L. Hu, A. S. Kuo and H. W. Liu, "A Study on Crack Tip Deformation and Ductile Fracture", Proceedings of 12th Annual Meeting of Soc. of Eng. Science, Austin, Texas (1975).

[9] W. L. Hu and H. W. Liu, "Characterization of Crack Tip Stresses and Strains - Small Scale Yielding and General Yielding", Proceedings of the Second Intl. Conference on Mechanical Behavior, Boston, Mass. (August 16-20, 1976).

FRACTURE TOUGHNESS AND THE STRETCHED ZONE
OF ULTRA-HIGH STRENGTH STEELS

Y. Kasai and N. Uehara
R & D Div. Daido Steel Co., Ltd.
2-30 Daido-cho, Minami-ku, Nagoya, Japan

ABSTRACT

Stereographic analyses of fracture surfaces of compact tension specimens were made on three grades of ultra-high strength steels heat-treated to yield strength level between 102-188 kg/mm^2. The fracture surface was separated into four regions, namely, a fatigue pre-cracked region, a stretched zone, a ductile fracture band characterized by small dimples, and an over-load fracture region with either cleavage or ductile nature. Plotting of the stretched zone height against the crack tip opening displacement calculated by

$$CTOD = A \cdot K_{IC}^2 / E \cdot \sigma_Y$$

showed that the experimentally obtained stretched zone height was bound by lines with slopes of 0.715 and 0.425, and best correlated with a line calculated by setting $A = (1-\nu^2)/2$, which indicates that the crack tip opening displacement can be best predicted by Thomason's model. The stretched zone height was correlated well with the fracture strain of the smooth tension bar. But the fracture toughness calculated by Hahn and Rosenfield's equation showed a deviation from the experimentally obtained fracture toughness.

INTRODUCTION

A stretched zone is observed between the fatigue pre-cracked and over-load fracture regions of the fracture toughness specimen [1-6]. A wider stretched zone is observed for a steel with higher fracture toughness as shown in Fig. 1 [1,2,5,7]. This is even observed in ultra-high strength steels, such as maraging and Ni-Cr-Mo steels, with yield strength of 180 kg/mm^2 or higher, and in some particular materials tested at $-180°C$ [8].

The stretched zone is usually considered the result of crack tip blunting since the calculated crack tip opening displacement (CTOD) from K_{IC} value [9,10] is close to the observed stretched zone width (SZW) [11,12]. This argument, however, is irrelevant. The SZW is the crack tip displacement along the crack propagation direction; CTOD is only the displacement normal to the crack plane.

Therefore, the comparison between SZW and CTOD is unreasonable.

The main purpose of the present study is to make a quantitative comparison between the dispacement in the stretched zone normal to the crack plane (stretched zone height, SZH) and the CTOD calculated from K_{IC} value. The SZH was determined by stereographic analysis using a scanning electron microscope.

Several analyses attempt to relate fracture toughness to the fracture ductility of the material [9,10,13-19]. The fact that K_{IC} is correlated with the size of the stretched zone, shown in Fig. 1, indicates the experimental support for these analyses. Therefore, the relation between the size of the stretched zone (SZH) and fracture ductility is also discussed in the study.

EXPERIMENTAL PROCEDURE

Materials and heat treatment

The chemical compositions of the three grades of steel used in the study are given in Table 1. JIS-SNCM 8 (equivalent to AISI 4340) was sampled from commercial heats, and 18Ni maraging and 10Ni-8Co steels were from 50 kg laboratory heats. Heat treatment conditions of these steels are

SNCM 8 ---------- oil-quenched from 850°C followed by tempering at 500, 550, and 600°C for 1 hour to give yield strength of 123.0-101.6 kg/mm^2

18Ni maraging --- solution-treated by air-cooling from 820°C. aged at 400°C for 40 minutes (σ_Y = 122.7 kg/mm^2), and at 480°C for 10 hours (σ_Y = 188.4 kg/mm^2).

10Ni-8Co -------- water-quenched from 790°C followed by aging at 500°C for 4 hours to give σ_Y = 118.1 kg/mm^2.

Fracture toughness test

Fracture toughness tests were conducted according to ASTM designation E399-72. Compact tension specimens were employed. Specimen thickness was 40 mm for SNCM 8, and 20 mm for both 18Ni maraging and 10Ni-8Co steels. All tests were conducted at room temperature. SNCM 8 steel tempered at 500°C was also tested at temperatures down to -175°C.

Fracture surface observations

The stereographic analysis method using a scanning electron microscope is proposed by Kimoto et al. [20]. Briefly, the height of the specific point from the fixed point is calculated from the difference in the distance of the specific point from the fixed point on the two photographs with tilt angles of 0° and θ°, respectively. The procedure is schematically depicted in Fig. 2.

The fixed point in this study was set on the fatigue pre-cracked surface far away from the stretched zone. A straight line was drawn through this point to the unstable fracture area along the crack propagation direction. The fracture surface profile was made by 15 to 20 measurements on the line. The profile thus obtained shows the section of the fracture surface in the X-Y plane. The displacement in the stretched zone in the load direction was determined by matching two profiles made from mating fracture surfaces. Magnification was x1000 and the tilt angle θ was 10 degrees in the present study.

EXPERIMENTAL RESULTS

Stereographic analysis of fracture surface

An example of the fracture surface and its profile, determined by the stereographic analysis, is shown in Fig. 3. The left half is a fatigue pre-cracked region, and the unstable propagation area (right half) mainly shows a quasi-cleavage nature with some dimples. The stretched zone is observed adjacent to the fatigue pre-cracked region. A narrow band characterized by dimples is also found between the stretched zone and the unstable fracture area. The fracture of this specific specimen is thought to proceed from the formation of the stretched zone by blunting of the fatigue pre-crack tip followed by ductile fracture, and, finally, to brittle fracture.

Matching the upper and lower profiles of the fracture surface reveals a more interesting feature. By matching the right halves (i. e. unstable fracture area) of the upper and lower fracture surfaces, not only the pre-cracked area and the stretched zone but also the dimpled band clearly show discrete separation between the upper and lower surfaces. A similar separation was observed in all the specimens tested in the study.

These observations lead to the creation of a new crack propagation model, Fig. 4, that refers to the Rice [21] and Thomason [10] models under the premise that the ductile fracture mechanism is the governing process of fracture at the crack tip. In this model, unstable fracture occurs after the following sequence. When a fatigue pre-cracked tip is loaded, a crack proceeds by stretched zone formation with a void nucleated ahead of the crack (stage ii). With more load applied, the tip of the stretched zone proceeds by increased crack tip opening displacement: a void grows and secondary voids nucleate (stage iii). When the coalescence occurs between the stretched zone tip and voids, final unstable fracture occurs. The SZH in this study is defined as the distance between the upper and lower tips of the fatigue pre-crack at the moment unstable fracture occurs.

Stretched zone width (SZW) and stretched zone height (SZH)

The plotting of SZW and SZH in Fig. 5 shows that a nearly 1 : 1 correspondence holds between these two characteristic dimensions of the stretched zone. The crucial point is that the SZH in this study is the sum of the crack opening displacement in the stretched zone and in the ductile fracture band prior to unstable fracture.

Stretched zone height (SZH) and crack tip opening displacement (CTOD)

The CTOD can be calculated by

$$CTOD = A \cdot K_{IC}^{2}/E \cdot \sigma_Y$$

Several values are proposed for a factor A. Thomason proposed $A = (1-\nu^2)/2$ for ductile materials [10]. The maximum and minimum values are 0.715 and 0.425, respectively [22,23].

The relationship between observed SZH and CTOD, calculated from experimentally obtained K_{IC}, is shown in Fig. 6. Experimentally obtained SZH is bound by the lines calculated by the slopes of 0.715 and 0.425, and best correlated with the line calculated by setting $A = (1-\nu^2)/2$.

Stretched zone height (SZH) and fracture ductility

Fracture surface observations of the various steels tested in the study suggest that the fracture toughness is governed by the material's ductility at the crack tip. As mentioned, various relations between the K_{IC} value and ductility have been proposed. The important problem common to those is how to estimate the local critical strain at the crack tip. In this study, however, the relation between SZH and fracture strain of smooth tension specimen was studied from the engineering point of view. Fig. 7 shows that the fracture strain of a smooth tension specimen is correlated with SZH, although the scatter is fairly wide.

DISCUSSION

Fractographic analysis

The important features of the present results follow: (1) a ductile fracture band was observed between the stretched zone and unstable fracture area, and (2) the sum of the opening displacement both in the stretched zone and in the ductile fracture band agreed numerically with the CTOD calculated from K_{IC} value.

The ductile fracture of the fatigue pre-cracked tip by the internal necking mechanism prior to unstable growth has been suggested by the other investigators [24,25], and fractographic observations revealed a small and shallow dimple zone between the stretched zone and large dimples in the unstable fracture zone [6]. These observations compare well with the present experimental observations, which leads to the crack propagation model, Fig. 4.

Krafft defined the 'process zone size (d_T)' as the distance between the crack tip and the plastic instability point ahead of the crack, and gave the relation between K_{IC} and σ_Y, σ_T, E and n [26,27]. Spitzig [1] has shown that the stretched zone width is essentially equal to the process zone size and is similar in magnitude to the average inclusion spacing. Birkle et al. [28] have also shown the agreement between inclusion spacing and process zone size. The calculated process zone size for steels employed in this study, however, does not agree with the stretched zone size. Therefore, it seems difficult to apply the tensile ligament instability fracture criterion to the fracture of materials where ductile fracture is the prevailing mechanism at the crack tip.

The other important finding of the present study is that the SZH is well correlated with the calculated CTOD from K_{IC}. Experimentally obtained CTOD (SZH) was best fitted by

$$CTOD = 0.455 \ K_{IC}^2/E \cdot \sigma_Y$$

The factor of 0.455 is equall to $(1-\nu^2)/2$, which indicates that CTOD can be best predicted by Thomason's model [10]. A similar analysis has been done by Broek on Al alloys [12]. He obtained a factor 0.4 for the relation between stretched zone depth and $K_{IC}^2/E \cdot \sigma_Y$, which compares well with the present result.

Fracture toughness and fracture ductility

Various models have been proposed which attempt to relate K_{IC} to ductility. Arguments differ about the procedure to estimate the critical crack tip strain and the kind of ductility measure to be employed, which greatly depends on the specimen configuration. From an engineering view point, the fracture strain obtained with a smooth tension specimen is the most convenient. Fig. 7 shows good correlation between the fracture strain obtained with the smooth tension bar and SZH, which is closely connected with K_{IC}. Therefore, the following equation given by Hahn and Rosenfield [9], which predicts K_{IC} from smooth tensile properties, was checked using the unpublished data generated by the present authors on materials other than those employed in this study.

$$K_{IC} = (2/3 \cdot E \cdot \sigma_Y \cdot n^2 \cdot \varepsilon_f)^{1/2}$$

The result shown in Fig. 8 shows good correlation between the measured and calculated K_{IC} values, and also shows a deviation from 1 : 1 correlation. One of the reasons for this deviation is the accuracy of determining n-value, as has been pointed out by Jones et al.[29]. Fig. 8 indicates that material properties which determine K_{IC} are σ_Y, E, n, and ε_f. Further investigation on the roles of these properties and the explicit incorporation of other parameters, such as inclusion spacing [30], will give a better fitting equation.

SUMMARY

Assuming that the stretched zone size observed on the fracture surface of the fracture toughness specimen is determined by material's ductility, SEM stereographic measurement of the stretched zone height was made on three grades of ultra-high strength steels. Results are summarized as follows:

(1) Matching the upper and lower profiles of the fracture surface revealed discrete opening in the ductile fracture band between the stretched zone and

unstable fracture region, as well as in the fatigue pre-cracked region and stretched zone.

(2) Crack opening displacement at the fatigue pre-crack tip, determined by matching upper and lower profiles of the fracture surface, was defined as stretched zone height (SZH). The SZH has a similar dimension to the stretched zone width (SZW) and agrees with the CTOD calculated from

$$CTOD = 0.455 \ K_{IC}^{2}/E \cdot \sigma_{Y}$$

(3) SZH is correlated with the fracture strain of the smooth tension bar. Hahn and Rosenfield's equation which relates K_{IC} to smooth tensile properties is shown basically an appropriate predictor, but needs refining to fit better with experimental data.

ACKNOWLEDGEMENT

The authors are indebted to Dr. S. Sawa, general manager, R & D Div., Daido Steel Co., Ltd. for permission to publish this study. They also wish to acknowledge Drs. T. Kato and S. Fukui for their fruitful and stimulating discussions. Finally, they wish to thank Mr. M. Saito for encouragement during the Study.

REFERENCES

(1) W. A. Spitzig, Trans. ASM, 61 (1968) 344

(2) R. C. Bates and W. G. Clark, Jr., ibid., 62 (1969) 380

(3) D. Elliot and H. Stuart, BISRA Open Rep. MG/C/49/'70 (1970)

(4) J. P. Tanaka, C. A. Pampillo and J. R. Low, Jr., Review of Developments in Plane Strain Fracture Toughness Testing, ASTM STP 463 (1970) 191

(5) A. J. Brothers, M. Hill, M. T. Parker, and W. A. Spitzig, Application of Electron Microfractography to Materials Research, ASTM STP 493 (1971) 3

(6) U. E. Wolff, ibid. (1971) 20

(7) N. Uehara, unpublished data, Daido Steel Co., Ltd.

(8) S. Fukui and N. Uehara, Tetsu-to-Hagane (J. I. S., Japan), 64 (1978) 841 (in Japanese)

(9) G. T. Hahn and A. R. Rosenfield, Applications Related Phenomena in Titanium Alloys, ASTM STP 432 (1968) 5

(10) P. F. Thomason, Int. J. Frac. Mech., 7 (1971) 409

(11) C. A. Griffis and J. W. Spretnak, Met. Trans., 1 (1970) 550

(12) D. Broek, Eng. Frac. Mech., 6 (1974) 173

(13) J. M. Barsom and J. V. Pellegrino, ibid., 5 (1973) 209

(14) H. W. Liu, Proc. 1st Intnl. Conf. Fracture, Vol. 1 (1965) 191

(15) J. S. Ke and H. W. Liu, Int. J. Frac. Mech., 7 (1971) 487

(16) D. P. Isherwood and J. G. Williams, Eng. Frac. Mech., 2 (1970) 19

(17) J. Malkin and A. S. Tetelman, ibid., 3 (1971) 151

(18) R. H. Sailors, T. & A. M. Rep. No. 367, Univ. Illinois (1973)

(19) D. Broek, Eng. Frac. Mech., 5 (1973) 55

(20) S. Kimoto, T. Suganuma, and T. Oshima, Shashin-Sokuryo (Photographic Measurement), 8 (1969) 8 (in Japanese)

(21) J. R. Rice and M. A. Johnson, Inelastic Behavior of Solids, McGraw-Hill (1970) 641

(22) J. R. Rice and G. F. Rosengren, J. Mech. Phys. Solids, 16 (1968) 1

(23) N. Levy, P. V. Marcal, W. J. Ostergren, and J. R. Rice, Int. J. Frac. Mech., 7 (1971) 143

(24) J. F. Knott and A. H. Cottrell, J. I. S. I., 201 (1963) 249

(25) R. C. Bates, W. G. Clark, and D. M. Moon, Electron Microfractography, ASTM STP 453 (1969) 192

(26) J. M. Krafft, Appl. Mat. Res., 3 (1964) 88

(27) J. M. Krafft, Nat. Symp. Fracture Mechanics, Lehigh Univ. (1967)

(28) A. J. Birkle, R. P. Wei, and G. E. Pellissier, Trans. ASM, 59 (1966) 981

(29) M. H. Jones and W. F. Brown, Jr., Review of Developments in Plane Strain Fracture Toughness Testing, ASTM STP 463 (1970) 63

(30) A. R. Rosenfield and G. T. Hahn, Colloque sur la Rupture des Materiaux, Grenoble, France (1972)

Table 1 Chemical composition of steels (w/o)

STEEL	C	Si	Mn	P	S	Cu	Ni	Cr	Mo	Co	Al	V	Ti
SNCM 8	0.39	0.33	0.86	0.029	0.028	0.18	1.79	0.68	0.20	—	—	—	—
18Ni MARAGING	0.003	0.02	0.01	0.004	0.003	0.01	18.50	0.01	5.15	8.55	0.099	—	0.87
10Ni-8Co	0.08	0.06	—	0.006	0.007	—	10.06	2.01	1.00	8.40	—	0.17	—

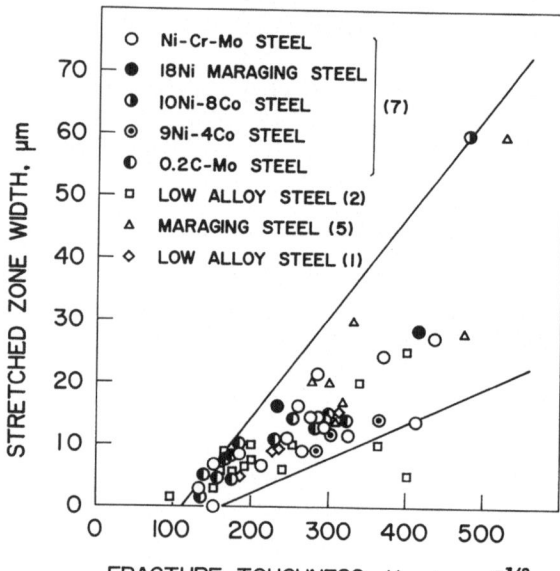

Fig. 1 Relation between fracture toughness and stretched zone
width.

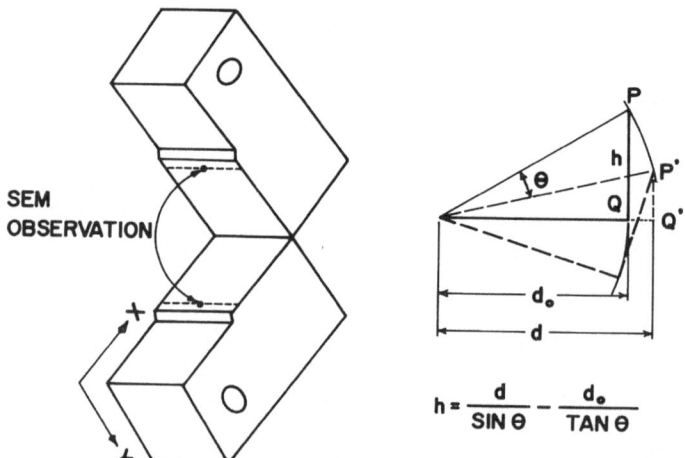

Fig. 2 Procedure of stereographic analysis of fracture surface
with scanning electron microscope (20).

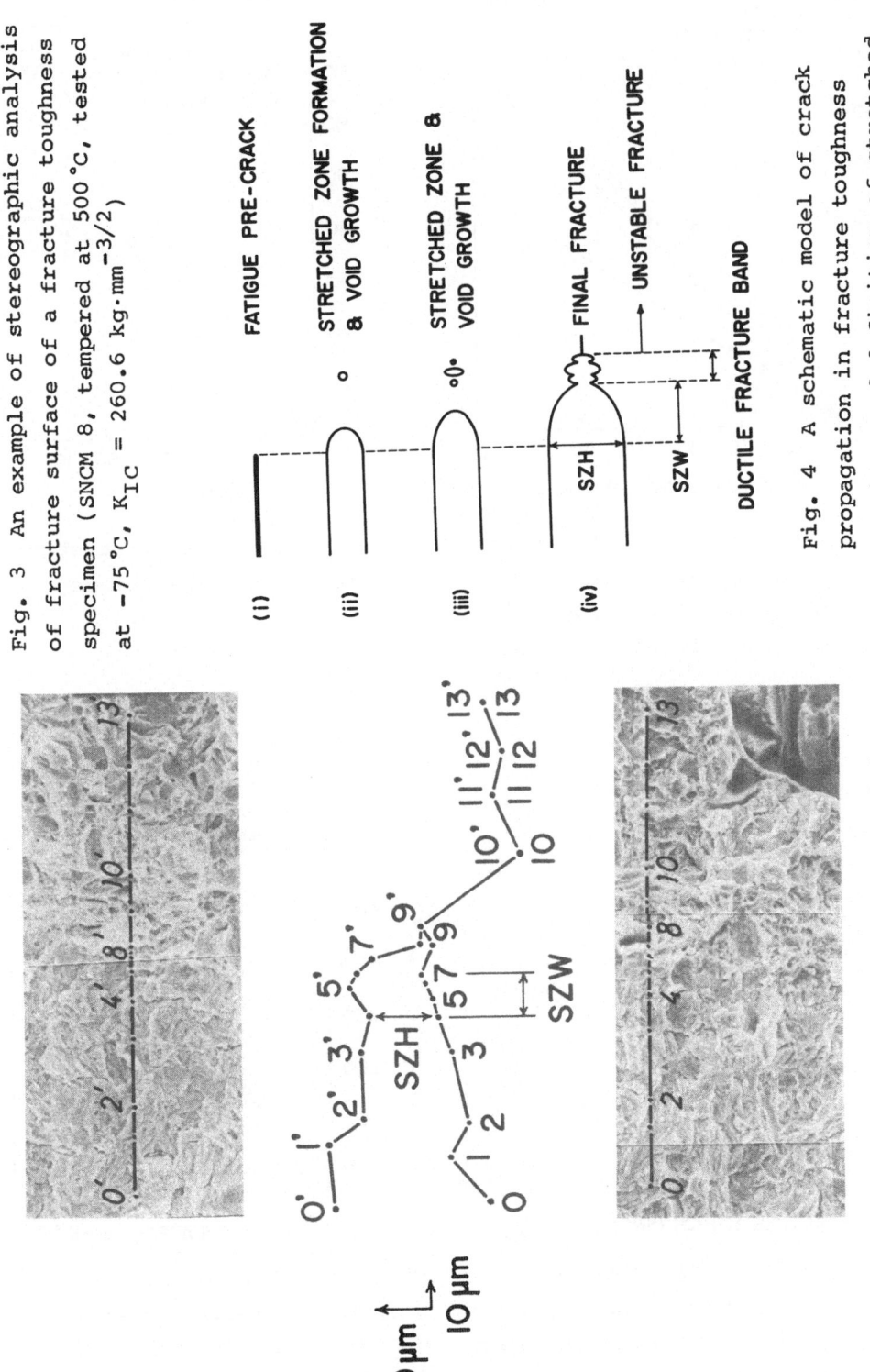

Fig. 3 An example of stereographic analysis of fracture surface of a fracture toughness specimen (SNCM 8, tempered at 500 °C, tested at −75 °C, K_{IC} = 260.6 kg·mm$^{-3/2}$)

(i) FATIGUE PRE-CRACK

(ii) STRETCHED ZONE FORMATION & VOID GROWTH

(iii) STRETCHED ZONE & VOID GROWTH

(iv) FINAL FRACTURE

UNSTABLE FRACTURE

SZH

SZW

DUCTILE FRACTURE BAND

Fig. 4 A schematic model of crack propagation in fracture toughness specimen and definition of stretched zone height.

208

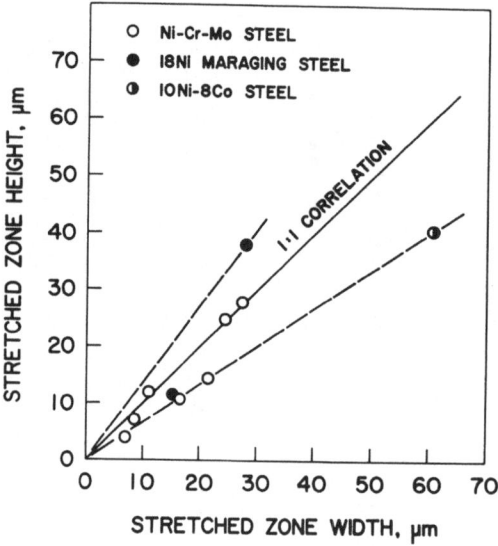

Fig. 5 Comparison of stretched zone width with stretched zone height of ultra-high strength steels of K_{IC} = 150-480 kg mm$^{-3/2}$.

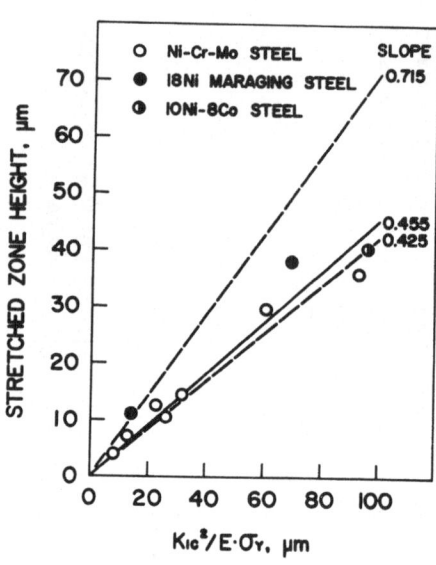

Fig. 6 Comparison of stretched zone height with calculated CTOD.

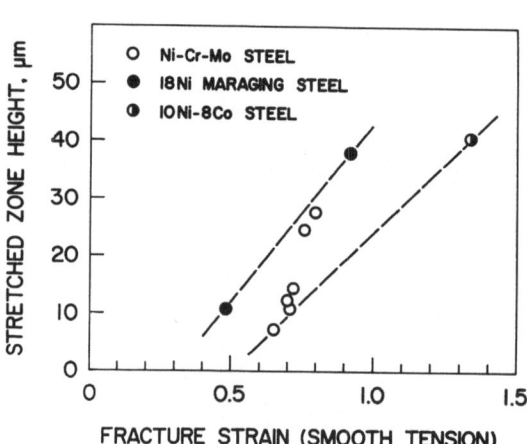

Fig. 7 Relationship between stretched zone width and fracture strain of smooth tension specimen.

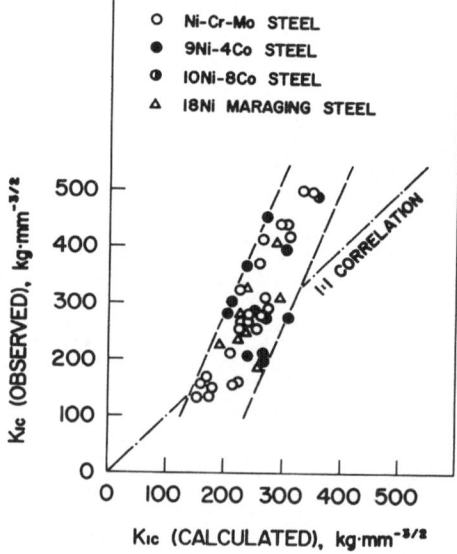

Fig. 8 Relationship between observed and calculated K_{IC} values of ultra-high strength steels.

ASSESSMENT OF NONDUCTILE FRACTURE INITIATION IN STRUCTURAL STEELS USING X-RAY DIFFRACTION

Y. Nakano and M. Katayama
Senior Researchers of Research Laboratories
Kawasaki Steel Corporation, 1 Kawasaki-cho, Chiba, Japan

ABSTRACT

A relation between fracture toughness J_{Ic} and the integral breadth ratio of the X-ray diffraction line profile, obtained from the fracture surface close to the precracked notch of a three point bend specimen, is obtained for four kinds of structural steel. The relation is strongly dependent upon the material strength. Using the material strength as a parameter, the relation between fracture toughness J_{Ic} (N/mm) and the integral breadth ratio of the X-ray diffraction line profile B/B_o is described as follows:

$$J_{Ic} = \exp[(B/B_o - 0.0006\sigma_{Yo} - 26.66)/136.71]$$

where σ_{Yo} is the yield stress of the material at room temperature in the unit of N/mm^2.

The effect of material strength on the plastic deformation at the notch tip is also discussed, using the integral breadth ratio of the X-ray diffraction line profile. For a given fracture toughness, the plastic zone at the notch tip is greater in a higher strength steel than in a lower one. It is concluded that nonductile fracture takes place in a higher strength steel after heavy plastic deformation concentrates in a small region at the notch tip, compared to a lower one. Crack opening displacement is a value corresponding to the plastic zone size at the notch tip, but it does not represent the extent of deformation in the plastic zone.

INTRODUCTION

Once nonductile fracture took place in steel structures or pressure vessels, the fracture surface is the only thing available for examination. To determine the cause of fracture quantitatively, information such as the applied stress, defect dimensions, temperature besides the fracture toughness of the material must be obtained. Among these, the fracture toughness is important as a characteristic of the material. Measuring the toughness of the failed material through a mechanical test, however, is not necessarily easy. Thus, the authors expect to develop a new technique, using the information from the fracture surface to determine the stress intensity factor at the tip of a defect when nonductile fracture takes place, that is, the fracture toughness of the material.

Optical and electron microscopies are typical techniques used to observe the fracture appearance. These methods give the qualitative information on the fracture surface, but not the quantitative.

X-ray diffraction, on the other hand, is a good technique for observing the fracture surface quantitatively as well as qualitatively. Research works in which X-ray diffraction technique is used as an engineering method observe the change in the microstructures of the material. Among them are the works on quantification of fatigue damage [1-4] and creep damage[5]. A trial to identify the fracture mode through X-ray diffraction technique [6] is also used. It is expected, therefore, that the X-ray diffraction technique can be successfully applied to the quantitative evaluation of nonductile fracture.

The present paper describes a quantitative evaluation of the fracture toughness of four kinds of structural steel, using X-ray diffraction from the fracture surface.

EXPERIMENTAL PROCEDURES

Materials

Four kinds of structural steel were used in the present experiment. They are a 75 mm thick, 590 N/mm^2 high tensile strength steel (HT60), a 25 mm thick, 785 N/mm^2 high tensile strength steel (HT80), a 30 mm thick, high tensile strength, hull structural steel (NK KD32), and a 40 mm thick, structural steel for welding purposes (JIS SM41C). The chemical composition and the mechanical properties of the steels are shown in Tables 1 and 2, respectively.

Fracture toughness test

The fracture toughness test was performed in a three point bend. The HT60 steel was tested using specimens with the thickness of 75 mm (original plate thickness), 50, 25, and 10 mm. The plate thickness was reduced by machining so that the center of the specimen thickness coincided with that of the original plate thickness. Other steels were tested, using specimens with the original plate thickness.

The dimensions of the three point bend specimens are shown in Fig. 1. The specimens were machined and fatigue-cracked in accordance with ASTM Standard E399-74. The ratio of the notch length to the specimen width was about 0.5.

The fracture toughness of the materials tested was determined as the J-integral value that was calculated using the following equation:

$$J = 2U_{total}/B(W-a) \qquad (1)$$

where U_{total} is the energy given as the area under the load-load point deflection curve prior to the nonductile fracture initiation or stable ductile crack initiation, whichever comes first. B, W, and a are the specimen thickness, the specimen width, and the notch length, respectively.

Static tensile test

Static tensile tests were performed to obtain the relation between the change in the integral breadth of the X-ray diffraction line profile and the plastic strain of the material. The specimen geometry is shown in Fig. 2.

X-ray measurement

An X-ray parallel beam method was applied to the quantitative evaluation of the change in microstructures in the fracture surface layer of the specimen created by nonductile fracture. The X-ray conditions are listed in Table 3. The X-ray irradiated area, with dimensions of 2 mm by 2 mm, was located at the center of the specimen thickness, and as close to the precracked notch front as possible in the bend specimens and in the area where the plastic strain was measured in the static tensile test specimens.

The effect of the fracture surface roughness on the X-ray measurement was ignored because the difference in the integral breadth of the X-ray diffraction line profile between the incidence angle ψ_o of 0° and that of 45° had been only three percent for the standard chromium powder sample [7].

EXPERIMENTAL RESULTS AND DISCUSSION

Fracture toughness test results

The energy consumed by the specimen to initiate a nonductile crack or a stable ductile crack, whichever came first, was used to calculate J_{Ic}. For HT80 and SM41C steels, an electrical potential method was used to detect the stable ductile crack initiation, while for HT60 and KD32 steels only specimens with no stable ductile crack initiation observed on the fracture surface were used for calculation of J_{Ic}.

All the values of J_{Ic} measured above satisfied Eq. 2, proposed by Paris [8].

$$B, a, W-a \geq 50(J_{Ic}/\sigma_f) \tag{2}$$

where σ_f is the flow stress at the test temperature of the material tested.

Figure 3 shows the plot of J_{Ic} vs. temperature.

Fracture toughness and broadening of the X-ray diffraction line profile

The microstructures of the material change with plastic deformation. The fracture surface layer of the specimen tested suffers from plastic deformation. It can be considered, therefore, that the extent of plastic deformation in the fracture surface layer corresponds to the stress intensity factor at the notch tip at the time of nonductile crack initiation, that is, the fracture toughness of the material tested.

212

Figure 3 shows the plot of fracture toughness J_{Ic} vs. the integral breadth ratio B/B_o of the X-ray diffraction line profile, where B is the integral breadth of the X-ray diffraction line profile from the fracture surface and B_o is that from the virgin material. Although there is some scattering, linear relation is observed between $\ln J_{Ic}$ and B/B_o. At a same J_{Ic} value, the value of B/B_o decreases in the order of HT80, HT60, KD32, and SM41C. In other words, the relation between J_{Ic} and B/B_o is strongly affected by the strength of the material.

Figure 5 shows the plot of the integral breadth ratio vs. the yield stress of the material at room temperature σ_{Yo} for three J_{Ic} values. The relation between B/B_o and σ_{Yo} can be approximately by three straight lines with the gradient of 0.0006.

Figure 6 shows the plot of the intercept of these straight lines with the axis of ordinates $(B/B_o - 0.0006\sigma_{Yo})$ vs. $\ln J_{Ic}$. From this figure, the following equation is obtained:

$$B/B_o - 0.0006\sigma_{Yo} = 136.71 \ln J_{Ic} + 26.66 \tag{3}$$

Thus, J_{Ic} (N/mm) can be calculated from the integral breadth ratio B/B_o of the X-ray diffraction line profile, measured on the nonductile fracture surface at the crack tip and the yield stress at room temperature σ_{Yo} (N/mm^2), as follows:

$$J_{Ic} = \exp[(B/B_o - 0.0006\sigma_{Yo} - 26.66)/136.71] \tag{4}$$

Since the following relation exists between J_{Ic} and the plane strain fracture toughness K_{Ic}:

$$K_{Ic} = \sqrt{J_{Ic}E/(1-\nu^2)}$$

where ν is the Poisson's ratio and E is the elastic modulus, Eq. 4 can be described as follows:

$$K_{Ic} = \sqrt{E \exp[(B/B_o - 0.0006\sigma_{Yo} - 26.66)/136.71]/(1-\nu^2)} \tag{5}$$

For an ordinary steel, Eq. 5 becomes

$$K_{Ic} = 475.72 \sqrt{\exp[(B/B_o - 0.0006\sigma_{Yo} - 26.66)/136.71]} \tag{6}$$

Fracture toughness and plastic strain

Since the integral breadth ratio of the X-ray diffraction line profile corresponds to the change in microstructures of the material due to plastic deformation, the relation between the integral breadth ratio and the plastic strain in the material is studied next.

Figure 7 shows the relation between the macroscopic plastic strain remaining in the static tensile test specimens that had been prestrained to certain amounts and the integral breadth ratio of the X-ray diffraction line profile. This relation was used to obtain the plastic strain corresponding to the integral breadth of the X-ray diffraction line profile from the nonductile fracture surface.

Figure 8 shows the plot of apparent plastic strain ε_{AP} obtained, as described above, vs. J_{Ic} on a semi-logarithmic graph. A linear relation exists between two parameters that is similar to the relation between J_{Ic} and B/B_o. Note, however, that the apparent plastic strain ε_{AP} corresponding to a given J_{Ic} value increases with the strength of material. In other words, the plastic deformation at the notch tip, required to give a certain amount of fracture toughness, increases with the strength of the material.

Plastic deformation is the greatest on the fracture surface, and it decreases sharply with the distance from the fracture surface [9]. To evaluate the plastic work done by the material under the X-ray irradiated area, the distribution of plastic strain ε_p from the fracture surface toward the inside must be determined. Now, suppose that the distribution of plastic strain from the fracture surface toward the inside is described by the following equation:

$$\varepsilon_p = \varepsilon_{Po}(1/1+r)^{1/(1+n)} \tag{7}$$

where ε_{Po} is the plastic strain on the fracture surface, r is the distance from the fracture surface, and n is the strain hardening coefficient. Though this equation is not empirically obtained, it does not differ very much from the empirically obtained distribution of plastic strain [9].

The plastic work done by the material under the X-ray irradiated fracture surface W is obtained by integrating the plastic strain energy by the volume under the X-ray irradiated fracture surface v as follows:

$$W = \int_v \int \sigma d\varepsilon_p dv \tag{8}$$

where σ is the stress.

Then, the plastic work done by the material under a unit X-ray irradiated area w is given by the following equation:

$$w = \int_r \int \sigma d\varepsilon_p dr \tag{9}$$

Since $\sigma = \sigma_o \varepsilon_p^{\,n}$ where σ_o is the material constant,

$$w = \int_r \int_o^{\varepsilon_P} \sigma_o \varepsilon_p^{\,n} d\varepsilon_p dr \tag{10}$$

Substituting Eq. 7 into Eq. 10,

$$w = (\sigma_o/1+n)\int_o^\infty [\varepsilon_{Po}/(1+r)]^{1+n} dr$$
$$= \sigma_o \varepsilon_{Po}^{\,1+n}/n(1+n) \tag{11}$$

Thus, the plastic work done by the material under a unit X-ray irradiated area w for $J_{Ic} = 10$ N/mm can be calculated as shown in Table 4. In the calculation, ε_{AP} was used instead of ε_p. The plastic work done is greater in a higher strength steel than in a lower one.

This fact can be understood by taking it into consideration that greater yield stress makes the plastic zone at the notch tip smaller (Fig. 9). In other words, nonductile fracture takes place in a higher strength steel after heavy plastic deformation concentrates in a small region at the notch tip, comparing to a lower strength steel.

Table 5 shows the values of J_{Ic}/σ_Y corresponding to J_{Ic} of 10 and 20 N/mm, where σ_Y is the yield stress of the material at the test temperature. It is well known that the J-integral value is related to crack opening displacement through flow stress. A relation has been obtained between J_{Ic}/σ_Y and crack opening displacement (COD, δ_c) as described by the following equation [10]:

$$J_{Ic}/\sigma_Y = 1.4\delta_c \tag{12}$$

Using the above equation, J_{Ic}/σ_Y can be converted to δ_c, as shown in Table 5. COD, corresponding to a given J_{Ic} value, decreases with the increase of strength. This fact is consistent with the dependency of the size of plastic zone at the notch tip on the material strength as described above, but it is different from that of the extent of plastic deformation. Therefore, COD is a value corresponding to the size of plastic zone at the notch tip but does not represent the extent of deformation in the plastic zone.

CONCLUSIONS

An X-ray observation of the nonductile fracture surface was applied to three point bend specimens of 590 and 785 N/mm^2 high tensile strength steels, a high tensile strength, hull structural steel, and a structural steel for welding purposes in order to evaluate the fracture toughness quantitatively.

The conclusions obtained follow:

1) Fracture toughness J_{Ic} (N/mm) could be related to the integral breadth ratio of the X-ray diffraction line profile B/B_o through the following equation:

$$J_{Ic} = \exp[(B/B_o - 0.0006\sigma_{Yo} - 26.66)/136.71]$$

where B is the integral breadth of the X-ray diffraction line profile obtained from the fracture surface, B_o is that from the virgin material, and σ_{Yo} is the yield stress at room temperature (N/mm^2).

2) For an ordinary steel, the plane strain fracture toughness K_{Ic} $(MPa\ m^{1/2})$ was given by the following equation:

$$K_{Ic} = 475.72 \sqrt{\exp[(B/B_o - 0.0006\sigma_{Yo} - 26.66)/136.71]}$$

3) For a given fracture toughness, the plastic zone at the notch tip was greater in a higher strength steel than in a lower one.

4) Nonductile fracture takes place in a higher strength steel after heavy plastic deformation concentrates in a small region at the notch tip, comparing to a lower strength steel.

5) Crack opening displacement is a value corresponding to the plastic zone size at the notch tip, but it does not represent the extent of deformation in the plastic zone.

REFERENCES

(1) S. Taira and K. Honda, "X-Ray Investigation on the Fatigue of Metals," Proc. 1st Int. Conf. Fracture, Sendai, Japan, 3, (1965), pp. 1581–1596.

(2) S. Taira, T. Goto, and Y. Nakano, "X-Ray Investigation of Low-Cycle Fatigue in an Annealed Low Carbon Steel," J. Soc. Mat. Sci. Japan, 17(183), (1968), pp. 1135-1139, (in Japanese).

(3) S. Taira, K. Hayashi, and K. Tanaka, "X-Ray Investigation on Fatigue Fracture of Notched Steel Specimen," J. Soc. Mat. Sci. Japan, 15(159), (1966), pp. 879-886, (in Japanese).

(4) M. Nagao, S. Kusumoto, and Y. Ito, "Macro-X-Ray Fractography - Fatigue Failure Analysis," J. Soc. Mat. Sci. Japan, 26(286), (1977), pp. 630-636, (in Japanese).

(5) S. Taira, E. Nakanishi, and Y. Kawabe, "Application of X-Ray Microbeam Technique to the Study of Creep," J. Soc. Mat. Sci. Japan, 14(147), (1965), pp. 1007-1013, (in Japanese).

(6) T. Goto, "A Fundamental Study on the Failure Analysis of Metal Components Using X-Ray Diffraction Technique," Mitsubishi Juko Giho, 11(3), (1974-5), pp. 328-340, (in Japanese).

(7) M. Katayama, Unpublished.

(8) P. C. Paris, Discussion on "The J-Integral as a Fracture Criterion," by J. A. Begley and J. D. Landes, Fracture Toughness, ASTM STP514, ASTM, (1972), pp. 21-22.

(9) S. Taira, J. Ryu, and Y. Komizo, "The Measurement of Local Plastic Strain Distribution near Cleavage Fracture Surfaces by the X-Ray Microbeam Diffraction Technique," Proc. 1973 Symp. Mech. Behavior of Materials, Mechanical Behavior of Materials, Kyoto, Japan, (1974), pp. 119-127.

(10) Y. Nakano, "Evaluation of Fracture Toughness of Structural Steels through J-Integral," J. Iron Steel Inst. Japan, 64(7), (1978), pp. 891-898, (in Japanese).

Table 1 Chemical Composition

(wt %)

Steel	C	Si	Mn	P	S	Cu	Ni	Cr	Mo	V
HT80	0.11	0.27	0.78	0.014	0.006	0.25	0.98	0.46	0.435	0.031
HT60	0.13	0.29	1.16	0.008	0.005	0.03	0.53	0.16	0.130	0.041
KD32	0.13	0.34	1.39	0.007	0.007	0.01	0.02	0.02	—	—
SM41C	0.15	0.24	1.16	0.019	0.007	0.01	0.02	0.01	—	—

Table 2 Mechanical Properties

Steel		Yield Stress (N/mm^2)	Tensile Strength (N/mm^2)	Elongation (%)	vTs (°C)
HT80, Q&T, t25		776	821	26	-110
HT60, Q&T, t75	1/4t	521	634	29	-51
	1/2t	530	655	28	-38
KD32, t30		314	475	27	-14
SM41C, Normalized, t40		302	473	39	-50

Table 3 X-Ray Conditions

Diffraction Plane	(211)
Target	Cr
Tube Voltage	30 kV
Tube Current	8 mA
Incidence Angle ψ_0	0°
Irradiated Area	2mm x 2mm

Table 4 Plastic Work Done by the Material under a unit X-Ray Irradiated Area Corresponding to J_{Ic} = 10 N/mm

Steel	HT80	HT60	KD32	SM41C
Temperature (°C)	-172	-158	-154	-120
σ_0 (N/mm^2)	105	85	61	53
n	0.13	0.15	0.19	0.24
ε_{PO}	0.16	0.102	0.035	0.04
w (N/mm)	90	36	5	3

Table 5 Values of J_{Ic}/σ_Y and COD Corresponding to J_{Ic} = 10 and 20 N/mm

J_{Ic} (N/mm)	J_{Ic}/σ_Y δ_c	Steel			
		HT80	HT60	SM41C	KD32
10	J_{Ic}/σ_Y(mm)	0.009	0.011	0.015	0.020
	δ_c (mm)	0.013	0.015	0.021	0.028
20	J_{Ic}/σ_Y(mm)	0.020	0.030	0.039	0.050
	δ_c (mm)	0.018	0.042	0.055	0.070

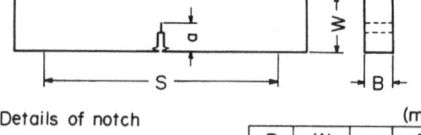

Details of notch

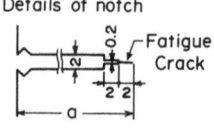

B	W	a	S
75	150	75	600
50	100	50	400
40	80	40	320
30	60	30	240
25	50	25	200
10	20	10	110

(mm)

Fig. 1 Geometry of Three Point Bend Specimen

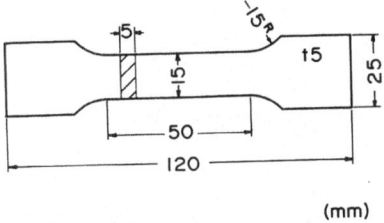

Fig. 2 Geometry of Static Tensile
Test Specimen

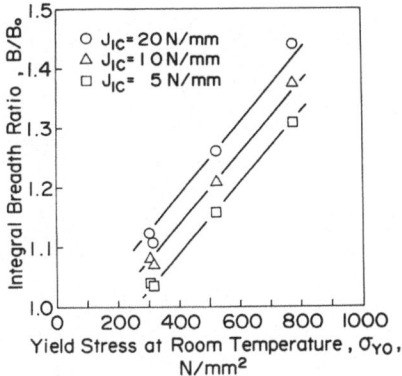

Fig. 5 Relation between Integral
Breadth Ratio and Yield Stress at
Room Temperature

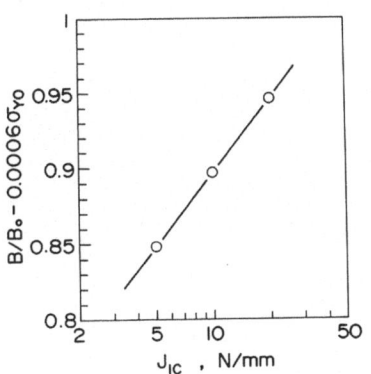

Fig. 6 Relation between
$(B/B_o - 0.0006\sigma_{Yo})$ and J_{Ic}

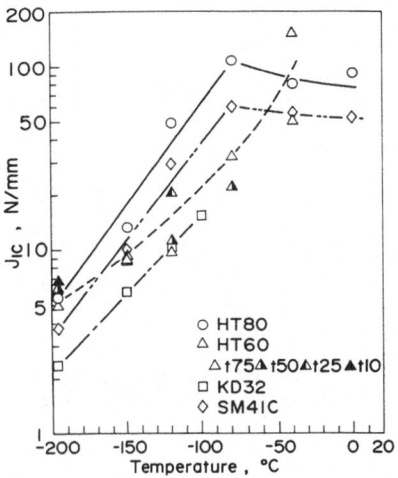

Fig. 3 Plot of J_{Ic} vs. Temperature

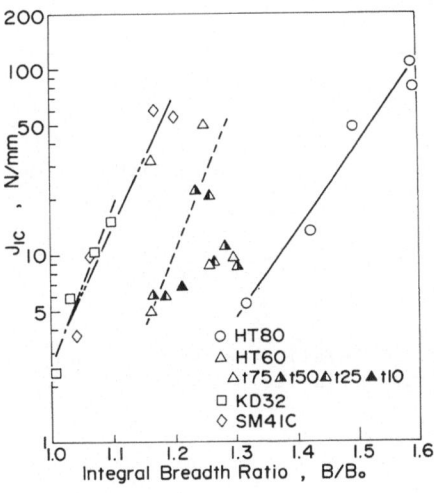

Fig. 4 Relation between J_{Ic} and
Integral Breadth Ratio

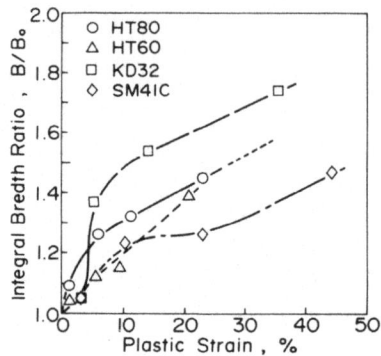

Fig. 7　Relation between Integral Breadth Ratio and Plastic Strain

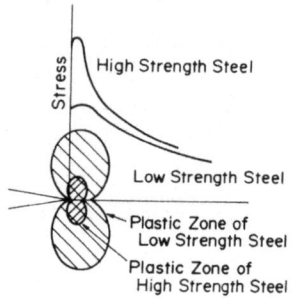

Fig. 9　Illustration of the Plastic Zone at the Notch Tip at a Certain Fracture Toughness Level for Different Strength Steels

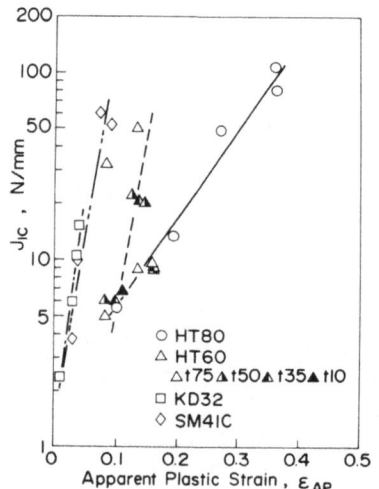

Fig. 8　Relation between J_{Ic} and Apparent Plastic Strain

V. MATERIAL EFFECTS ON FRACTURE TOUGHNESS

THE INFLUENCE OF MICROSTRUCTURAL VARIATIONS ON THE FRACTURE PROCESS OF DUAL PHASE STEEL

T. Kunio*, M. Shimizu**, K. Yamada***
J.K. Kim**** and H. Hirusawa****
*Prof., **Assoc. Prof., ***Assist. Prof., and ****Graduate Students
of Mechanical Engineering
Keio University, Hiyoshi 3-14-1, Kohoku-ku, Yokohama 223, Japan

ABSTRACT

A study has been made of the strength and fracture of the dual phase microstructure, in which the martensitic phase encapsulated islands of the ferritic phase in association with the cleavage cracking of ferrite grains.

It was found that the final fracture occured in a brittle manner, starting from the Griffith crack which consisted of the cleavage crack in the ferrite grains. Furthermore, the effects of the ferrite grain sizes on the Griffith crack were discussed.

INTRODUCTION

In recent years, considerable interest has been aroused in dual phase steel, whose microstructures consist of a mixture of the two phases of martensite and ferrite [1]-[12]. The studies of the dual phase structure utilizing martensite and ferrite aim at the development of high strength materials without the loss of ductilities [3][6]. The main role of martensite in the dual phase microstructure is related to the overall strength level, while the ferrite is expected to contribute to the high ductilities. However, the experimental results do not always agree with such a simple expectation.

The mechanical properties of these microstructures were associated with the interaction effects of these two phases, rather than the individual properties of martensite and ferrite [1][3]. While the strength is a function of the volume fraction of martensite [2][3][5][6], the ductilities are strongly influenced by the morphology of the two phases even for a given volume fraction.

The morphology of the two phases is characterized as follows: in one mor-

phology ferrite encapsulates islands of the martensite (FEM microstructure);
in the other, martensite encapsulates islands of ferrite (MEF microstructure).
It has been recognized in the MEF microstructure that the fracture occurs in a
brittle manner, accompanying cleavage cracking of the ferrite grains in contrast to the ordinary ductile fracture of the FEM microstructure [1].

This unusual behavior of cleavage cracking of the ferrite grain in the
MEF microstructure was reasonably explained by the enhancement of internal
stresses associated with the phase transformation of martensite and also with
the plastic constraint of the martensite during the loading process [1][5][11]
[12]. However, questions arise about the manner in which the size of ferrite
grains influences the cleavage cracking of the present MEF microstructure and
also the way in which the final fracture is related to the present cleavage
cracking of ferrite grains.

In an attempt to answer the questions, the MEF microstructures with various grain sizes of ferrite were employed, and the influence of microscopic
internal stresses was maintained as constant by careful heat treatments.

EXPERIMENTAL PROCEDURE

The material is a low carbon steel of 0.20%C. This plain carbon steel was
machined into a cylindrical shape of a tensile specimen with 10mm in gauge
length and 9mm in diameter, after austenitizing at various temperatures between
1000°C and 1300°C to obtain a wide range of ferrite grain sizes. The specimens
were then heat-treated in the conditions as shown in Fig.1.

These careful heat-treatments gave the specimens the same morphology of
MEF microstructure as that shown in Fig.2. The hardness of martensite and
ferrite and the volume fraction of martensite in these microstructures were
kept constant, but the ferrite grain sizes varied from 28 to 93 μm. The details
of the metallurgical properties of these MEF microstructure were given in
Tab.1. This table indicates that the effects of grain sizes on the mechanical
properties of the MEF microstructure might have been extracted from those of
the other microstructural parameters. All of the specimens were electro-polished
and then subjected to tensile loads with the strain rate of 1×10^{-4}/sec.

Both the initiation and the stable growth of the crack, involving the
ferrite cleavage cracking , were examined by optical and scanning electron
microscopes on the surfaces and longitudinal cross sections of the specimens.
Examination of the increasing rate of the number of cracks during the stable
growth was made by counting all the cleavage cracks on the several longitudinal
sections of each specimen, which was unloaded from the stress or strain levels
at any instance of stable crack growth.

The fraction of brittle fractured facets, was made by the line counting method [13].

RESULTS AND DISCUSSIONS

Characteristics of stress-strain behavior of the four different grain sized specimens were given in Fig.3. This figure shows that the fracture stress strongly depends upon the ferrite grain size, though no appreciable differences appear in the nature of the fracture process of the four types of specimens mentioned above. Details will be given in the following paragraphs.

The fracture stress and strain increase with the decrease of the ferrite grain size, while the proof stress is not depedent on the difference in the grain sizes. The grain size independent characteristic of the proof stress was immediately recognized from the stress -strain relationship in Fig.3, where the elastic behavior of bulk specimens was characterized by a single curve of the stress-strain relationship.

Characteristics of the tensile behavior were replotted against the reciprocals of the square root of the grain sizes $d^{-1/2}$, shown in Fig.4. Changes of reduction in area against the grain size clearly indicate that the brittle behavior in the present dual phase steel is dependent on the grain size. This brittle behavior is characterized by the cleavage cracking of ferrite grains in the MEF microstructure, seen in Fig.5, in which typical cleavage facets of the ferrite grains appeared on the fractured surface. In addition, the result of fracture surface analysis shows that the fraction of the cleaved ferrite grain exceeded the volume fraction of ferrite in the matrix, shown in Fig.6. This might be explained by the formation of the fractured surface, resulting from the coalescence of neighbouring cleaved grains which were three dimensionally distributed in the volume of the matrix. Fig.7 indicates the cleavage crack of ferrite grain inside the series C specimen, which was unloaded from the stress $\sigma=1009$MPa just before the fracture and was examined by sectioning the specimen. The final growth of the crack leading to the fracture of the specimen was caused by the coalescence of these cracks, which were nucleated independently and which increased their number with the increases in the applied stress. Fig.8 shows that a common tendency of the number of cleavage cracking versus applied strain was observed for all kinds of specimens and also that the critical number of the cracking at fracture depends on grain size. Furthermore, Fig.9 indicates that the critical fraction of cleaved ferrite grains at the onset of fracture is larger for a large grain sized specimen. Since the fraction of cleaved ferrite grains could be related to the size of the crack in the specimen, the large grain sized specimen might have the large critical crack length. On the other hand, from the experimental fact that the hardness of martensite were almost the same for all kind of specimens, it is considered that no appreciable difference between them might exist in the magnitude of the crack growth resistance at the onset of the growth of the critical crack.

It can be concluded from these considerations, together with the evidence shown in Fig.10 which shows clear grain size dependence of the critical stress for the initiation of cleavage crack, that the size of the ferrite grain could be directly related to the magnitude of the length of the critical crack. This might explain why the fracture stress determined from the critical condition for the crack growth had the grain size dependence.

The above discussion may imply that the fracture stress of the present dual phase microstructure is expressed by the Griffith equation [14]:

$$\sigma_f = \sqrt{\frac{G_{IC}E}{\pi(1-\nu^2)c}} \qquad ---------- \quad (1)$$

If this equation is applicable in the present case, the critical crack length can be calculated as follows. The plain strain fracture toughness G_{IC} was obtained approximately as 3000 J/m^2 from the following equation [14]:

$$G_{IC} = \frac{(1-\nu^2)K_{IC}^2}{E} \qquad ---------- \quad (2)$$

where the K_{IC} value was experimentally obtained [15] from the ASTM method [16] for the martensite-ferrite dual phase microstructure. Then, the substitutions of G_{IC}, Young's Modulus E, Poisson's ratio ν, and the tensile fracture stress σ_f brought about the expected length of the Griffith cracks, 2c. The results obtained were given in Tab.2. This table shows that the calculated values of the Griffith crack length were several times larger than the ferrite grain sizes of the specimen. This might suggest that the Griffith crack was not composed of the cleavage crack of a single grain but composed of the several coalesced cleavage cracks, the number of which would depend on the grain size of the specimen. The number of cleaved ferrite grains was calculated from the following equation on the basis of the morphology of the present microstructure where the ferrite grain is encapsulated by the martensite of the thickness t (MEF microstructure) as shown in Fig.2.

$$L_C = Nd_f + (N-1)2t \qquad ---------- \quad (3)$$

L_C: calculated Griffith crack length

d_f: diameter of mean ferrite grain

t: thickness of the martensite surrounding the ferrite grain

N: number of the ferrite grains

Results obtained were also tabulated in Tab.2. The number of cleaved ferrite grains might be observed in the specimen just before the critical crack growth, when the fracture of the specimen follows the Griffith criterion. Then, an examination of the Griffith crack length, consisting of the coalesced cleavage cracking of several ferrite grains at the fatal growth, was made on the two kinds of specimen: one is unloaded from the stress just before the fracture : the other is the half part of the fractured specimen. In the former specimens, such cracks was not detected through a careful examination of the longitudinal section of the specimen. However, in the latter specimens, the cracks having a few grains in the length were observed. The maximum length of these observed cracks were 5~7 grains for the series A specimen, 4~6 grains for series B, 3~4 grains for series C, and 2~3 grains for series D specimens. All these cleaved ferrite grains were coalesced with each other by the cracking of the adjacent martensite as shown in Fig.11. The number of cleaved ferrite grains associated with the Griffith crack might be larger for the specimen at the onset of fracture than that observed in the half part of fractured specimen.

Accordingly, a fairly good agreement was obtained between the prediction and the experimental result. Thus, it may be concluded that the fracture strength of the present specimen was well characterized by the Griffith criterion.

CONCLUSIONS

A study has been made of the strength and fracture of a dual phase microstructure in which the martensite encapsulated islands of ferrite in association with the cleavage cracking of ferrite grains. The results obtained are summarized as follows:

1. Both the initiation stress for cleavage cracking and the fracture stress increase linearly with respect to the reciprocals of the square root of the grain sizes.
2. The final fracture of the specimen occurred in a brittle manner, which was characterized by the Griffith criterion. The Griffith crack consisted of the several number of coalesced ferrite grains in the MEF microstructures.

An estimation of this Griffith crack length, made by the calculation, showed a fairly good agreement with the experimental result.

226

REFERENCE

(1) T. Kunio, M. Shimizu, K. Yamada, and H. Suzuki, "An Effect of the Second Phase Morphology on the Tensile Fracture Characteristics of Carbon Steels", Eng. Fract. Mech., 7, (1975), 411 pp.

(2) H.W. Hayden and S. Floreen,"The Influence of Martensite and Ferrite on the Properties of Two-Phase Stainless Steels Having Microduplex Structure", Met. Trans., (1970), 1955 pp.

(3) I. Tamura, Y. Tomota, A. Akao, M. Ozawa, and S. Kanatani, "The Strength and Ductility of Two-Phase Iron Alloys", J. Iron and Steel Inst. of Japan, 59-3, (1973), 454 pp.

(4) K. Minakawa, M. Shimizu and T. Kunio, "Relation between Fracture Toughness and Microstructural Variables", Trans. Japan Soc. of Mech. Eng., 41-351, (1975), 3033 pp.

(5) Y. Tomita, S. Oki, and K. Okabayashi, "Static Tensile Properties of Mixed Structure of Martensite and Residual Ferrite in Ni-Cr-Mo Steels Containing Low Carbon", J. Iron and Steel Inst. of Japan, 63-8, (1977), 1321 pp.

Y. Tomita, S. Oki, and K. Okabayashi, "Static Tensile Properties of Mixed Structure of Martensite and Residual Ferrite in Ni-Cr-Mo Steels Containing Medium Carbon", J. Iron and Steel Inst. of Japan, 64-1, (1978), 78 pp.

(6) R.G. Davies, "Influence of Martensite Composition and Content on the Properties of Dual Phase Steels", Met. Trans., 9A, (1978), 671 pp.

(7) R.G. Davies, "Influence of Silicon and Phosphorous on the Mechanical Properties of Both Ferrite and Dual Phase Steels", Met. Trans., 10A, (1979), 113 pp.

(8) Y. Soyama and M. Tagaya, "Lower Yield Point of the Steel Containing Hard Patches in Free Ferrite Matrix", Japan Soc. Material Sci., 14-142, (1965), 542 pp.

Y. Soyama, "Stress-Strain Diagram of the Steel Constituted of Hard Patches and Free Ferrite", Japan Soc. Material Sci., 15-148, (1966), 17 pp.

(9) B. Karlsson and B.O. Sundström, "Inhomogeneity in Plastic Deformation of Two-Phase Steels", Mater. Sci., Eng., 16, (1974), 161 pp.

(10) R.L. Carins and J.A. Charles, "Mechanical Properties of Steels with Controlled Martensite-Ferrite Microstructures", Journal Iron and Steel Inst., 205, (1967), 1051 pp.

(11) J.K. Kim, H. Hirusawa, K. Yamada, M. Shimizu, and T. Kunio, "The Effect of the Second Phase on Ductile-Brittle Transition Behavior of Carbon Steels with Martensitic-Ferritic Combined Microstructure", Trans. Japan Soc. of Mech. Eng., 45-393, (1979), 415 pp.

(12) J.K. Kim, M. Shimizu, and T. Kunio, "Effect of Internal Stress on Fracture Behavior of Carbon Steel with Duplex Microstructure", Preprint of 53th Annual Meeting of JSME, 784-1, (1978), 61 pp.

(13) R.T. Dehoff and F.N. Rines, "Quantitative Microscopy", McGraw-Hill, (1968), 402 pp.

(14) G.R. Irwin, "Fracture Mechanics", Structural Mechanics, Proc. 1st Sympo. on Naval Structural Mechanics, Pergamon, (1960), 557 pp.

(15) H. Ide, K. Minakawa, M. Shimizu, and T. Kunio, "A Study on the Stable Crack Growth and the Evaluation of Fracture Toughness of the Material Having Microstructural Heterogeneity", Preprint of 53th Annual Meeting of JSME, 784-1, (1978), 64 pp.

(16) "ASTM E399-70T", ASTM STP 463, (1970),249 pp.

227

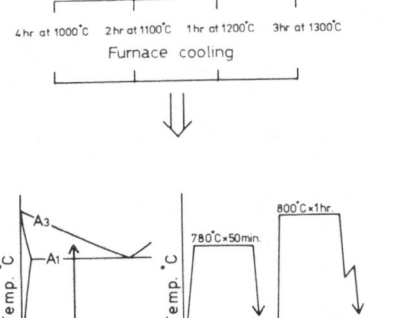

Fig. 1 Heat treatments of the

specimen

Tab. 1 Metallurgical properties

Series	Ferrite grain size d μm	Martensite volume fraction V_M %	Micro-Vickers hardness (25g, 100 indentation)		Hardness ratio	Connectivity %
			Martensite	Ferrite		
A	28	55	679	200	3.4	95
B	38	55	686	201	3.4	96
C	60	59	665	199	3.3	97
D	93	59	689	203	3.4	98

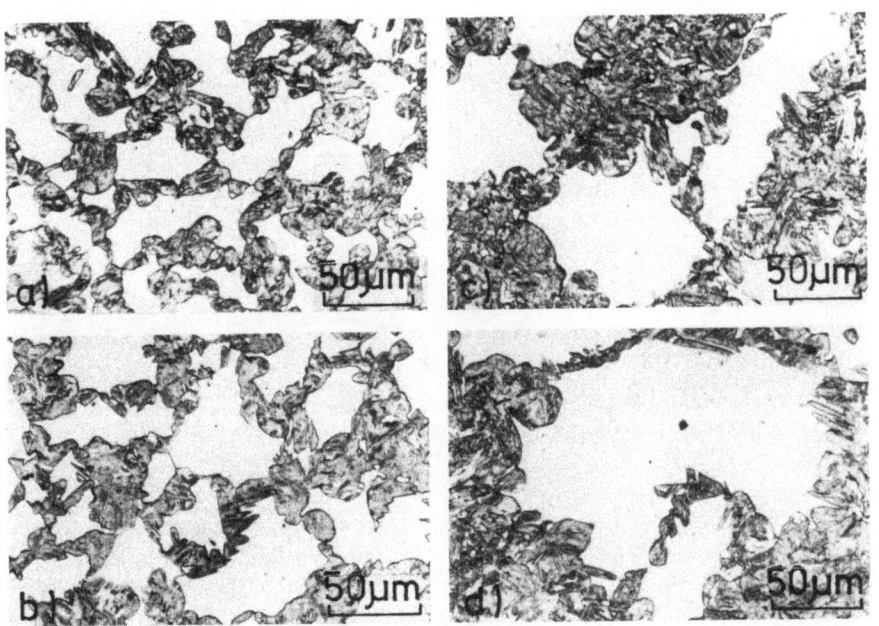

Fig. 2 Typical microstructures of dual phase steels

(a) Series A, (b) Series B, (c) Series C, (d) Series D

228

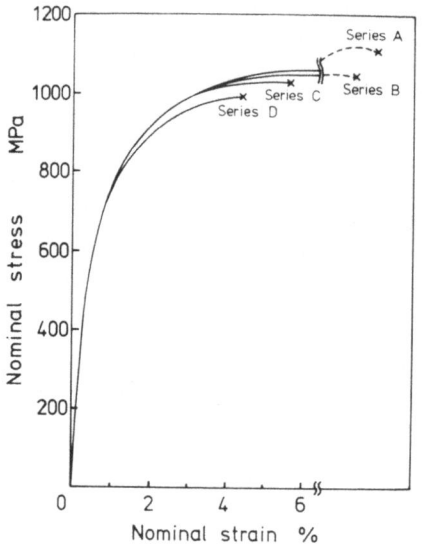

Fig. 3 Stress-strain relationship
of dual phase steels

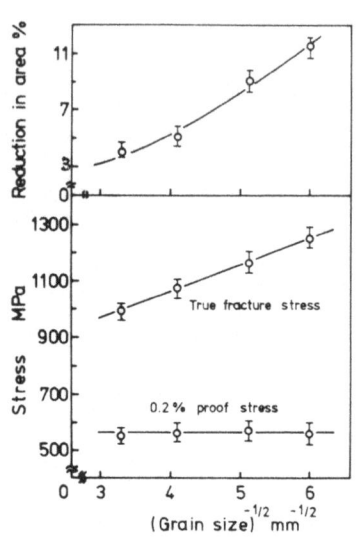

Fig. 4 Mechanical properties of
dual phase steels

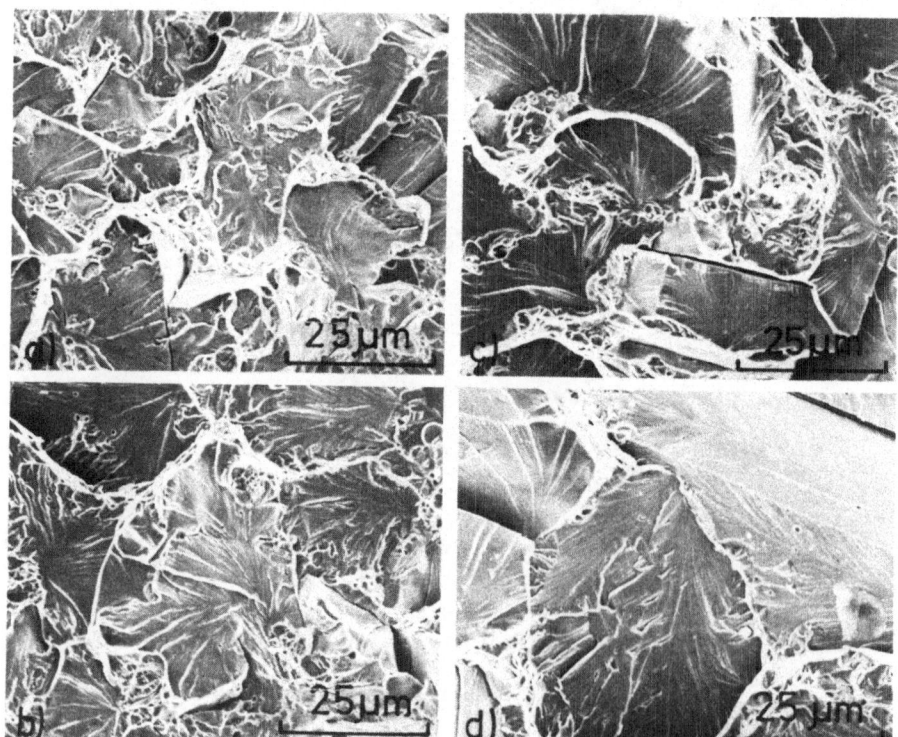

Fig. 5 Scanning electron micrographs of typical

fracture surface of (a) series A, (b) series B,
(c) series C, and (d) series D specimens, respectively

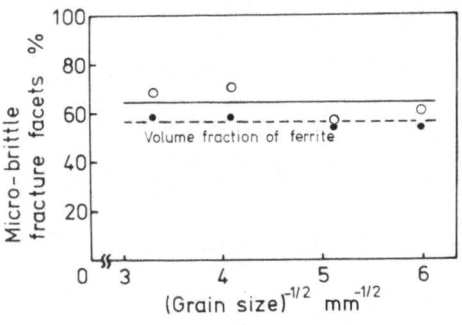

Fig. 6 Relationship between the
fraction of cleavage facets
and the volume fraction of
ferrite as a function
of (grain size)$^{-\frac{1}{2}}$

Fig. 7 Example of cleavage crack in
the ferrite grain formed
inside the specimen
(Series C ; unloaded from 1009 MPa)

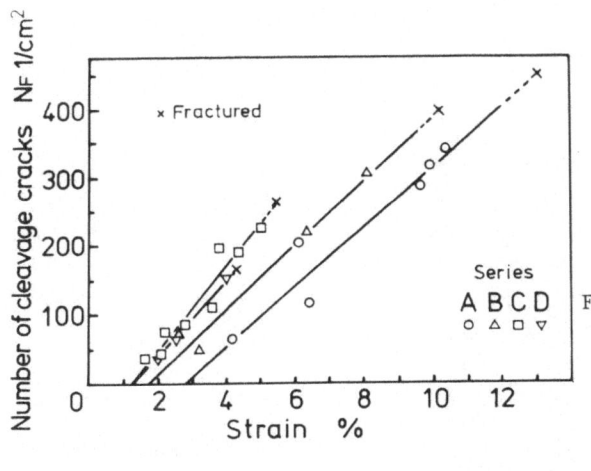

Fig. 8 Increasing rate of the
number of cleavage
cracks as a function
of applied strain

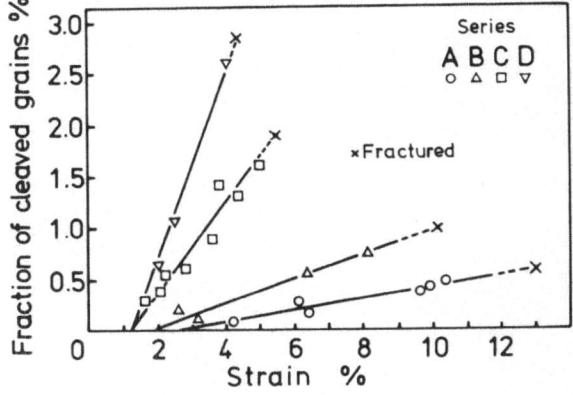

Fig. 9 Fraction of cleaved
ferrite grains as a
function of applied
strain

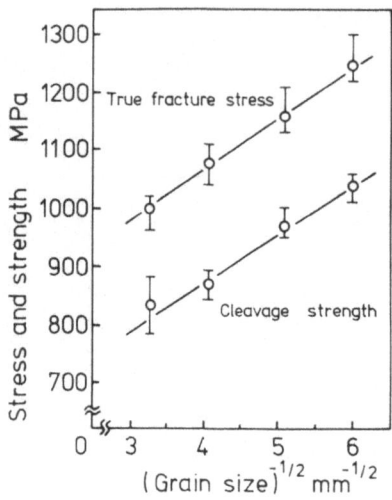

Fig. 10 Influence of ferrite grain size on the critical stress for cleavage cracking and the fracture stress

Tab. 2 Estimated number of cleaved ferrite grains calculated from Griffith equation

Series	Ferrite grain size d_f μm	Thickness of second phase t μm	Fracture stress σ_f MPa	L_c (=2c) μm	N
A	2.8	4.2	1250	270	8 (7.8)
B	3.8	5.7	1170	312	7 (6.6)
C	6.0	10.6	1080	364	5 (4.8)
D	9.3	16.0	990	432	4 (3.8)

L_c : Critical crack length (Griffith crack length 2c)
N : Number of cleaved ferrite grains
( ) : Calculated value of the number of cleaved ferrite grains from the equation (3)

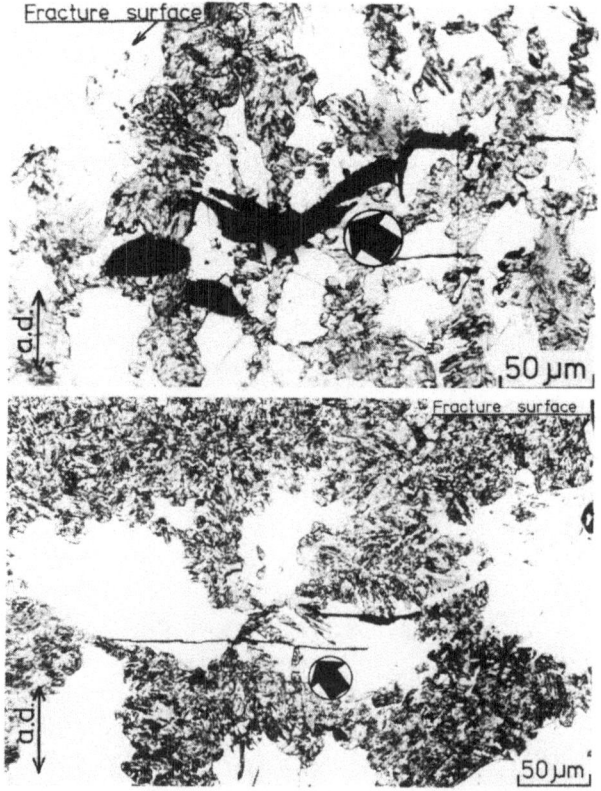

Fig. 11 Typical coalesced microcracks observed on the longitudinal sections near fracture surface of tensile specimens

(above: series A, below: series B)

J_{Ic} CALCULATIONS FOR PRESSURE VESSEL STEEL FROM MICROVOID KINETICS[*]

D. A. Shockey
Assistant Department Director

L. Seaman
Senior Research Engineer

R. L. Burback
Physicist

D. R. Curran
Department Director

Shock Physics and Geophysics, SRI International, Menlo Park, CA   94025 USA

ABSTRACT

The tensile fracture process in A533B steel at 366 K was studied at the microlevel and found to consist of the nucleation, growth, and coalescence of voids. A computational model of the process was constructed, and the void size distributions, measured in the necked regions of tensile bars, were used to quantify the model. The model was then incorporated into a finite-difference Lagrangian stress wave code to simulate the fracture toughness testing of a hypothetical center-cracked panel. Calculated values for J_{Ic}, the critical crack opening angle, and the critical crack opening displacement only slightly exceeded values typically measured, demonstrating that a route now exists for calculating fracture toughness from microvoid kinetics data. Potential applications of such models include monitoring radiation embrittlement in nuclear reactor components from tests on small surveillance specimens, predicting failure in structural elements that do not already contain a macrocrack, and explaining variations in toughness and strength caused by variations in inclusion size distribution.

INTRODUCTION

The toughness of the steel used in nuclear reactor pressure vessels can be degraded by continual neutron irradiation from the reactor core. To ensure

---

[*] A more detailed account of this work will be submitted to the Journal of Pressure Vessel Technology.

that the fracture toughness does not reach an unsafe level, small surveillance specimens are placed in the reactor at start-up and removed periodically for fracture toughness testing.

A shortcoming of this monitoring procedure is that the tests, usually Charpy impact tests, do not yield a number that can be used to calculate allowable loads or crack sizes. Thus the present research program was undertaken with the goal of establishing a method for obtaining more useful fracture toughness data from small reactor surveillance specimens.

Our approach was to establish the microphenomenology of the tensile rupture process in A533B steel at operating temperatures, to construct and quantify a computational model of the process based on observations and measurements of microfailure features, to calibrate the model and verify its reliability by using it in conjunction with stress codes to compare fracture behavior in tensile specimens of several geometries, and finally to demonstrate its ability to calculate fracture toughness by simulating a standard fracture toughness test.

TENSILE FAILURE PHENOMENOLOGY

The phenomenology of tensile failure in A533B steel at reactor operating temperature (561 K) was investigated [1] by examining fracture surfaces and polished sections of fractured and partially fractured Charpy and round bar tensile specimens tested at upper shelf temperatures (355 K).[*] Fracture surfaces showed ductile hemispherical dimples in a bimodal size distribution (Fig. 1) and polished cross sections showed spherical voids (Fig. 2). We concluded that tensile fracture occurred by nucleation, growth, and coalescence of voids.

Nucleation takes place predominantly at MnS and Al_2O_3 inclusions, which are typically 5 to 10 μm in diameter, and grow plastically and independently of one another as the tensile strain is increased, maintaining a roughly spherical shape. As adjacent growing voids approach one another, a population of much smaller voids nucleates at the 0.2-μm-diameter Fe_3C particles, which are homogeneously dispersed throughout the material along the interfaces of the bainite laths. The submicron voids form on rather well-defined surfaces connecting the larger voids and grow to radii of about a micron before coalescing with neighboring voids by plastic impingement. Thus the mechanism of coalescence of larger voids is the nucleation, growth, and coalescence of smaller voids.

---

[*] Extensive fracture testing performed elsewhere [2] demonstrated that no change in fracture mode or in energy absorbed occurs in A533B steel between 355 K and 561 K.

VOID KINETICS DATA

To obtain quantitative expressions describing the microfailure process, we measured the size distributions of inclusions, statistical void populations, and fracture surface dimples and correlated these measurements with the imposed strain field. The photograph in Fig. 3 shows the voids intersecting the plane of polish through a round bar tensile specimen. The voids were counted and measured. These surface data were converted via a statistical transformation to obtain void radius per unit volume for each zone indicated in the figure. The cumulative void size distributions regularly increase in size and number with proximity to the fracture plane from the inclusion size distribution to the dimple size distribution. No voids were observed in regions of plastic strain less than 11%, suggesting that 11%, the strain at the onset of necking, is also the threshold strain for nucleation. Void growth rates were deduced from these data by drawing (horizontal) lines of constant void number across the distribution lines and plotting against strain the points where the lines intersect the size distribution curves.

THE COMPUTATIONAL MODEL

The ductile fracture model consists of analytical expressions for void nucleation, growth, and coalescence and an algorithm for gradually reducing the specimen strength as the voids develop. The model was constructed to describe as closely as possible experimental observations and measurements.

Void nucleation was presumed to occur by decohesion of the matrix from inclusions. Thus, material is assumed to contain no voids until it experiences a strain of 11%, at which time it acquires a population of voids identical to the inclusion size distribution (Fig. 3) which was fitted to the exponential form:

$$N_g = N_o \exp(-R/R_n)$$

where N_g is the number of inclusions (voids) with radii greater than R, N_o is the total number per cm^3, and R_n is the characteristic size parameter of the distribution. The voids are presumed to be distributed uniformly throughout each computational cell, so that the failing material can be treated as homogeneous and isotropic. The experimental observations indicate nucleation is primarily a function of plastic shear strain $\bar{\varepsilon}^P$ and suggest the following bilinear nucleation model:

$$N_o = N_1(\bar{\varepsilon}^P - \varepsilon_1) \quad , \quad \varepsilon_1 < \bar{\varepsilon}^P < \varepsilon_2$$

$$N_o = N_1(\varepsilon_2 - \varepsilon_1) + N_2(\bar{\varepsilon}^P - \varepsilon_2) \quad , \quad \varepsilon_2 < \bar{\varepsilon}^P$$

where N_1 and N_2 are material specific void densities, and ε_1 and ε_2 are strain thresholds determined from the data. The results of the tensile tests suggest values of 5×10^6 cm^{-3} and 1×10^5 cm^{-3} for N_1 and N_2, respectively, and 0.11 and 0.13 for ε_1 and ε_2.

Once nucleated, voids grow gradually by plastic flow, elastic strain, and thermal expansion. The plastic growth law

$$V_v = V_o \exp \left[-T_1 \frac{P_s}{Y} (\bar{\varepsilon}^P - \bar{\varepsilon}_o^P) \right]$$

where P_s is the average stress in the solid material and T_1 is a dimensionless coefficient determined from a plot of void volume versus strain, was obtained from measurements on a series of notched and unnotched tensile bars [3]. The stress triaxiality (P_s/Y) at the specimen centers was calculated using Bridgman's [4] equation. The measured final void volume V_v, inclusion volume V_o, final and nucleation plastic strains $\bar{\varepsilon}^P$ and $\bar{\varepsilon}_o^P$, and the final yield strength were also obtained for the material at the center of the specimen. This equation is very similar to the theoretically derived Rice-Tracey [5] growth law for stress triaxialities less than one.

The elastic expansion of the void is given by

$$\Delta V_v = -V_v \Delta P_s (1-3K/4G)/K$$

which was derived from Love's [6] expression for radial motion of an infinite sphere under external load. This expansion is included to provide a physically accurate stiffness for porous material, although its influence is usually small. The thermal expansion of the void is simply the ratio of the temperature factors

$$1 + \Gamma \rho_{so} E/K$$

at the beginning and end of the strain increment, where Γ is the Grüneisen ratio, E is internal energy, and ρ_{so} is the initial solid density.

Combining the expressions for growth by plastic flow, elastic strain, and thermal expansion yields the following expression for the final void volume at the end of a strain increment:

$$V_{v3} = V_{v1} \left[\frac{1 + \Gamma \rho_{so1} E/K}{1 + \Gamma \rho_{so3} E/K} \right] \left[1 - \frac{\Delta P_s}{K} \left(1 + \frac{3K}{4G} \right) \right] \exp \left[-\frac{T_1 P_s}{Y} \Delta \bar{\varepsilon}^P \right]$$

As voids form and grow in the material, the load-bearing capacity of the specimen decreases. This reduction in strength is an important consequence of void development and must be accounted for in the failure model. This is accomplished by treating the material as consisting of a solid phase and a void phase. The stresses in the solid intact material are calculated from a Mie-Grüneisen equation of state and then reduced by the ratio of gross density to solid density to obtain the average stresses in the void-containing material.

A detailed model of the void coalescence was not attempted. Instead, we assumed that void growth continues until a relative void volume of about 0.01, indicated by the round-bar tension experiments, is reached, at which point coalescence and specimen failure occur.

CONFIRMATION OF THE MODEL

The model was written as a subroutine DFRACTS, incorporated into the Lagrangian, two-dimensional stress wave propagation computer code TROTT $\begin{bmatrix} 7 \end{bmatrix}$ and used to compute the failure behavior in a smooth and in a severely notched tensile bar. Comparisons of measured and computed force-displacement curves and relative void volume as a function of location in the specimens were made to test the reliability of the model at two levels of stress triaxiality. Pressure, deviator stress, and void volume were computed from the imposed strains at each cell of the subroutine DFRACTS. Because the relations between these variables are nonlinear, the solution is obtained by iteration of the following steps:

- Compute the deviator stresses in the solid from the deviator strains.
- Divide the imposed volume change into many increments, each small enough so that rapid convergence to a pressure can be obtained.
- For each volume change increment, compute the number of voids nucleated, estimate the solid pressure, and compute the plastic and elastic void growth. Reestimate the pressure and repeat the growth calculation until the solid and void volume change approximates the imposed volume change satisfactorily.
- Transform the computed solid stresses and pressure to gross stresses and pressure.

Iterative simulations were made first of a smooth tensile bar test, varying the stress-strain curve and the void growth coefficient until the computed load-deflection curve and void size distributions agreed with observation. Then a simulation of a circumferentially notched* round bar tensile test (which produces a significantly higher stress triaxiality and hence a different environment for void development) was performed, using the stress-strain relation and void growth coefficient indicated by the smooth bar iterative simulations. Good agreement was obtained between computed and measured load-deflection curves and rupture strains.

---

*The specimen had a 1.27-cm-dia. gross section and a 0.897-cm-dia net section. The notch had a 60° included angle with 0.025 cm root radius.

SIMULATION OF A FRACTURE TOUGHNESS TEST

To test the ability of the model to produce values for standard fracture toughness parameters, we performed a computational simulation of the center-cracked panel shown in Fig. 4(a). The panel dimensions and crack length were chosen to give plane strain conditions and J-controlled initiation and growth; that is, the half-crack length, ligament and panel thickness exceeded 200 J_{Ic}/σ_o [8,9]. For economy a complete simulation using the coarse computational grid of Fig. 4(b) was performed first to establish the boundary motions of a small butterfly-shaped region at the crack tip. This smaller region was then subdivided into much smaller cells, Fig. 4(c), and the failure behavior of this region was computed by applying the boundary motions indicated in the first simulation. These motions, governed by the equations for the conservation of momentum, lead to strain changes in each cell, from which stresses are computed with a stress-strain relation that includes standard elastic and plastic and work-hardening relations as well as the provisions for nucleation and growth of voids.

Table I presents the computed values for three fracture toughness parameters at the point where the void volume in one cell has reached 1% (the onset of macrocrack growth) and compares the values with values reported by Shih et al. [10,11].

Table I

COMPARISON OF COMPUTED AND MEASURED TOUGHNESS PARAMETERS

	Present Computation	Experimental Measurements [10,11]
Fracture toughness, J_{Ic} (MJ/m^2)	0.45	0.23 - 0.43
Crack opening, δ_c (cm)	0.084	0.035- 0.072
Crack opening half-angle (degrees)	54	*

The fracture resistance predicted by the microfracture model is slightly above the range of values measured experimentally, demonstrating that fracture toughness parameters can be computed from microvoid kinetics models. Such a model provides a route for obtaining toughness data from substandard size fracture specimens and in cases where no well-defined macrocrack exists. In addition, a microvoid model offers a route for understanding the influence of inclusion size distribution on toughness and thereby a way to link microstructure with continuum properties. Future work should aim at modeling void coalescence and calculating stable crack growth.

---

*Measured values of this quantity are quite high and not well defined during the initial stages of crack extension. A constant value of 11° to 17° is approached after the crack has advanced several mm. [11]

ACKNOWLEDGMENTS

Financial support for this work was provided by EPRI. Drs. T. U. Marston and R. L. Jones served as technical monitors. K. C. Dao performed the experimental work.

REFERENCES

1. D. A. Shockey, L. Seaman, K. C. Dao, and D. R. Curran, "Kinetics of Void Development in Fracturing A533B Tensile Bars," accepted for publication in Journal of Pressure Vessel Technology.

2. T. U. Marston et al., "Fracture Toughness of Ferritic Materials in Light Water Nuclear Reactor Vessels," MML-75-152, Combustion Engineering, Inc., Chattanooga, TN., October 1975.

3. D. A. Shockey, K. C. Dao, L. Seaman, R. L. Burback, and D. R. Curran, Computational Modeling of Microstructural Fracture Processes in A533B Pressure Vessel Steel, Final Technical Report to Electric Power Research Institute, Palo Alto, California, on Contract RP 1023-1 September 1979.

4. P. W. Bridgman, "The Stress Distribution at the Neck of a Tension Specimen," Transactions American Society Metals, 32 (1944), pp. 553-572.

5. J. R. Rice and D. M. Tracey, "On the Ductile Enlargement of Voids in Triaxial Stress Fields," Journal of the Mechanics and Physics of Solids, 17, (1969), pp. 201-217.

6. A. E. H. Love. A Treatise on the Mathematical Theory of Elasticity. London: Cambridge University Press, (1906), p. 142.

7. L. Seaman and D. R. Curran, TROTT Computer Program for Two-Dimensional Stress Wave Propagation. SRI International Final Report for U.S. Army Ballistic Research Laboratory, Aberdeen Proving Ground, Maryland, (August 1978).

8. C. F. Shih, W. R. Andrews, M. D. German, R. H. Van Stone, and J.P.D. Wilkinson, Methodology for Plastic Fracture, (July 13, 1978). Electric Power Research Institute, Report No. SRD-78-116.

9. R. M. Meeking and D. M. Parks, "On Criteria for J-Dominance of Crack Tip Fields in Large-Scale Yielding," Presented at ASTM symposium on Elastic-Plastic Fracture, Atlanta, Georgia, 1977.

10. C. F. Shih, "An Engineering Approach for Examining Crack Growth and Stability in Flawed Structures," presented at OECD-CSNI Specialist Meeting, Washington University, St. Louis, MO, Sept. 25-27, 1979.

11. C. F. Shih, et al. "Crack Initiation and Growth Under Fully Plastic Conditions: A Methodology for Plastic Fracture," EPRI Special Report NP-701-SR, T. U. Marston, editor, Electric Power Research Institute, Palo Alto, CA 94304 (February 1978).

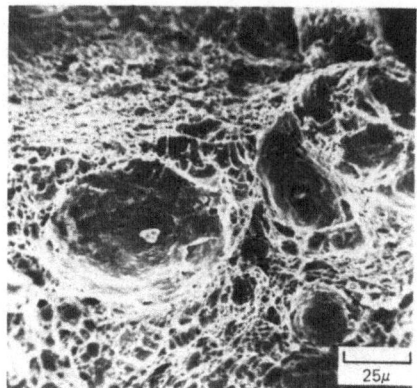

MP-6110-25A

Figure 1 · Scanning electron micrograph of the fracture surface
of a Charpy specimen of CDB A533B steel tested at
366 K showing ductile dimples in two size ranges.
Several of the larger dimples contain Al_2O_3 particles
that may have served as their nuclei.

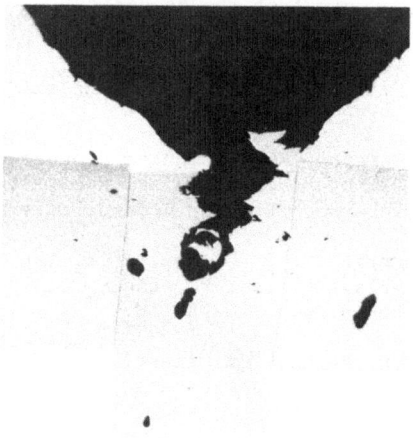

MP-6110-9

Figure 2   Polished cross section through notch base of
interrupted Charpy impact specimen showing
incipient ductile crack and isolated voids near
the crack tip.

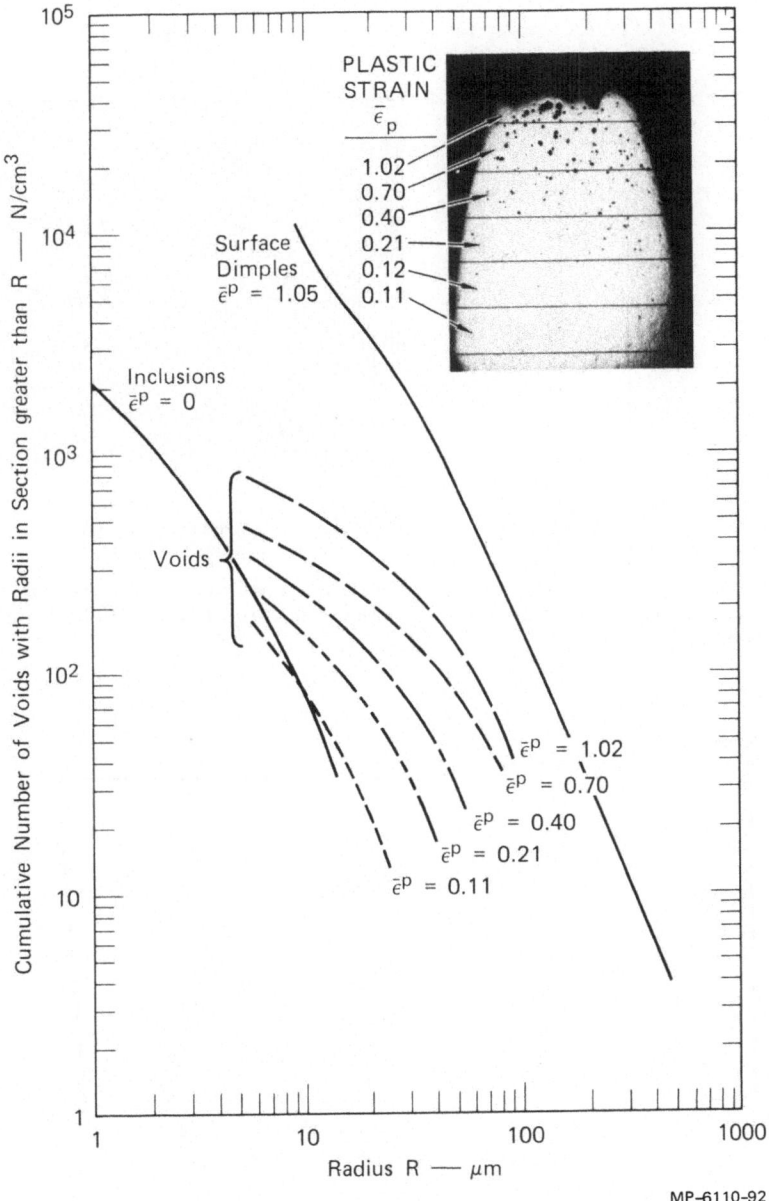

Figure 3   Size distributions of inclusions, voids, and surface dimples in a smooth tensile bar of A533B steeel, heat CDB, at various plastic strains, $\bar{\epsilon}^P$.

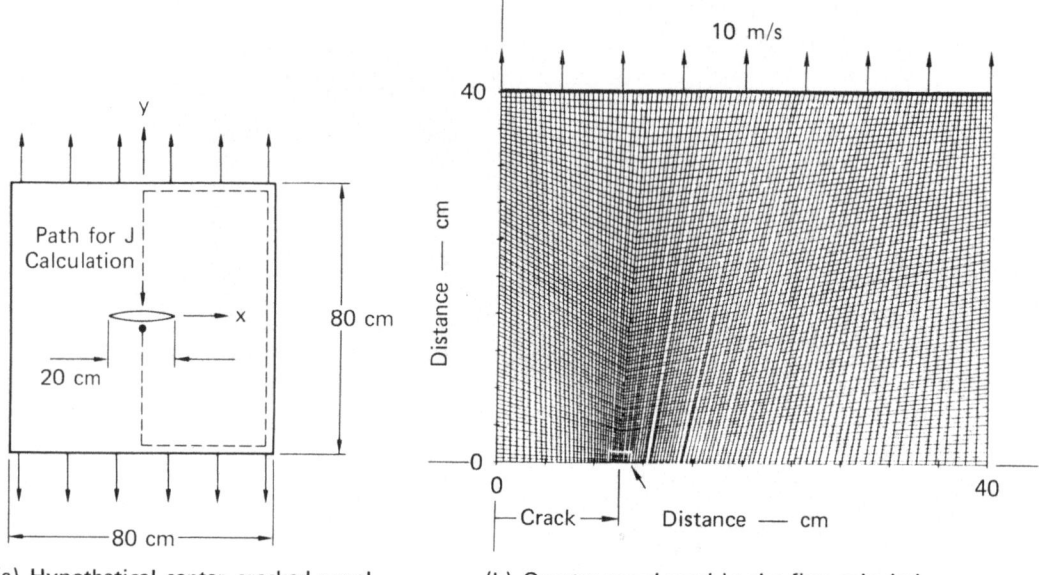

(a) Hypothetical center-cracked panel test simulated with the microfracture model to obtain J_{Ic}.

(b) Quarter panel used in the first calculation. Min. cell size = 0.16 cm, 100 cells by 82 cells.

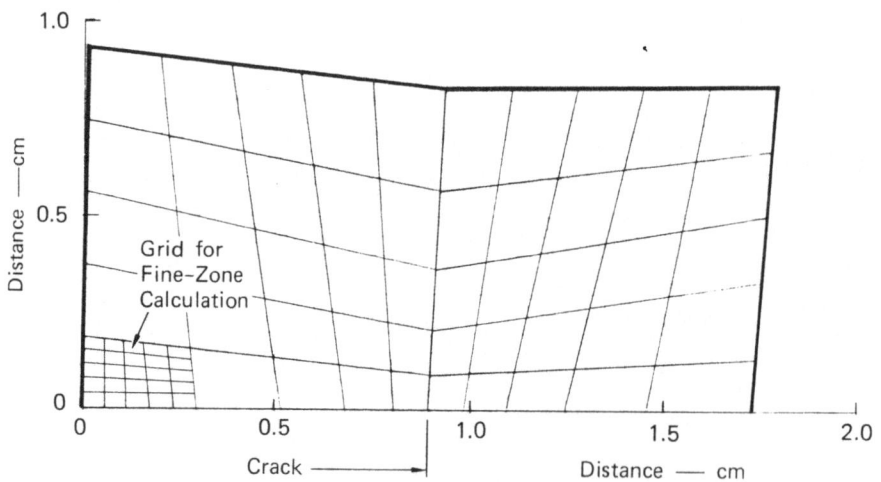

(c) Second calculation of the subregion around the crack tip. Min. cell size = 0.015 cm, 50 cells by 25 cells.

MA-6110-93

Figure 4   Specimen and cell geometries for the center-cracked panel simulation.

FRACTURE TOUGHNESS OF TEMPER EMBRITTLED
2 1/4Cr-1Mo AND 3.5 Ni-Cr-Mo-V STEELS

T. Iwadate, Dr. Eng
Engineer, Research Laboratory
The Japan Steel Works, Ltd., Chatsu-machi 4, Muroran, Japan

T. Karaushi
Engineer, Industrial Machine Design Section
The Japan Steel Works, Ltd., Chatsu-machi 4, Muroran, Japan

J. Watanabe, Dr. Eng
General Manager, Material Research Laboratory
The Japan Steel Works, Ltd., Chatsu-machi 4, Muroran, Japan

ABSTRACT

Fracture toughness K_{Ic} values of temper embrittled 2 1/4Cr-1Mo and 3.5 Ni-Cr-Mo-V steels were measured at the wide temperature range including the upper shelf, and they were compared with the one of un-embrittled steels. The effect of temper embrittlement on fracture toughness was investigated, and its behavior was analyzed. A method to predict fracture toughness of temper embrittled materials from the Charpy V-notch test results was also presented.

INTRODUCTION

2 1/4Cr-1Mo pressure vessel steel and 3.5 Ni-Cr-Mo-V steam turbine rotor steel have a propensity toward temper embrittlement [1], [2]. Therefore, these materials may be embrittled during heat treatment in manufacturing and/or during long term services, in which Charpy V-notch impact transition curve is shifted to higher temperature. Yet, many of these structural components are operated at temperatures in the upper shelf toughness range of materials because of recent improvement in material toughness. So, in order to analyze the safety of these structural components during the service, the fracture toughness behavior of the materials in the entire temperature range, especially the upper shelf, must be known in relation to the extent of temper embrittlement.

For this purpose, the effect of temper embrittlement on fracture toughness K_{Ic} of a 2 1/4Cr-1Mo pressure vessel steel and a 3.5 Ni-Cr-Mo-V rotor shaft steel was investigated, and their fracture toughness behavior was analyzed. Furthermore, a method for predicting the fracture toughness K_{Ic} of temper embrittled materials from their Charpy V-notch impact test results was developed successfully on the basis of the results of this test.

MATERIALS

The materials investigated are a 2 1/4Cr-1Mo pressure vessel steel and a 3.5 Ni-Cr-Mo-V steam turbine rotor steel, both having the potential for temper embrittlement. Chemical compositions of these materials are presented in Table 1. The 2 1/4Cr-1Mo steel was taken out of a pressure vessel used for approximately 7 years at temper embrittlement temperature range. Tests were carried out for two conditions: (1) as received (temper embrittled) and (2) de-embrittled by heating the steel to 595°C and water quenched. The 3.5 Ni-Cr-Mo-V steel sample was obtained through a forging and was tested in (1) as received and (2) temper embrittled conditions (by Gould's step cooling [3]). The mechanical properties of these materials are summarized in Table 2, and Charpy V-notch impact transition curves are shown in Fig. 1.

EXPERIMENTAL PROCEDURE

Fracture toughness K_{Ic} values were measured by K_{Ic} and J_{Ic} tests. The K_{Ic} tests were performed in accordance with ASTM Method E399-78 using 2T-CT specimens. In the higher toughness temperature range, J_{Ic} tests were conducted in accordance with the method recommended by the ASTM E24 Task Group [4], except for the use of an actual blunting line to determine J_{Ic}. The test specimens used were 1T-CT specimens, modified so displacements could be measured at the centerline of loading. The fatigue crack length to specimen width ratio, a/w, was about 0.6, when the final precracking stress intensity factor (Kmax) was kept less than 25 MPa-m$^{1/2}$.

Fracture toughness J was calculated using the equation corrected for the tension loading component [4].

$$J = \frac{1 + \alpha}{1 + \alpha^2} \frac{2A}{Bb} \tag{1}$$

where, $\alpha = 2[(a/b)^2 + (a/b) + 1/2]^{1/2} - 2(a/b + 1/2)$,

    A = area under the load-displacement curve taken at the displacement of interest,

    B = specimen thickness, and

    b = remaining uncracked ligament.

Corresponding K_{Ic} values of the steels were obtained by converting J_{Ic} to K_{Ic} using Equation (2).

$$K_{Ic} = \left(\frac{J_{Ic} E}{1 - \nu^2} \right)^{1/2} \tag{2}$$

where E is Young's modulus and ν is Poisson's ratio.

At the upper shelf temperatures where the stable crack growth was observed, the multiple specimen resistance curve test technique was used. At the
transition temperatures where the fast fracture occurred at maximum load by an
instability, more than three specimens were tested at the same temperature.
The lowest value among the specimens with no crack extension was defined as J_{Ic}
[5], [6].

The orientation of K_{Ic} and J_{Ic} test specimens were T-L for the 2 1/4Cr-1Mo
steel and C-R for the 3.5 Ni-Cr-Mo-V steel per ASTM Method E399-78.

RESULTS AND DISCUSSION

Fracture toughness behavior

Fracture toughness K_{Ic} and J_{Ic} tests were performed in the wide temperature range from -196°C to 427°C. Figs. 2 through 4 show J versus Δa R-curves
obtained for the 2 1/4Cr-1Mo and 3.5 Ni-Cr-Mo-V steels at the upper shelf temperature region. At the transition temperature range, no R-curve was developed
because no or small stable crack extension was observed. All J values needed
to develop the R-curves met the following size criterion [4].

$$a, B \geq 25 \ J/\sigma_Y \tag{3}$$

where σ_Y is the average of the yield and the ultimate tensile strength. These
figures indicate a significant decrease of fracture toughness J_{Ic} with increasing temperature. The same tendency was reported by Logsdon and Begley [7] for
other steels.

Fig. 5 shows the relationship between the slope of J resistance curves
(dJ/da) and test temperature, which serves as a measure of the resistance to
stable crack growth at various test temperatures. The slope of all materials
decreased with increasing temperature. The effect of temper embrittlement on
dJ/da was not observed.

The fracture toughness K_{Ic} versus temperature curves of the 2 1/4Cr-1Mo
pressure vessel steels and 3.5 Ni-Cr-Mo-V rotor steels are illustrated in Figs.
6 and 7, respectively. These indicate that K_{Ic} versus temperature behavior is
remarkably affected by temper embrittlement, similar to the Charpy V transition
curves (Fig. 1). A significant decrease of fracture toughness, especially in
the transition range, is demonstrated.

In the upper shelf region, K_{Ic} values converted from J_{Ic} decrease with
increasing temperature and this is more remarkable for the 2 1/4Cr-1Mo pressure
vessel steel. Fracture toughness at the temperature where upper shelf fracture
behavior is first experienced is lower for the temper embrittled steels than
for the unembrittled steels. However, beyond the temperature where the upper

shelf of the embrittled steel is first observed, K_{Ic} values coincide with both the temper embrittled and unembrittled materials. This fact has a practical importance: temper embrittlement of a material will not damage the safety of a component, if it is loaded at temperatures higher than the upper shelf temperature of the embrittled material.

In order to explain this fracture toughness behavior, the stretched zone width created at the tip of the crack of the specimens was measured using a scanning electron microscope. The specimen was inclined at an angle of 30° to the fatigue crack surface to obtain the accurate stretched zone width [8].

Fig. 8 shows an example of the stretched zone, and its widths at 165°C and 300°C show only a minor difference. The critical stretched zone width SZWc, beyond which a stable crack initiates, was measured on the specimens tested at the upper shelf region, and its result is shown in Fig. 9. The relationship between the stretched zone width and its corresponding J of the 3.5 Ni-Cr-Mo-V steels is presented in Fig. 10.

These figures indicate that the decrease in fracture toughness with increasing temperature in the upper shelf may be caused by the decrease of the flow stress when the temperature is increased. This is supported by the following two reasons: first, the critical stretched zone width SZWc decreases with increasing temperature (Fig. 9); second, the material's resistance to stretching also decreases with increasing temperature as illustrated by change of slope in Fig. 10.

Fig. 10 also shows that SZWc is closely related to fracture toughness at the transition temperature range [8]. Unembrittled 3.5 Ni-Cr-Mo-V steel B tested at -40°C showed a very narrow SZWc.

Correlation between K_{Ic} and CVN

The correlation between the upper shelf fracture toughness K_{Ic} and the Charpy V-notch impact energy at the corresponding temperature was proposed empirically by Rolfe and Novak [9].

$$(\frac{K_{Ic}-us}{\sigma_{0.2}})^2 = 0.6478(\frac{CVN-us}{\sigma_{0.2}} - 0.0098) \tag{4}$$

where $K_{Ic}-us$ is the upper shelf fracture toughness, in MPa-m$^{1/2}$, CVN-us is in Joule and $\sigma_{0.2}$ is in MPa.

When the upper shelf temperature was defined as the temperature where CVN specimens first experience zero-percent brittle fracture, a strong correlation between K_{Ic} and CVN energy was found, shown in Fig. 11 [6], [7], [10] - [13]. Fig. 12 shows the relationship between the fracture toughness and CVN energy in the whole upper shelf region. This figure indicates that no general correlation was observed between $K_{Ic}-us$ and CVN-us, because at the temperature range investigated, CVN result is still in transition range [7].

Fig. 13 and Fig. 14 show the fracture toughness of embrittled and non-embrittled 2 1/4Cr-1Mo and 3.5 Ni-Cr-Mo-V steels [7], [10], [14]. Data points are normalized using K_{Ic}/K_{Ic-us} and the excess temperature, where K_{Ic-us} is K_{Ic} at the initial temperature of the Charpy upper shelf fracture behavior, and the excess temperature is the test temperature minus FATT. A master curve was obtained for 2 1/4Cr-1Mo and 3.5 Ni-Cr-Mo-V steels, shown in Fig. 13 and Fig. 14, respectively.

The fracture toughness K_{Ic} versus temperature curves of temper embrittled 2 1/4Cr-1Mo pressure vessel and 3.5 Ni-Cr-Mo-V rotor steels, including the upper shelf region, can be predicted using a method mentioned below [10].

(1) Perform the CVN test and obtain temperature versus energy and fracture appearance curves.

(2) Determine the 50 percent FATT and 100 percent ductile fracture appearance temperature. The latter temperature is called the upper shelf temperature in this paper.

(3) Perform the tension test at the upper shelf temperature and obtain the yield strength.

(4) Obtain the K_{Ic-us} from Fig. 11 or Eq. (4).

(5) Draw predicted K_{Ic} versus temperature curves using the master curves (Fig. 13 and Fig. 14), K_{Ic-us} and FATT.

This prediction method could be applied with successful results, if a specific grade of steel were selected, because the effect of the strain rate on crack toughness depends upon the chemistry and strength of the steel.

CONCLUSIONS

In order to analyze the safety of 2 1/4Cr-1Mo pressure vessels and 3.5 Ni-Cr-Mo-V steam turbine rotor shafts, fracture toughness K_{Ic} of temper embrittled steels was measured in a wide range of temperatures including the upper shelf.

1. K_{Ic} versus temperature behavior is remarkably affected by temper embrittlement, similar to Charpy V transition curves. A significant decrease in fracture toughness, especially in the transition range, is demonstrated.

2. Fracture toughness of the embrittled materials is lower than toughness of the unembrittled materials up to the temperature where the upper shelf fracture behavior is first experienced in the embrittled materials. However, beyond this temperature, no noticeable difference is found in the toughness of both materials. Consequently, temper embrittlement of a material will not damage the safety of a component, if it is loaded at

temperatures higher than the upper shelf temperature of the embrittled material.

3. In the wide range of temperatures, stretched zone formation preceding the stable crack initiation plays a dominant role on fracture toughness behavior.

4. Based on the fracture toughness data obtained, a prediction method of fracture toughness K_{Ic} from Charpy V-notch test results was developed.

REFERENCES

[1] L. G. Emmer, C. D. Clauser, and J. R. Low, Jr., "Critical Literature Review of Embrittlement in 2 1/4Cr-1Mo Steel", WRC Bulletin, No. 183, Welding Research Council, (May, 1973).

[2] R. Narayan and M. C. Murphy, "A Review of Temper Embrittlement as It Affects Major Steam Turbine and Generator Rotor Forgings", Journal of the Iron and Steel Institute, Vol. 211, (1973), pp. 493-501.

[3] G. C. Gould, "Long Time Isothermal Embrittlement in 3.5 Ni, 1.75 Cr, 0.50 Mo, 0.20 C Steel", ASTM STP 407, (1968), pp. 90-105.

[4] G. A. Clarke, W. R. Andrews, J. A. Begley, J. K. Donald, G. T. Embley, J. D. Landes, D. E. McCabe, and J. H. Underwood, "A Procedure for the Determination of Ductile Fracture Toughness Values Using J Integral Techniques", Journal of Testing and Evaluation, JTEVA, Vol. 7, No. 1, (Jan., 1979), pp. 49-56.

[5] T. Iwadate, J. Watanabe, K. Sakabe, T. Nakata, and T. Karaushi, " Comparison of the Measured K_{Ic} with K_{Ic} Converted from J_{Ic}", Proceedings of the JSME Autumn Meeting, No. 760-13, (1976), pp. 55-60, (in Japanese).

[6] H. Tsukada, T. Iwadate, Y. Tanaka, and S. Ono, "Static and Dynamic Fracture Toughness Behavior of Heavy Section Steels for Nuclear Pressure Vessels", To be presented at the Fourth International Conference on Pressure Vessel Technology, London, (May, 1980).

[7] W. A. Logsdon and J. A. Begley, "Upper Shelf Temperature Dependence of Fracture Toughness for Four Low to Intermediate Strength Ferritic Steels", Engineering Fracture Mechanics, Vol.9, (1977), pp. 461-470.

[8] H. Kobayashi, K. Hirano, N. Nakamura, and H. Nakazawa, "A Fractographic Study on Evaluation of Fracture Toughness", Proceedings of the Fourth International Conference on Fracture, Vol. 3, (1977), pp. 583-592.

[9] J. M. Barsom and S. T. Rolfe, "Correlation between K_{Ic} and Charpy V-Notch Test Results in the Transition Temperature Range", ASTM STP 466, (1970), pp. 281-302.

[10] T. Iwadate, T. Karaushi, and J. Watanabe, "Prediction of Fracture Toughness K_{Ic} of 2 1/4Cr-1Mo Pressure Vessel Steels from Charpy V-Notch Test Results", ASTM STP 631, (1977), pp. 493-506.

[11] J. A. Begley and P. R. Toolin, "Fracture Toughness and Fatigue Crack Growth Rate Properties of a Ni-Cr-Mo-V Steel Sensitive to Temper Embrittlement", International Journal of Fracture, Vol. 9, No. 3, (1973), pp. 243-253.

[12] M. Kashihara, S. Kawaguchi, R. Yanagimoto, S. Sawada, and K. Tsuchiya, "Development of Gigantic Forgings with Ultra-Heavy Section Exceeding 2,500 mm Diameter from Ingots Weighing over 400 Tons", Proceedings of 8th

International Forgemasters Meeting, Kyoto, (1977).

[13] J. Watanabe, T. Iwadate, and T. Karaushi, "Recent Development of the Strength Evaluation of Large Forgings", Proceedings of 8th International Forgemasters Meeting, Kyoto, (1977).

[14] T. Iwadate, J. Watanabe, K. Ohnishi, S. Saikudo, Y. Ohshio, T. Tsukikawa, S. Yamamoto, M. Nakano, and H. Ueyama, "Fracture Analysis of 2 1/4Cr-1Mo Steel Test Vessel and Its Correlation to Acoustic Emission", To be presented at the Fourth International Conference on Pressure Vessel Technology, London, (May, 1980).

Table 1   Chemical composition of materials investigated (wt.%).

Material	C	Si	Mn	P	S	Ni	Cr	Mo	V	As	Sn	Sb
2$\frac{1}{4}$Cr - 1Mo[a]	0.20	0.21	0.70	0.022	0.013	0.30	2.58	1.01	0.01	0.028	0.022	0.0045
3.5Ni-Cr-Mo-V	0.29	0.32	0.36	0.010	0.008	3.70	1.80	0.48	0.14	0.008	0.013	0.0031

a Taken out of a pressure vessel after 7 years service.

Table 2   Mechanical properties of materials investigated at 25°C.

Material		0.2% Yield strength MPa	Tensile strength MPa	Elongation in 50 mm %	Reduction of area %	Charpy V-notch Energy absorption J	FATT[c] °C
2$\frac{1}{4}$ Cr - 1Mo	A	427	605	31.1	67.5	27	95
	B[a]	412	600	32.0	69.9	127	-40
3.5Ni-Cr-Mo-V	A[b]	735	857	23.3	58.9	8	174
	B	735	874	22.0	62.1	64	25

a   Water quenched from 595°C.
b   Gould's step-cooling heat treatment.
c   50% shear fracture appearance transition temperature.

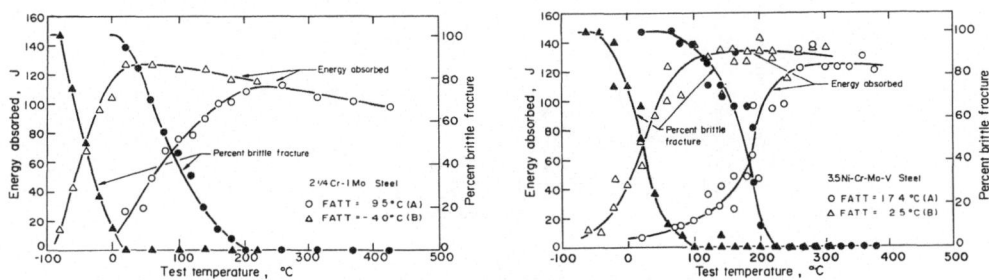

Fig. 1   Charpy V-notch impact transition curves of 2 1/4Cr-1Mo pressure vessel steels and 3.5 Ni-Cr-Mo-V rotor steels.

248

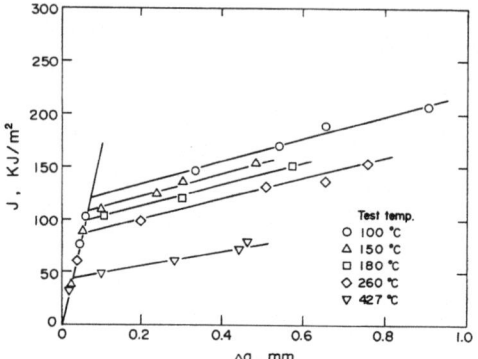

Fig. 2   J resistance curves for
2 1/4Cr-1Mo steel A (FATT=95°C)

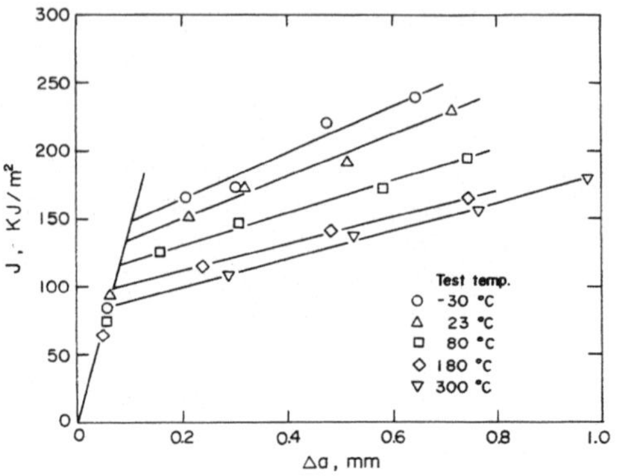

Fig. 3   J resistance curves for
2 1/4Cr-1Mo steel B (FATT=-40°C)

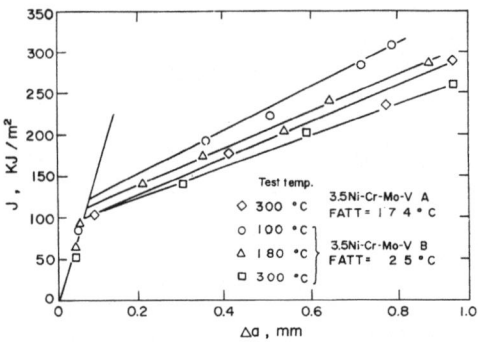

Fig. 4   J resistance curves for
3.5 Ni-Cr-Mo-V steels.

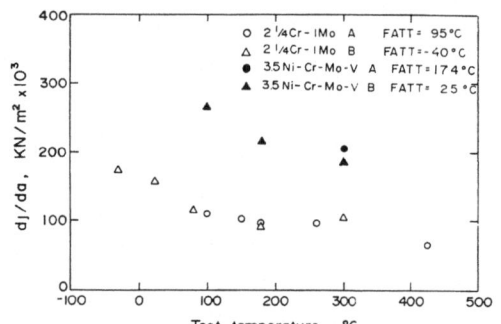

Fig. 5   dJ/da versus temperature
behavior.

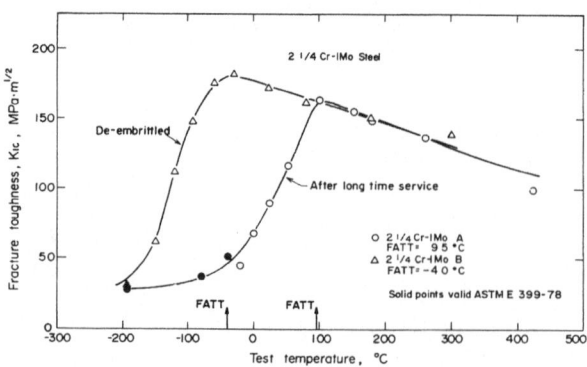

Fig. 6   Fracture toughness versus temperature behavior
of 2 1/4Cr-1Mo pressure vessel steels.

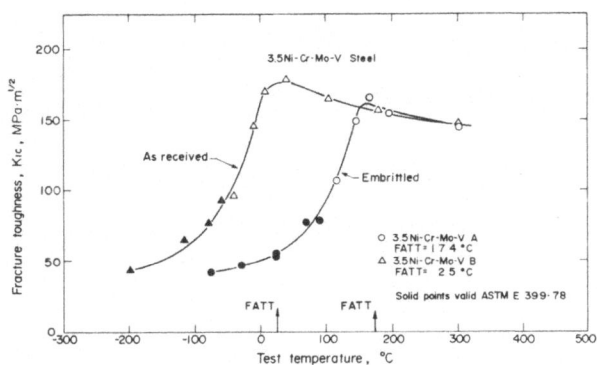

Fig. 7 Fracture toughness versus temperature behavior of
3.5 Ni-Cr-Mo-V rotor steels.

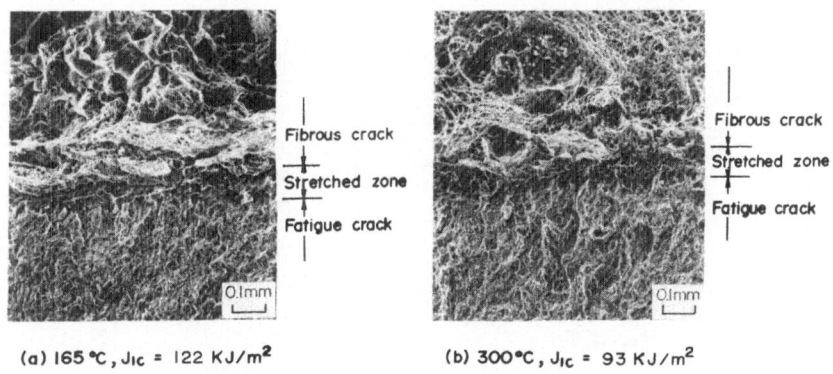

(a) 165 °C, J_{IC} = 122 KJ/m^2 (b) 300°C, J_{IC} = 93 KJ/m^2

Fig. 8 Stretched zone at upper shelf temperatures of
3.5 Ni-Cr-Mo-V rotor steel A (FATT = 174°C).

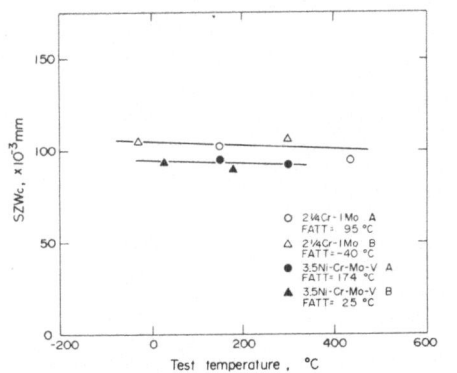

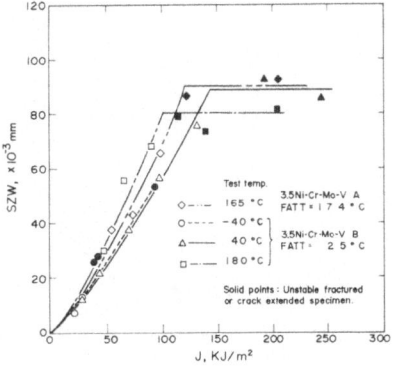

Fig. 9 Stretched zone width (SZWc)
versus temperature behavior.

Fig. 10 Relationship between stretched
zone width (SZW) and J of
3.5 Ni-Cr-Mo-V rotor steels.

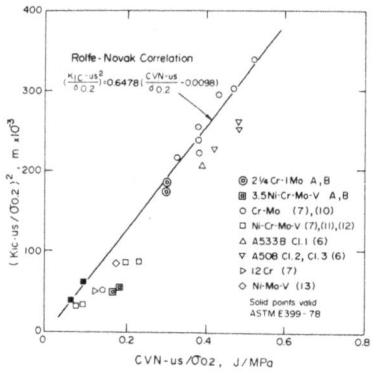

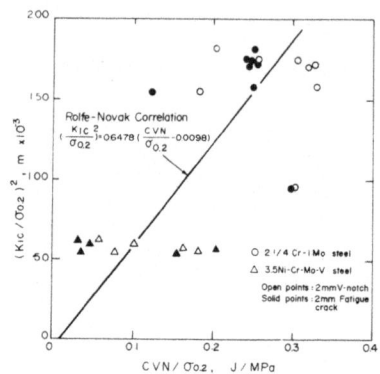

Fig. 11   Relationship between K_{Ic} and CVN energy at the upper shelf temperature where CVN specimens first experience 100 percent ductile fracture.

Fig. 12   Relationship between K_{Ic} and CVN energy at the upper shelf temperatures.

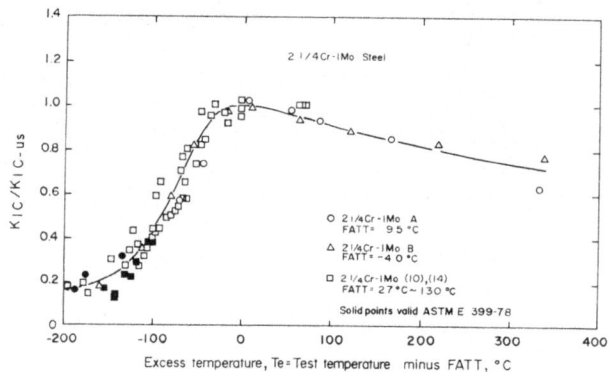

Fig. 13   Relationship between excess temperature and K_{Ic}/K_{Ic-us} of 2 1/4Cr-1Mo pressure vessel steels.

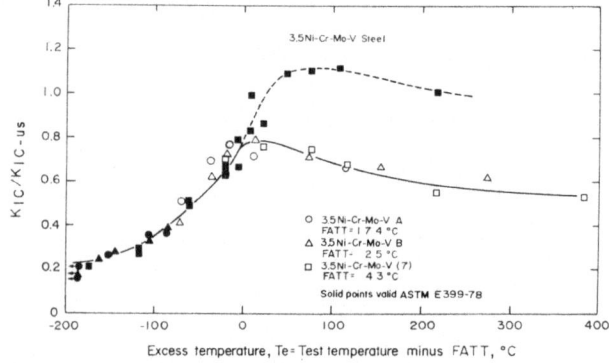

Fig. 14   Relationship between excess temperature and K_{Ic}/K_{Ic-us} of 3.5 Ni-Cr-Mo-V rotor steels.

VI. CREEP AND FATIQUE

FRACTURE MECHANICS OF CREEP CRACK GROWTH
IN METALLIC MATERIALS

Kiyotsugu Ohji*, Keiji Ogura**, Shiro Kubo*** and Yasuyuki Katada***
*Professor, **Associate Professor and ***Research Associate
Department of Mechanical Engineering, Osaka University
Yamada-kami, Suita, Osaka 565 Japan

ABSTRACT

Analysis of creep crack growth was made, based on the nonlinear singular stress field near a creep crack tip. The creep crack growth rate of metallic materials was expressed as a function of both a modified J-integral and the amount of crack growth. Experimental evidences were given. The invalidity of the linear elastic stress intensity factor and the net section stress as a correlation parameter of creep crack growth rate was substantiated. The size dependence of the creep crack growth rate and the concept of "confined creep deformation" were discussed.

INTRODUCTION

Creep crack growth is one of the most important and interesting problems which remain undeveloped in the field of applied fracture mechanics. The undeveloped state of this problem is mainly due to the highly nonlinear nature of metallic creep. Some papers [1-3] claimed apparent success in correlating the creep crack growth rate with the linear elastic stress intensity factor, while others [4-6] proposed s strong correlation between net section stress and creep crack growth rate. However, neither of these two parameters theoretically represents the real mechanical state in the vicinity of a creep crack tip, and accordingly, they cannot claim general validity in the correlation of creep crack growth rate.

In this paper, the authors present a new, nonlinear, fracture mechanics approach to the creep crack growth problem, which is based on the nonlinear, singular stress field predominant in the vicinity of a creep crack tip. Experimental evidences are given to support this approach. The size dependence of creep crack growth rate and a state of "confined creep deformation" are discussed in terms of the approach's structural applications.

CREEP CRACK GROWTH ANALYSIS BASED ON THE PLASTIC SINGULAR STRESS FIELD [7]

For simplicity, only a Mode III creep crack growth analysis in the steady-state creep condition is described in this paper. This analysis can easily be extended to the Mode I creep crack growth problem [7].

The Norton-type, steady-state creep law is assumed in this analysis.

$$\dot{\gamma} = \dot{\gamma}_c (\tau/\tau_c)^n \tag{1}$$

where $\dot{\gamma}$ is the creep rate of shear srain, τ is the applied shear stress, and $\dot{\gamma}_c$, τ_c, and n are material constants. The singular stress field in the vicinity of a crack tip is obtained by applying the principle of "the elastic analogue in the creep stress analysis" to the plastic singular stress field solution given by Hutchinson [8].

$$\left.
\begin{array}{c}
\tau = \left\{ \dfrac{2n}{\pi(n+1)} \dfrac{\tau_c^{\,n}}{\dot{\gamma}_c} J' \right\}^{1/(n+1)} r^{-1/(n+1)} \tilde{\sigma}_e(\theta) \\[2em]
\displaystyle \max_{0 \le \theta \le \pi} \tilde{\sigma}_e(\theta) = \tilde{\sigma}_e(0) = 1
\end{array}
\right\} \tag{2}$$

Here (r, θ) are the two-dimensional polar coordinates with the origin at the crack tip, as shown in Fig.1, and $\tilde{\sigma}_e(\theta)$ is a characteristic function which determines the distribution of τ in the vicinity of the crack tip with respect to θ. J' denotes the modified J-integral, which is defined as

$$J' = \int_\Gamma \{ W' dy - T_i \frac{\partial \dot{u}_i}{\partial x} dc \} \tag{3}$$

where

$$W' = \int_0^{\dot{\varepsilon}_{ij}} \sigma_{ij} d\dot{\varepsilon}_{ij} \tag{4}$$

Γ is a line contour surrounding the crack tip; T_i is the traction vector; $\dot{u}_i$ is the displacement rate vector, and dc is an element of arc length along Γ, as indicated in Fig.1. σ_{ij} and $\dot{\varepsilon}_{ij}$ are the stress and the strain rate tensors, respectively. Eq.(2) indicates that J' uniquely characterizes the singular stress or strain rate field predominant in the vicinity of the crack tip under the steady-state creep condition.

Fracture criterion can be formally expressed in terms of creep damage η, which is defined as zero in the initial state and as unity at fracture. The cumulative creep damage rate $\dot{\eta}$ may be given in a generalized form [9] as

$$\dot{\eta} \doteq A \tau^b \tag{5}$$

where A and b are material constants. From Eqs.(2) and (5), the cumulative damage rate at a distance of r on the crack line is given as

$$\dot{\eta} = A \left\{ \frac{2n}{\pi(n+1)} \frac{\tau_c^n}{\dot{\gamma}_c} \left(\frac{J'}{r} \right) \right\}^{\alpha} , \qquad \alpha = b/(n+1) \tag{6}$$

In order to take the effect of the stress gradient into account when calculating the creep crack growth rate by using Eq.(6), a characteristic length ρ_s is introduced. This quantity may be taken as a characteristic constant which represents the structural properties of the material under investigation. The crack is assumed to extend stepwise by this characteristic length ρ_s, when the value of creep damage at the distance of ρ_s ahead of the current crack tip is accumulated to a critical value, i.e., unity.

Referring to Fig.2, the point at a distance of $k\rho_s$ (k: a positive integer) ahead of the initial crack tip is designated as the k-point. η_{kj} and $\dot{\eta}_{kj}$ are defined as an accumulated value of creep damage and its cumulative rate, respectively, at the k-point when the current crack tip is located at the $(j-1)$-point.

The time interval necessary for the crack to grow from the $(j-1)$-point to the j-point is denoted by t_{pj}. Crack growth at the k-point occurs when the accumulated damage value of the k-point results in unity. Then, the following two equations are obtained.

$$\sum_{j=1}^{k} \dot{\eta}_{kj} t_{pj} = 1 \tag{7}$$

$$t_{pj} = (1-\eta_{jj})/\dot{\eta}_{jj} \tag{8}$$

The crack growth rate dl/dt may be defined as

$$dl/dt = \rho_s/t_{pj} \tag{9}$$

From Eqs.(6) through (9), the following equations are obtained.

$$\sum_{j=1}^{k} (1-\eta_{jj})/(k-j+1)^{\alpha} = 1 \tag{10}$$

$$\frac{dl}{dt} = A \left\{ \frac{2n}{\pi(n+1)} \frac{\tau_c^n}{\dot{\gamma}_c} \right\}^{\alpha} \frac{\rho_s^{1-\alpha} (J')^{\alpha}}{1-\eta_{jj}} \tag{11}$$

By solving Eq.(10) for $(1-\eta_{jj})$ in a successive manner and substituting it into Eq.(11), we can obtain an explicit expression of dl/dt. However, since this reduction of dl/dt for the explicit expression is difficult to do in a rigorous manner, we may use an approximate expression for the summation involved in Eq.(10). According to a literature survey on the values of n and b, $\alpha = b/(n+1) < 1$ might be the general case which prevails in metallic creep fracture. In this case, assuming that $(1-\eta_{jj})$ changes only slightly when j is much greater than unity, we may have,

$$1-\eta_{jj} \fallingdotseq 1/\sum_{i=1}^{j} i^{-\alpha} \fallingdotseq 1/\int_{1}^{j} x^{-\alpha}\,dx \fallingdotseq (1-\alpha)/j^{1-\alpha}, \qquad (\ j \gg 1\ ) \qquad (12)$$

From this equation, we may write,

$$1-\eta_{ii} = (1-\eta_{jj})\,(j/i)^{1-\alpha}, \qquad (\ j \gg 1\ ) \qquad (13)$$

Substituting Eq.(13) into Eq.(10),

$$(1-\eta_{jj})j^{1-\alpha}\sum_{j=1}^{j} i^{\alpha-1}\,(j-i+1)^{-\alpha} = 1 \qquad (14)$$

The summation involved in Eq.(14) is approximated by the gamma function as

$$\sum_{i=1}^{j} i^{\alpha-1}\,(j-i+1)^{-\alpha} \fallingdotseq \int_{0}^{1} x^{\alpha-1}\,(1-x)^{-\alpha}\,dx = \Gamma(\alpha)\,\Gamma(1-\alpha), \qquad (\ j \gg 1\ )$$

$$(15)$$

By using Eqs.(11), (14), (15) and

$$j \fallingdotseq (l-l_0)/\rho_s,$$

the crack growth rate is given as

$$\frac{dl}{dt} = A\,\Gamma(\alpha)\,\Gamma(1-\alpha)\left\{\frac{2n}{\pi(n+1)}\frac{\tau_c^{\,n}}{\dot{\gamma}_c}\right\}^{\alpha}(l-l_0)^{1-\alpha}(J')^{\alpha}, \qquad (\ l-l_0 \gg \rho_s\ ) \qquad (16)$$

Note that when $\alpha < 1$, the crack growth rate is affected by $(l-l_0)$ as well as by J'. According to Eq.(16), the effect of $(l-l_0)$ is estimated as $(l-l_0)^{1-\alpha}$, and this may not have a serious effect on dl/dt, since the value of $(1-\alpha)$ may be much smaller than unity.

DEPENDENCE OF CREEP CRACK GROWTH RATE ON SPECIMEN SIZE

In this section, the dependence of dl/dt on specimen size is examined on the basis of the analytical results derived in the preceding section.

A series of proportional specimens with different sizes are considered. A representative length of the specimen, e.g., a width of the specimen, is denoted by a^*. The modified J-integral J' and the stress intensity factor in linear fracture mechanics K_{el} of the specimen may be written in terms of a^* as

$$J' = (\dot{\gamma}_c/\tau_c^{\,n})\,\tau_n^{\,n+1}\,a^*\,f_1(l/a^*) \qquad (17)$$

$$K_{el} = \tau_n\sqrt{a^*}\,g_1(l/a^*) \qquad (18)$$

where τ_n is the net section stress, and f_1 and g_1 are functions of $l/a*$ only.

For a series of proportional specimens, τ_n/τ_{n0} is a function of $l/a*$, where the suffix "0" indicates the initial value of the respective quantity. Then, Eqs.(17) and (18) can be rewritten in terms of τ_{n0} as

$$J' = (\dot{\gamma}_c/\tau_c^{\,n})\,\tau_{n0}^{\,n+1}\; a* f_2(l/a*) \tag{19}$$

$$K_{el} = \tau_{n0}\sqrt{a*}\; g_2(l/a*) \tag{20}$$

where f_2 and g_2 are functions of $l/a*$ only. From Eqs.(19) and (20), the following relation is also obtained.

$$J' = (\dot{\gamma}_c/\tau_c^{\,n})K_{el0}^{\,n+1}\,(a*)^{1-(n+1)/2}f_2(l/a*)/\{g_2(l_0/a*)\}^{n+1} \tag{21}$$

Using these relations, Eq.(16) can be written as

$$\frac{dl}{dt} \propto \left[J_0' \frac{f_2(l/a*)}{f_2(l_0/a*)} \right]^{\alpha} (a*)^{1-\alpha} (l/a*-l_0/a*)^{1-\alpha} \tag{22}$$

$$\frac{dl}{dt} \propto (\tau_{n0})^b\; a* \left(\frac{l}{a*} - \frac{l_0}{a*} \right)^{1-\alpha} \{ f_2(l/a*) \}^{\alpha} \tag{23}$$

$$\frac{dl}{dt} \propto (K_{el0})^b(a*)^{1-b/2} \left[\frac{l}{a*} - \frac{l_0}{a*} \right]^{1-\alpha} \{f_2(l/a*)\}^{\alpha}/\{g_2(l_0/a*)\}^b \tag{24}$$

Then the size dependence of the proportional specimens can be obtained as follows.

For constant J_0' tests,

$$dl/dt \propto (a*)^{1-\alpha} \qquad\qquad \text{for a fixed } J'$$

For constant τ_{n0} tests,

$$dl/dt \propto a* \qquad\qquad \text{for a fixed } \tau_n \tag{25}$$

For constant K_{el0} tests,

$$dl/dt \propto (a*)^{1-b/2} \qquad\qquad \text{for a fixed } K_{el}$$

Note that, according to this analysis, a size dependence may theoretically exist in the $(dl/dt)-J'$ correlation when $\alpha = b/(n+1) < 1$. This size dependence may be estimated from Eq.(25) as $(a*)^{1-\alpha}$.

In usual metallic materials, however, the exponent $(1-\alpha)$ is estimated as very small in comparison with unity. This will be shown later in this paper and means that the size dependence in the $(dl/dt)-J'$ correlation may not be serious, as long as creep specimens of the usual size range are used. However, when the laboratory test results are used for the design of large-scale structures, this factor should be taken into account.

EXPERIMENTAL EVIDENCES AND DISCUSSION

In conjunction with the theoretical efforts described in the first part of this paper, several types of creep crack growth experiments were conducted by the authors' research group. The results of these experiments consistently indicated the validity of the foregoing analysis. In this section, some of these experimental evidences are presented.

Using the size dependence of the creep crack growth rate derived from the preceding analysis in a series of proportional specimens may be helpful for checking the validity of the analysis. "Deep"(D) and "Shallow"(S) center-cracked(CN) plate specimens were used in this study [10-12]. The "deep" and "shallow" cracks were selected because the crack growth behavior of the "deep" crack [9] is considerably different from that of the "shallow" crack [13]. Proportional (two-dimensional, thickness: 1.5mm) specimens of three different sizes (L, M, and S) were prepared; the proportional factor between any two consecutively sized specimens was 2. Fig.3 shows the geometries of the speci-mens. The material used was the SUS 304 stainless steel (equivalent to AISI 304 stainless steel). The experiments were conducted under a condition of constant initial net section stresses, σ_{n0} = const., at a temperature of 923K (650°C). The values of J' were experimentally determined by using the formu-lae developed by the authors' group [14].

Figs. 4 through 6 show the correlations of creep crack growth rate with the three mechanical parameters, J', σ_n, and K_I, selected in this study. A comparison of these three figures indicates that J' is the best correlation parameter among the three selected, which was expected from the analysis. The K_I-correlation curves showed a remarkable size dependency, resulting in the widest spread of experimental data. The σ_n-correlation seems much better than the K_I-correlation. However, a definite trend toward size dependece and also a difference of slopes in the correlation curves between the deep and the shal-low cracks are observed in Fig.5. In the case of the J'-correlation curves, practically no size dependence was observed and the spread of the experimental data was narrowest, particularly when the initial transient region in each cor-relation curve was neglected.

The values of n and b for this material were 7.4 and 7.1, respectively. Therefore, $\alpha = b/(n+1) = 0.85 < 1$. According to Eq.(25), a size dependence of proportional specimens in the (dl/dt)-J' correlation is estimated approximately as $(a*)^{1-\alpha} = (a*)^{0.15}$. This results in an 1.23 factor for the maximum range of the specimen sizes used in this experiment. The factor of 1.23 is quite consistent with the experimental observations in Fig.4, in which the size dependence of this order was not discernible in the $(dl/dt)-J'$ correlation when compared with a width of the scatter band observed. The slope of the correlation curves in Fig.4 is slightly smaller than unity. This is also compatible with the theoretical prediction of $\alpha = b/(n+1) = 0.85$.

The size dependence in the $(dl/dt)-\sigma_n$ correlation is quantitatively examined. The experiments were conducted, keeping the initial net section stress value σ_{n0} constant. According to Eq.(25), the size dependence may be compensated by plotting $(dl/dt)/(2W)$ against σ_n, taking $a*$ as the specimen width $2W$. Fig.7 shows the results of this adjustment by using the experimental results of the "deep" center-cracked proportional specimens. This adjustment is quite successful for eliminating the size dependence. A similar method for eliminating the size dependence on the basis of Eq.(25) may also be applicable to the $(dl/dt)-K_I$ correlation. Strong evidence of this was observed.

The idea of using proportional specimens was also adopted by Koterazawa and Mori [15], and Taira et al. [16]. The superiority of J' over σ_n and K_I was substantiated in these investigations.

THE STRESS FIELD AND MODIFIED J-INTEGRAL NEAR A CRACK TIP UNDER THE CONDITION OF CONFINED CREEP DEFORMATION [17]

As seen from Eq.(2), J' uniquely provides the intensity of the singular stress field near a crack tip when the Norton-type, steady-state creep law is selected as the constitutive equation. When the elastic strain prevails, except in a region near the crack tip, however, the intensity of the singular stress field near the crack tip may be related to the elastic stress intensity factor K_{el} or the elastic J-integral J_{el}, as in the case of small-scale yielding. Assuming that (1) the total strain can be divided into the elastic and creep strains, (2) the crack does not grow, and (3) the definition of modified J-integral, J', is valid near the crack tip even under the elastic-creep condition, then during incipient creep deformation, the J'-value near the crack tip is given by

$$J' = J_{el}/\{(n+1)t\} \tag{26}$$

where t denotes loading time. J' decreases and approaches its steady-state value J'_{st} as the creep deformation increases. As a measure of this transition period, a characteristic time t_{tr} is defined, which is proportional to J_{el}/J'_{st}. This transition should be taken into account when predicting the initial phase of creep crack growth in low ductility materials.

CONCLUSIONS

A nonlinear fracture mechanics approach to the creep crack growth problems, which has been developed by the authors' research group, was presented. Several of the important conclusions derived from this investigation follow:

(1) Based on the plastic singular stress field predominant in the vicinity of a creep crack tip, an equation of crack growth rate was derived, which explicitly represented the influences of various mechanical and material parameters.

(2) Under usual metallic creep conditions, creep crack growth rate was expressed as a function of the modified J-integral, J', and the amount of crack growth.

(3) On the basis of this analysis, the dependence of the creep crack growth rate on specimen size observed in experiments using a series of proportional specimens was discussed. Experimental evidences which supported these predictions were demonstrated.

(4) Methods of compensating the foregoing size dependence were proposed.

(5) Experimental and analytical data showed that among stress intensity factor K_{el}, net section stress σ_n, and modified J-integral J', J' was the best correlation parameter of the creep crack growth rate, and that under moderate creep conditions both the stress intensity factor and the net section stress failed their general validity in the correlation of creep crack growth rate.

(6) During incipient creep deformation, where the elastic strain prevailed except in a region near the crack tip, the value of the modified J-integral was very large, compared to its steady-state value.

REFERENCES

(1) M.J.Siverns, and A.T.Price, "Crack Propagation under Creep Conditions in a Quenched $2\frac{1}{4}$ Chromium 1 Molybdenum Steel", Int. J. Fracture, 9-2, (1973), pp.199-207.

(2) J.L.Kenyon, G.A.Webster, J.C.Radon, and C.E.Turner, "An Investigation of the Application of Fracture Mechanics to Creep Cracking", Int. Conf. on Creep and Fatigue in Elevated Temperature Applications, Philadelphia 1973, (1974), pp.156.1-156.8.

(3) R.Koterazawa, and Y.Iwata, "Fracture Mechanics and Fractography of Creep and Fatigue Crack Propagation at Elevated Temperature", Trans. ASME, Ser.H, 98-4, (1976), pp.296-304.

(4) C.B.Harrison, and G.N.Sandor, "High-Temperature Crack Growth in Low-Cycle Fatigue", Engng. Fracture Mech., 3-4, (1971), pp.403-420.

(5) S.Taira, and R.Ohtani, "Creep Crack Propagation and Creep Rupture of Notched Specimens", Int. Conf. on Creep and Fatigue in Elevated Temperature Applications, Philadelphia 1973, (1974), pp.213.1-213.7.

(6) R.D.Nicholson, and C.L.Formby, "The Validity of Various Fracture Mechanics Methods at Creep Temperatures", Int. J. Fracture, 11-4, (1975), pp.595-604.

(7) S.Kubo, K.Ohji, and K.Ogura, "An Analysis of Creep Crack Propagation on the Basis of the Plastic Singular Stress Field", Engng. Fracture Mech., 11-2, (1979), pp.315-329.

(8) J.W.Hutchinson, "Singular Behavior at the End of a Tensile Crack in a Hardening Material", J. Mech. Phys. Solids, 16-1, (1968), pp.13-31.

(9) K.Ohji, K.Ogura, and S.Kubo, "Creep Stress Analysis of a Notched Body under Longitudinal Shear and Its Application to Crack propagation Problems", Proc. 1974 Symp. on Mech. Behavior of Materials, Kyoto 1974 (Soc. Materials Sci., Japan), 1, (1974), pp.455-466.

(10) K.Ohji, K.Ogura, S.Kubo, Y.Katada, T.Katsuhara, and N.Iwanaga, "A study on Fracture Mechanics Parameter of Creep Crack Growth Using Proportional Deep- and Shallow-Cracked Plate Specimens", Trans. Japan Soc. Mech. Engrs., 45-394(A), (1979), pp.550-558, (in Japanese).

(11) K.Ohji, K.Ogura, S.Kubo, and Y.Katada, "A Study on Governing Mechanical Parameters of Creep Crack Growth Rate by Using Notched Round Bar Specimens with Proportional Geometry", Proc. 21st Japan Congr. Materials Res., Soc. Materials Sci., Japan, (1978), pp.99-104.

(12) K.Ohji, K.Ogura, S.Kubo, and Y.Katada, "An Apllication of Nonlinear Fracture Mechanics to Creep Crack Growth Problems (An Evaluation of Experimental Data in Terms of Modified J-Integral)", Preprint of Japan Soc. Mech. Engrs., No.780-9, (1978), pp.123-128, (in Japanese).

(13) K.Ohji, K.Ogura, and S.Kubo, "An Analysis of the Mode Ⅲ Creep Crack Propagation in a Semi-Infinite Body", Trans. Japan Soc. Mech. Engrs., 45-396(A), (1979), pp.836-842, (in Japanese).

(14) K.Ohji, K.Ogura, and S.Kubo, "Estimates of J-Integral in the general Yielding Range and Its Application to Creep Crack Problems", Trans. Japan Soc. Mech. Engrs., 44-382, (1978), pp.1831-1838, (in Japanese).

(15) R.Koterazawa, and T.Mori, "Applicability of Fracture Mechanics Parameters to Crack Propagation under Creep Condition", Trans. ASME, Ser.H, 99-4, (1977), pp.298-305.

(16) S.Taira, R.Ohtani, and T.Kitamura, "Application of J-Integral to High-Temperature Crack Propagation, Part I-Creep Crack Propagation", Trans. ASME, J. Engng. Materials and Technology, 101-2, (1979), pp.154-161.

(17) K.Ohji, K.Ogura, and S.Kubo, "The Stress Field and Modified J-Integral near a Crack Tip under the Condition of Confined Creep Deformation", submitted to J. Soc. Materials Sci., Japan, (in Japanese).

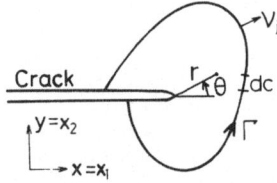

Fig.1 A line contour in modified J-integral

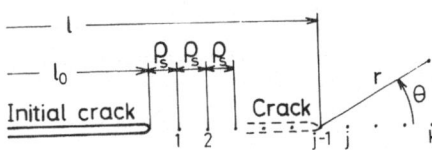

Fig.2 Procedure of creep crack growth analysis

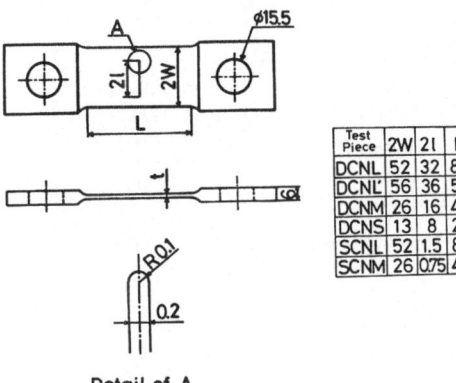

Test Piece	2W	2l	L	t
DCNL	52	32	88	
DCNL′	56	36	50	
DCNM	26	16	44	1.5
DCNS	13	8	22	
SCNL	52	1.5	88	
SCNM	26	0.75	44	

Detail of A

Fig.3 Geometry of proportional center-cracked(CN) plate specimens (mm)

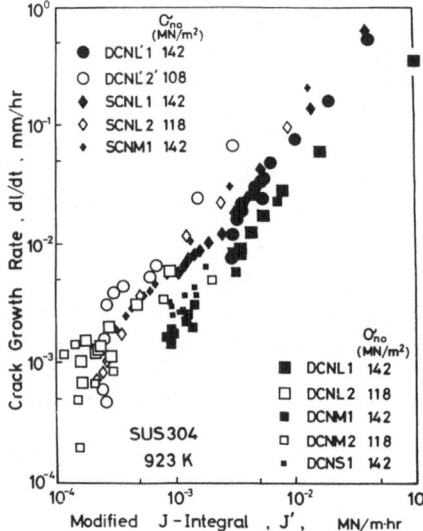

Fig.4 Correlation of creep crack
growth rate with the modified *J*-integral

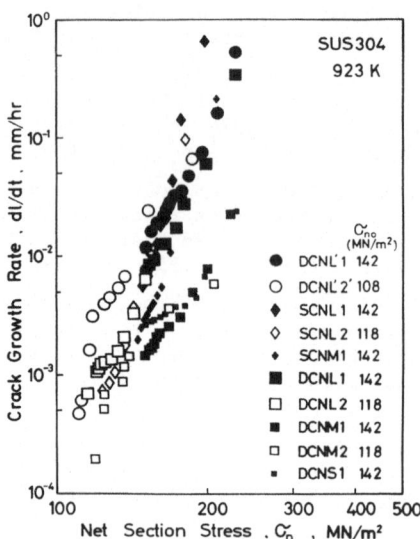

Fig.5 Correlation of creep crack
growth rate with the net
section stress

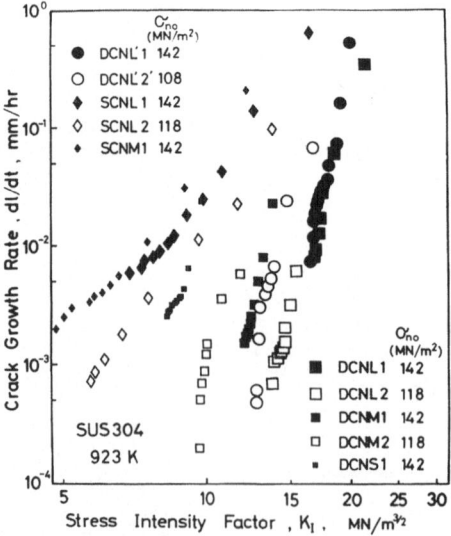

Fig.6 Correlation of creep crack
growth rate with the stress
intensity factor

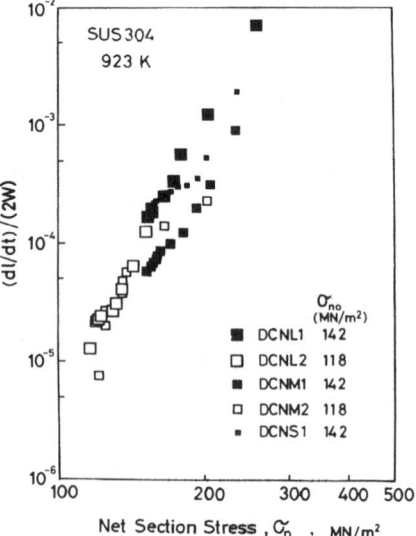

Fig.7 Compensation of size dependence
in the net section stress correlation

A NONLINEAR FRACTURE MECHANICS APPROACH
TO CRACK PROPAGATION IN THE CREEP-FATIGUE INTERACTION RANGE

R. Ohtani
Associate Professor of Engineering Science
Kyoto University, Yoshida, Sakyo-ku, Kyoto, Japan

T. Kitamura
Central Research Institute of Electric Power Industry
Komae, Tokyo, Japan

K. Yamada
Graduate Student, Kyoto University, Kyoto, Japan

INTRODUCTION

High temperature fatigue tests were conducted with center cracked thin-walled cylindrical specimens in a closed-loop servo hydraulic testing machine. The applicability of nonlinear fracture mechanics to the time-dependent as well as the cycle-dependent fatigue crack propagation was determined, using creep J-integral and fatigue J-integral.

METHODS OF EVALUATION OF J-INTEGRAL

Methods of evaluation of the J-integral range for high temperature fatigue crack were examined, supposing that the strain in the cracked body are composed of the time independent elastic-plastic strain and the time dependent creep strain, and the former strain in the vicinity of a crack is represented by the fatigue J-integral range and the latter near tip strain by the creep J-integral range [1, 2].

The fatigue J-integral range is equivalent to the cyclic J-integral proposed by Dowling [3], and is applied to the low cycle fatigue crack propagation at room temperature [3∿6] as well as at elevated temperature [1, 2, 7]. Fig. 1(a) shows the graphical method of estimating the fatigue J-integral range for center cracked plates. Since the Dowling's method is not suitable for evaluating more accurate values of the J-integral for shallow cracks, the load versus crack center opening displacement loop ($P - V$ loop) is used instead of the load versus load point displacement loop ($P - \delta$ loop) [7]. The crack center opening displacement (CCOD), V, is the crack opening displacement at the center of the notch. The initial value, V_0, of the specimen used in the present study is shown in Fig. 2. The fatigue J-integral range, ΔJ_f, is then given by [7]

$$\Delta J_f = \frac{\Delta K_{eff}^2}{E'} + \frac{S_p}{Bb} \quad , \tag{1}$$

where ΔK_{eff} is the effective stress intensity factor range, $E' = E$ for plane stress and $E' = E/(1 - v^2)$ for plane strain, in which E is the Young's modulus and v is the Poisson's ratio, S_p is the hatched area inside the $P - V$ hysteresis loop, B is the specimen thickness, and b is the ligament length. If a tension going curve of the $P - V$ hysteresis loop can be represented by the following equation of a power form:

$$V_p = A P^n \quad , \tag{2}$$

the second term of the right-hand side of Eq. (1), S_p/Bb, being the plastic component of ΔJ_f, becomes [1, 2, 7]

$$\frac{S_p}{Bb} = \frac{1}{Bb} [\int_0^{V_p} P \, d V_p - \frac{1}{2} P^* V_p] = \frac{n-1}{n+1} \Delta\sigma_{net} V_p \quad , \tag{3}$$

where $\Delta\sigma_{net} = P^*/2Bb$.

On the other hand, the creep J-integral is a modification of the J-integral where the strain field is replaced by the strain rate field, and is applied to the creep crack propagation [8∿10] as well as the time-dependent low cycle fatigue crack propagation [1, 2, 7, 11, 12]. In Fig. 1(b), the method of evaluation of the creep J-integral, $\dot{J}$, for a center crack subjected to constant load is illustrated with the simple estimating equation proposed by Ohji and coworkers [13]:

$$\dot{J} = \frac{\alpha - 1}{\alpha + 1} \sigma_{net} \dot{V} \cong \sigma_{net} \dot{V} \quad , \tag{4}$$

where α is the creep exponent in a power law of stress versus creep rate, σ_{net} is the net section stress, and $\dot{V}$ is the CCOD rate. The form of the equation is similar to that of Eq. (3).

In the case of the creep-fatigue interaction range for rectangular stress waveforms, the fatigue J-integral range given by Eq. (1) is applicable at the ascent of stress in every cycle, and the creep J-integral given by Eq. (4) is adopted in the stress hold period, as shown in Fig. 1 (c). In addition, if the magnitude of the creep J-integral is constant during half cycle under tensile stress, the creep J-integral range, ΔJ_c, becomes [1, 2]

$$\Delta J_c = \int_0^{1/2v} \dot{J} \, dt \cong \frac{1}{2Bb} \int_0^{V_c} P \, d V_c = \frac{S_c}{2Bb} = (\sigma_{net})_{max} V_c \quad , \tag{5}$$

where ν is the frequency, S_c is the rectangular hatched area in Fig. 1(c), and $(\sigma_{net})_{max} = P_{max}/2Bb$.

In Fig. 1(d) illustrated is another example of the graphical method of estimating ΔJ_f and ΔJ_c for a fatigue crack under push-pull condition of a triangular CCOD waveform or any other waveforms in which the elastic-plastic strain and the creep strain grow simultaneously. In order to partition the CCOD into time independent component, V_p, and time dependent component, V_c, a half cycle rapid tension is introduced at several stages during fatigue test, to be obtained the $P - V_p$ curve. This is the same method as adopted in the strainrange partitioning approach to the high temperature low cycle fatigue [14, 15]. Assuming that the creep component of CCOD, V_c, has a functional relation with the load, P, as $V_c = f(P)$, the creep J-integral range, ΔJ_c, can be expressed by the following [1, 2].

$$\Delta J_c \cong \frac{1}{2Bb} \int_0^{V_c} P \, dV_c = \frac{1}{2Bb} \int_0^{P_{max}} P f'(P) \, dP$$

$$= \frac{1}{2Bb} \left\{ [P f(P)]_0^{P_{max}} - \int_0^{P_{max}} f(P) \, dP \right\} = \frac{S_{c1} - S_{c2}}{2Bb} . \qquad (6)$$

EXPERIMENTAL PROCEDURES

The material tested is a 0.16% carbon steel of a 25 mm diameter hot rolled bar [1, 2]. Fig. 2 shows the shape and the size of the specimen used in the experiment. The specimen is a thin-walled cylinder with a notch at the center of the gage section. The notch was introduced by drilling and filing. After machining, the specimens were annealed at 900°C for one hour.

The fatigue crack propagation tests were conducted in a closed-loop servo hydraulic testing machine with an induction heating equipment. All tests were carried out in an air environment. By attaching the tips of a pair of quartz glass bars of the extensometer to the upper and lower edges of the notch [2], the CCOD, V, was measured by an LVDT and recorded as $V - t$(time) curves or P(load) $- V$ curves by an X-Y pen recorder or a synchroscope. During the tests, the crack length was measured by a travelling microscope of the magnification of about 30.

TEST RESULTS AND DISCUSSIONS

Fig. 3 shows the relationship between crack propagation rate, dl/dN, for various frequencies and total J-integral range, ΔJ_T. The total ΔJ_T is the sum of the fatigue ΔJ_f and the creep ΔJ_c. As the frequency decreases, the fatigue propagation rate increases to approach to the creep crack propagation rate.

Here, the creep crack propagation rate data represented by cross marks in Fig. 3 show the dl/dt [mm/h] versus J [kgf/mm·h] relation.

When ΔJ_T is divided into two components, ΔJ_f and ΔJ_c, most of the data can be classified into two groups: The one holds the $dl/dN - \Delta J_f$ relation (Fig. 4) and the other is expressed by the $dl/dN - \Delta J_c$ relation (Fig. 5) [2]. It is found from Fig. 4 that the $dl/dN - \Delta J_f$ relation for high frequency fatigue at 400°C agree well with those for strain controlled fatigue at 20, 100, and 300°C. The test results on 316 stainless steel at room temperature, 600 and 650°C [7], and those on A533B steel [3] and 0.04% carbon steel [5, 6] at room temperature obtained by other investigators are shown by straight lines in Fig. 4. All the data are fairly close with each other.

Fig. 5 indicates that dl/dN correlates well with ΔJ_c for time-dependent fatigue tested at low frequencies and high temperatures. Static creep crack propagation data are also plotted in Fig. 5 as the $dl/dt - \dot{J}$ relation. It is evident that, when dl/dN and ΔJ_c for the time-dependent fatigue are converted into $dl/dt = 2\nu(dl/dN)$ and $\dot{J} = 2\nu \cdot \Delta J_c$, the $dl/dt - \dot{J}$ relation for fatigue coinsides with that for creep. Also represented by straight lines are the fatigue data on 316 stainless steel [7] and Hastelloy X [16], and the creep data on 304 stainless steel [8]. The correlation between crack propagation rate and J-integral appears to be insensitive to the test temperature and the sort of material.

The data indicated by the symbols, ◉ and ▲, in Figs. 4 and 5, show the crack propagation rate at intermediate frequencies in the creep-fatigue interaction range. It is found that the propagation rate, dl/dN, is nearly equal to or just a little higher than both the cycle-dependent fatigue crack propagation rate, $(dl/dN)_f$, (see Fig. 4) and the time-dependent one, $(dl/dN)_c$ (see Fig. 5). The result suggests that the crack propagation rate, dl/dN, can be given as a sum of $(dl/dN)_f$ and $(dl/dN)_c$ [1, 2].

It is worthy of note that the time-dependent fatigue crack propagation rate, $(dl/dN)_c$, is more than ten times as large as the cycle-dependent one, $(dl/dN)_f$, for the same magnitude of J-integral. The difference is the same for the 316 stainless steel shown in Fig. 6 [7]. The same tendency is found in a $\Delta\varepsilon_{pp} - N_{pp}$, and a $\Delta\varepsilon_{cc} - N_{cc}$ relation of smooth specimens in high temperature low cycle fatigue obtained by the strainrange partitioning approach [15, 17]. That is, the fatigue failure life, N_{cc}, is shorter than N_{pp} for the same magnitude of characteristic strain ranges, $\Delta\varepsilon_{pp}$ and $\Delta\varepsilon_{cc}$. Of two laws on fatigue crack propagation rate corresponding to other partitioned strainranges, $\Delta\varepsilon_{pc}$ and $\Delta\varepsilon_{cp}$, the law for $\Delta\varepsilon_{pc}$ appears to coinside with the $dl/dN - \Delta J_f$ relation for $\Delta\varepsilon_{pp}$, and the law for $\Delta\varepsilon_{cp}$, with the $dl/dN - \Delta J_c$ relation for $\Delta\varepsilon_{cc}$, as will be reported elsewhere.

ACKNOWLEDGEMENT

The authors sincerely acknowledge their indebtness to the late Dr. Shuji Taira, formely Professor of Engineering, Kyoto University, for his continual guidance and encouragement which have been given in doing the investigation.

REFERENCES

(1) S. Taira, R. Ohtani, T. Kitamura, and K. Yamada, "J-Integral Approach to Crack Propagation under Combined Creep and Fatigue Condition", J. of Soc. Mat. Sci., Japan, 28-308, (1979), pp. 414-420, (in Japanese).

(2) R. Ohtani, "Crack Propagation Under Creep-Fatigue Interaction", to be presented in the Proceeding of Int. Conf. on Engineering Aspect of Creep, to be held in Sheffield, 15-19 September, 1980.

(3) N.E. Dowling, "Geometry Effects and the J-Integral Approach to Elastic-Plastic Fatigue Crack Growth", ASTM STP 601, (1976), pp. 19-32.

(4) N.E. Dowling, "Crack Growth During Low-Cycle Fatigue of Smooth Axial Specimens", ASTM STP 637, (1977), pp. 97-121.

(5) S. Taira, K. Tanaka, and S. Ogawa, "Fatigue Crack Propagation Law under Elastic-Plastic and General Yield Conditions", J. of Soc. Mat. Sci., Japan, 26-280, (1977), pp. 93-98, (in Japanese).

(6) S. Taira, K. Tanaka, and T. Hoshide, "Evaluation of J-Integral Range and Its Relation to Fatigue Crack Growth Rate", Proc. of the 22nd Japan Cong. on Mat. Res., Soc. of Mat. Sci., Japan, (1979), pp. 123-129.

(7) S. Taira, R. Ohtani, and T. Komatsu, "Application of J-Integral to High-Temperature Crack Propagation, Part II — Fatigue Crack Propagation", ASME J. of Engng. Mat. and Tech., 101-2, (1979), pp. 162-167.

(8) S. Taira, R. Ohtani, and T. Kitamura, "Application of J-Integral to High-Temperature Crack Propagation, Part I — Creep Crack Propagation:, ibid., pp. 154-161.

(9) K. Ohji, K. Ogura, and S. Kubo, "Creep Stress Analysis of a Notched Body Under Longitudinal Shear and its Application to Crack Propagation Problems", Proc. 1974 Symp. on Mech. Behav. of Mat., Soc. of Mat. Sci., Japan, (1974), pp. 455-466.

(10) J.D. Landes and J.A. Begley, "A Fracture Mechanics Approach to Creep Crack Growth", ASTM STP 590, (1976), pp. 128-148.

(11) M.P. Harper and E.G. Ellison, "The Use of the C^* Parameter in Predicting Creep Crack Propagation Rates", J. Strain Analysis, 12-3, (1977), pp. 167-179.

(12) R. Koterazawa and T. Mori, "Crack Propagation under Intermittent Overloading at Elevated Temperatures", J. of Soc. Mat. Sci., Japan, 27-303, (1978), pp. 1178-1184, (in Japanese).

(13) K. Ohji, K. Ogura, and S. Kubo, "Estimates of J-Integral in General Yielding Range and its Application to Creep Crack Problems", Trans. JSME, 44-382, (1978), pp. 1831-1838, (in Japanese).

(14) S.S. Manson, "New Directions in Materials Research Dictated by Stringent Future Requirements", Proc. Int. Conf. on Mech. Behav. of Mat., Spec. Vol., Soc. of Mat. Sci., Japan, (1972), pp. 3-60.

(15) G.R. Halford, M.H. Hirschberg, and S.S. Manson, "Temperature Effects on the

268

Strainrange Partitioning Approach for Creep Fatigue Analysis", <u>ASTM STP 520</u>, (1973), pp. 658-669.

(16) S. Taira, R. Ohtani, T. Yonekura, M. Osada, and T. Kitamura, "Time Dependent Fatigue Crack Propagation at Elevated Temperature", to be contributed to Trans. JSME.

(17) K. Hirakawa and K. Tokimasa, "Effect of Environment on Partitioned Strain-Life Relations of SUS304 Stainless Steel", <u>J. of Soc. Mat. Sci., Japan</u>, <u>28</u>-308, (1979), pp. 386-392, (in Japanese).

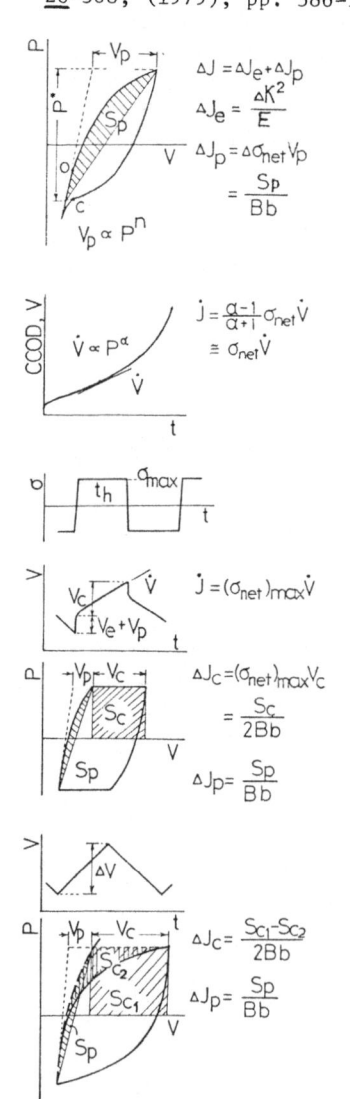

Fig.1 Methods of estimation of fatigue J-integral, ΔJ_f, creep J-integral, $\dot{J}$, and creep J-integral range, ΔJ_c.

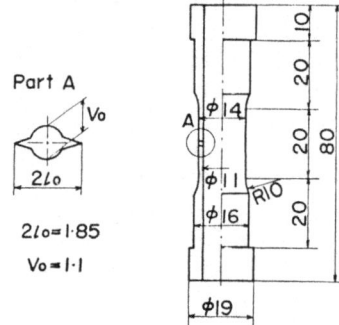

Fig.2 Specimen (Dimensions in mm).

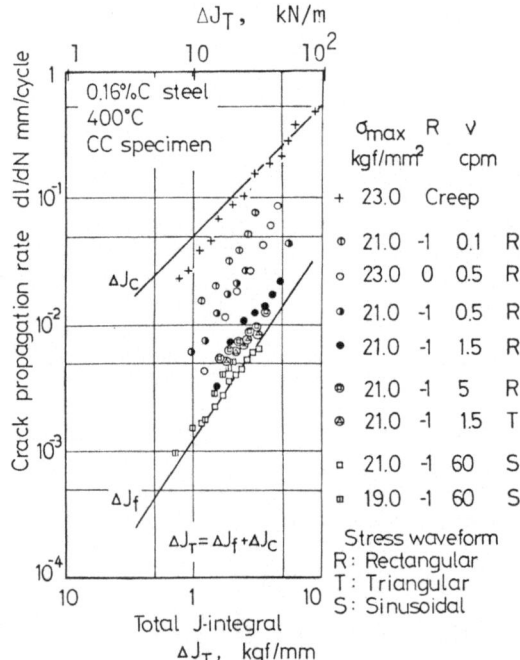

Fig.3 Crack propagation rate versus total J-integral range diagram.

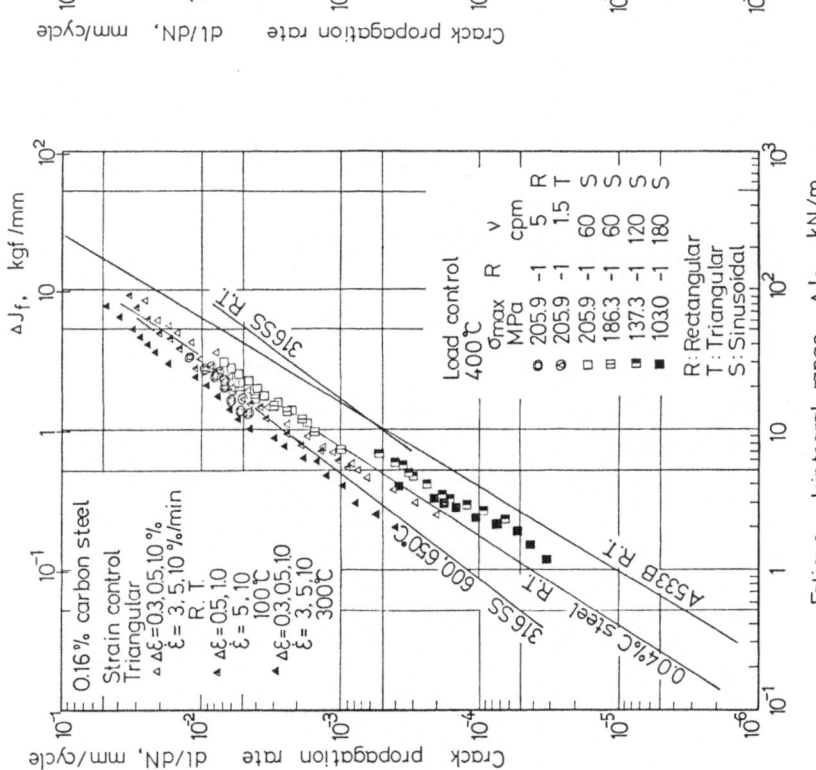

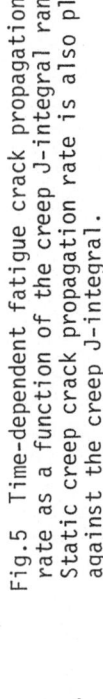

Fig.5 Time-dependent fatigue crack propagation rate as a function of the creep J-integral range. Static creep crack propagation rate is also plotted against the creep J-integral.

Fig.4 Cycle-dependent fatigue crack propagation rate as a function of the fatigue J-integral range.

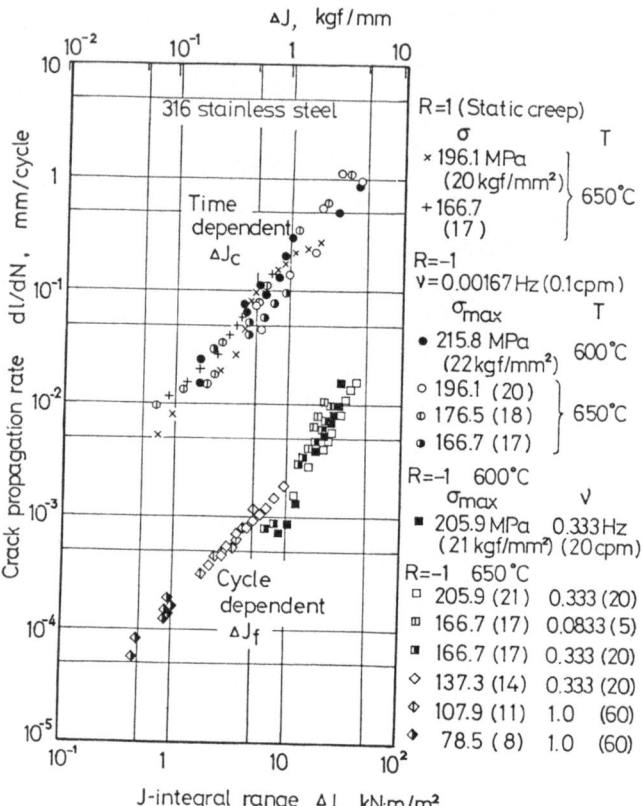

Fig.6 Relationship between dl/dN and ΔJ_f, and dl/dN and ΔJ_c of 316 stainless steel at 600 and 650°C.

A CONSIDERATION OF FATIGUE LIMITS AND NOTCH SENSITIVITY BASED ON SUCCESSIVE OBSERVATIONS OF THE CRACK INITIATION PROCESS

Hironobu NISITANI
Professor of Faculty of Engineering,
Kyushu University, Fukuoka, 812, Japan

Ken-ichi TAKAO
Associate Professor of Faculty of Science and
Engineering, Saga University, Saga,840, Japan

1. INTRODUCTION

It is well known that ferrous metals have distinct fatigue limits while most nonferrous metals do not and that the notch sensitivity in fatigue is different from metal to metal. For example, Al-alloy is more sensitive than low carbon steels[1][2]. These two phenomena seem to be independent. However, they are closely related to each other; that is, both phenomena are correlated by the behavior of microcracks initiated from surfaces.

In the present study, rotating bending fatigue tests were carried out for an age-hardened Al-alloy and low carbon steels, and fatigue processes were observed successively on the replicas of the surfaces of both smooth and notched specimens, with special attention being given to the starting point of fatigue cracking. Based on these observations,the physical backgrounds of the existence of fatigue limits of low carbon steels and of the difference in notch sensitivity between Al-alloy and low carbon steel became clear.

2. MATERIALS, SPECIMENS, AND TESTING PROCEDURES

Materials used are an age-hardened Al-alloy A2017BE-T4(0.2 % offset yield strength of 308 MPa,UTS of 443 MPa)and annealed low carbon steels[steel A(0.13 %C), LYS of 203 MPa, UTS of 327 MPa; steel B(0.17%C), LYS of 176 MPa, UTS of 318 MPa; steel C(0.17%C), LYS of 211 MPa, UTS of 379 MPa]. Steels were reannealed in a vacuum at 600°C for an hour after machining. Dimensions of the specimen are shown in Fig.1. Since the fatigue notch factor K_f for the specimen having a notch root radius ρ of 5 mm is about 1.06, this type of specimen(Fig.1, ρ=5 mm) can be used as a smooth specimen. The fine shallow notch was introduced in order to localize the severely damaged region and to easily follow the starting point of final fracture.

Before testing, most of the specimens were electro-polished to remove more than 20 μm of the work hardened layers from the surface.

The Ono type rotating bending fatigue machine was used.

The notched surfaces of specimens were replicated at pertinent intervals during stress repetitions to observe the fatigue process successively by using optical and electron microscopes(TEM). Fractographs including the starting point of final fracture were taken by using SEM.

The specimen, made of prestrained steel C and having the fine parallel lines drawn at 45° to the specimen axis on the surface, were also prepared to measure the crack opening displacement COD of the microcracks at the maximum and minimum stresses during one stress cycle. The COD was obtained by measuring the displacement of parallel lines on the electron micrographs($\times$10,000) of the microcrack, as shown in Fig.10(a).

3. RESULTS AND DISCUSSION

3.1. The initiation processes of fatigue cracks in Al-alloy and low carbon steel

Figure 2 shows the change in the surface state of an Al-alloy, due to the repetitions of the stress(stress amplitude σ_a=196 MPa) which are about 50% higher than the fatigue strength at 10^7 cycles. Figure 3 shows the slip band of an Al-alloy subjected to static extension, after about 0.5 N_i cycles of σ_a(=196 MPa), where N_i means the number of cycles necessary to form a fatigue crack the size of a grain.

These figures indicate that a slip band of an Al-alloy initiates in a fairly small region(less than 4μm at $N=10^4$) compared to the grain size(about 100 μm), and it gradually grows as the number of cycles N increases. However, the matrix other than the slip band is almost unchanged from the original surface state(The cause of these phenomena may be the work softening). For this reason, an Al-alloy is more notch sensitive than usual metals in which the finite region comparable to the grain size is disrupted as a whole(e.g., see Fig.5). Under static extension, the slip band opens up easily into a microcrack, even when the size of the slip band is much smaller than that of a grain and even under the total strain of 0.3 % . Therefore, it is difficult to distinguish a slip band from a microcrack, or the initiation process from the propagation process[3]. Thus, the fatigue process of an Al-alloy can be subdivided into Forsyth's well known stage I and stage II crack propagation processes[4].

Figure 4 shows the change in the surface state of steel A, due to the repetitions of the stress(σ_a=196 MPa) which are about 15 % higher than the fatigue limit. Figure 5 shows electron micrographs of the central parts in Fig.4. Figure 6 shows the region of steel B which is to become a crack in the future under static extension, produced by about 0.3 N_i cycles at σ_a(=176 MPa).

As seen from these figures, the depth or contrast of shade in the same region which is to become a crack increases with increasing N on the optical micrographs. Observations by TEM show that the grain boundary which is to become a crack is being disrupted with the nearby ferrite matrix[5].Although the size of the region is larger than that of Al-alloy in Fig.3, it does not open up even under the static extension of total strain of 4%. That is, the crack has not initiated at this stage. After the initiation of the crack(Fig.4, $N=4\times10^5$), it propagates into neighboring grains due to the concentration of stress and strain at the crack tips. If the static extension is applied at the beginning of this process, the crack opens up easily[6].

Based on the above observations and a series of experiments carried out by the authors[3][5][7][8][9], the crack initiation processes of Al-alloy and low carbon steel are illustrated schematically in Fig.7.

The fatigue process of Al-alloy is thought to follow Forsyth's[Fig. 7 (a)]. However, the fatigue process of low carbon steel is different from that of Al-alloy, and it is subdivided into two entirely different processes, i.e.,an initiation process and a propagation process. Until the initiation of a crack, fatigue damage may be accumulated gradually as a whole at the definite region [shaded area in Fig.7(b)] whose dimensions are closely related to the grain size. The characteristics of crack propagation may not be included in this process[3][5][8]. In low carbon steel, cracking along the slip band in a grain also goes through the same process as described above[7]. Moreover, the initiation process is also observed when the stress of the fatigue limit is reversed more than 10^7 cycles[3][5]. In this case, the maximum crack observed on the specimen surface is the crack which propagates a little after the initiation and then stops propagating, as shown in the next section. In metals other than Al-alloy which do not have a definite fatigue limit(annealed copper[10]for example), the general crack initiation process is analogous to the case of low carbon steel [Fig.7]. However, once initiated, the crack do not stop propagating in the metal as they do in low carbon steel. This is attributable to the difference in work hardening and strain aging properties of metals[10][11].

Figure 8 shows the SEM fractographs,including the starting points of final fracture. The initiation region of the cracks is fairly smooth, and at present, the difference in the crack initiation processes of the two materials is difficult to clarify from the observation of the fracture surface alone.

3.2. Fatigue limit of carbon steel and non-propagating microcracks

Figure 9 shows the non-propagating microcrack observed on the surface of a smooth unbroken specimen produced by 10^7 cycles at σ_a(=167 MPa), slightly below the fatigue limit of steel A(σ_{wo}=172 MPa). The starting point of cracking is shown by arrows. The crack shown in Fig.9 is observed not only on the other low carbon steels[7], but also on the medium or high carbon steels[8].

Previously,one of the authors carried out experiments on the coaxing effect of carbon steels, and found out that the starting point of final fracture by the step loading was the non-propagating microcrack produced under the initial stress level,i.e.,the fatigue limit of the materials[10][11]. From this, the fatigue limit of a carbon steel may be defined as the limiting stress that produces the non-propagating microcrack.

Figure 10 shows crack opening displacements of the non-propagating and propagating microcracks in prestrained steel C, having almost the same size. The latter crack was produced under the stress about 10%higher than the fatigue limit[12].

The non-propagation of the microcrack described above is closely related to the crack closure phenomenon[13]. Figure 10 shows that the opening and closing behavior of the non-propagating microcrack is much different from that of the propagating microcrack,i.e.,the shape near the tip of the former shows the "cusp" type on the surface; on the other hand, the latter shows "blunt" type. The COD near the tips of the non-propagating crack is much smaller than that of the propagating crack, under the maximum stresses during the respective stress cycles.The non-propagating microcracks are almost closed at or near their tips, even under the maximum stress in a stress cycle. This behavior has also been recognized in the case of the so-called non-propagating crack emanating from a sharp notch[14][15]. Such a phenomenon of non-propagation of cracks is thought to be due to the plastic deformation, left in the wake of the propagating crack [13][15]and due to the change in material properties,such as work hardening and strain aging at or near the crack tips[10][11].

3.3. Notch sensitivity in fatigue of Al-alloy and low carbon steel

Figure 11 shows the relation between the stress gradient $\chi [=(1/\sigma_{max})(d\sigma/dz)_{z=0}$, z is the distance from the surface]and the maximum reversed stress at the crack initiation limit $K_t\sigma_{wl}$,in Al-alloy and low carbon steel[2]. In this figure, σ_{wl} means the maximum nominal stress that causes about the same damage at the notch root as the one at the surface of a smooth specimen after 10^7cycles of the stress of the fatigue limit. Therefore, a horizontal line of $K_t\sigma_{wl}/\sigma_{w0}=1$ means that at the fatigue limit, the maximum stress at the notch root is equal to the fatigue limit of a smooth specimen.

Figure 12 shows the relation between the fatigue notch factors of the specimens having the same form and size in the two materials.

These figures indicate that Al-alloy is more notch sensitive than low carbon steel. This seems to be mainly due to the fact that the thickness of the surface layer affecting the crack initiation is much smaller for Al - alloy than for steel C[2][9] (This means that crack initiation is controlled by the stress at a point in the former and by the mean stress over the finite thickness in the latter). The initiation process of a crack is different from metal to metal as described in Sec.3.1,and this apparently causes the difference in the thickness of the surface layer affecting the crack initiation from metal to metal.

Figure 13 shows the crack propagation curves for the notched specimens of the materials used and was obtained from successive observations. This shows almost the same trends as those described in Sec.3.1, irrespective of the notch geometries.

4. CONCLUSIONS

(1) The crack initiation process of Al-alloy may correspond to Forsyth's stage I of crack propagation. However, the process of low carbon steels is the one in which the definite region, whose sizes are closely related to the grain size, turns to the free surface by the repetitions of slip, gradually and as a whole. This initiation process is different from the later propagation process.

(2) The fatigue limit of carbon steel is the maximum reversed stress at which the fatigue crack the size of a grain propagates into neighboring grains and then becomes a non-propagating microcrack. Such a crack is almost closed at or near its tip even under the maximum tensile stress in one cycle of the reversed stress under which the crack is formed.

(3) Al-alloy is more notch sensitive in fatigue than low carbon steel. This is due to the fact that the initiation processes of fatigue cracks are different in the two materials.

REFERENCES

[1] R.L.Templin, "Fatigue of Aluminum", Proc. of American Society for Testing and Materials, 54(1954),pp.641–699.

[2] H.Nisitani and T.Goto, "Fatigue Notch Sensitivity of an Aluminum Alloy", Trans. of the Japan Society of Mechanical Engineers, 42-361 (Sept.,1976), pp.2666–2672(in Japanese).

[3] H.Nisitani and K.Takao,"Successive Observations of Fatigue Process in Carbon Steel, 7:3 Brass and Al-alloy by Electron Microscope", Trans. of the Japan Society of Mechanical Engineers, 40-340(Dec.,1974), pp.3254–3266(in Japanese).

[4] P.J.E.Forsyth,"A Two Stage Process of Fatigue Crack Growth", Proc. of the Crack Propagation Symposium, Cranfield, Vol.1 (1962), pp.76–94.

[5] H.Nisitani and K.Takao,"Successive Observations of Fatigue Process in Low Carbon Steel by Electron Microscope",Proc.of the International Conference on Mechanical Behavior of Materials, Kyoto, vol.2 (1972), pp.153–164.

[6] G.C.Smith, "The Initial Fatigue Crack", Proc.Roy.Soc.,Ser.A,242(1957),pp. 189–197.

[7] K.Takao, H.Nisitani, and H.Sakaguchi, " Influence of Strain Aging on Crack

276

Initiation Process in Rotating Bending Fatigue of Low Carbon Steel", Pre-print of the Society of Materials Science, Japan, Symposium on Fatigue , (1979), to be published(in Japanese).

[8] H.Nisitani and I.Chishiro,"Non-Propagating Micro-Cracks of Plain Specimens and Fatigue Notch Sensitivity in Annealed or Heat-Treated 0.5%C Steel", Trans. of the Japan Society of Mechanical Engineers, 40-329(Jan.,1974),pp. 41—52(in Japanese).

[9] H.Nisitani and M.Kage,"Rotating Bending Fatigue of Electropolished Speci-mens with Transverse Holes—Observation of Slip Bands and Non-propagating Cracks Near the Holes", Trans.of the Japan Society of Mechanical Engineers, 39-323(July,1973),pp.2005—2012(in Japanese).

[10] H.Nisitani and S.Nisida,"Correlation between the Existence of Fatigue Limit and Non-Propagating Micro-Crack", Bull. of the Japan Society of Mechanical Engineers, 17-1o3(Jan.,1974),pp.1—11.

[11] H.Nisitani and T.Ikenaga,"Relation between Coaxing Effect and Micro-Cracks of Carbon Steel", Science of Machine, 27-8(Aug.,1975), pp.995—1000 (in Japanese).

[12] K.Takao and H.Nisitani,"Opening and Closing Behavior of Non-propagating Micro-Cracks in Rotating Bending and Torsional Fatigue Tests of Carbon Steels", Journ. of the Society of Materials Science, Japan, 28-311(Sept., 1979), pp.873—879 (in Japanese).

[13] W.Elber,"Fatigue Crack Closure Under Cyclic Tension",Engng.Fracture Mech., 2-1(Jan.,1970), pp.37—45.

[14] H.Nisitani and K.Takao,"Behavior of a Tip of Non-Propagating Fatigue Crack During One Stress Cycle",Engng.Fracture Mech.,6-2(Apr.,1974),pp.253—260.

[15] H.Nisitani and M.Kage,"Observation of Crack Closure Phenomena at the Tip of a Fatigue Crack by Electron Microscopy", Advances in Research on the Strength and Fracture of Materials, ICF IV, Pergamon Press, vol.2 B (1978), pp.1099—1108.

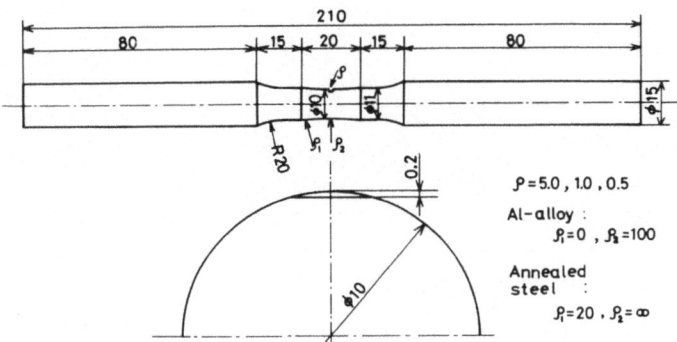

Fig.1   Dimensions of a specimen and details of a notched part

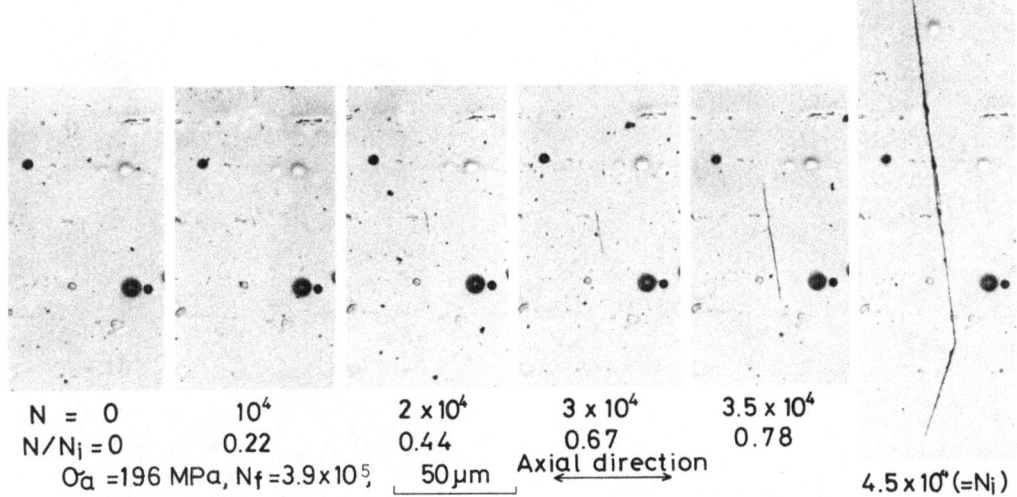

N = 0 10^4 2×10^4 3×10^4 3.5×10^4
N/N$_i$ = 0   0.22   0.44   0.67   0.78
σ_a =196 MPa, N$_f$=3.9$\times$10^5,   50 μm   Axial direction 4.5×10^4(=N$_i$)

Fig.2   The change in surface of Al-alloy

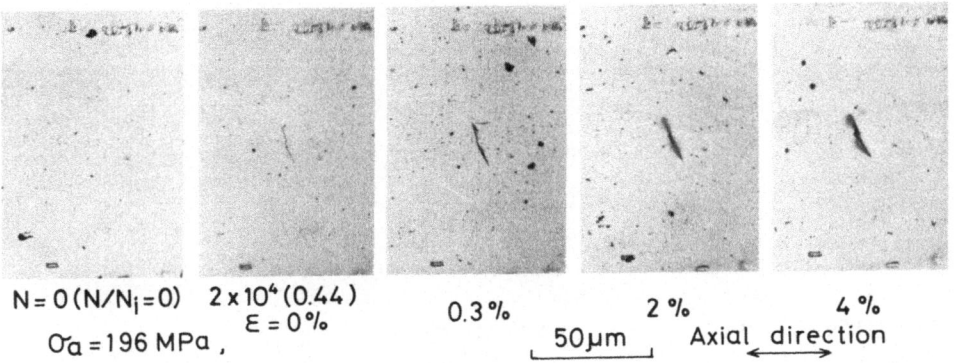

N = 0 (N/N$_i$=0) 2$\times$10^4(0.44)   0.3%   2%   4%
σ_a =196 MPa ,   ε = 0%   50μm   Axial direction

Fig.3   Behavior of the slip band of Al-alloy under static extension

278

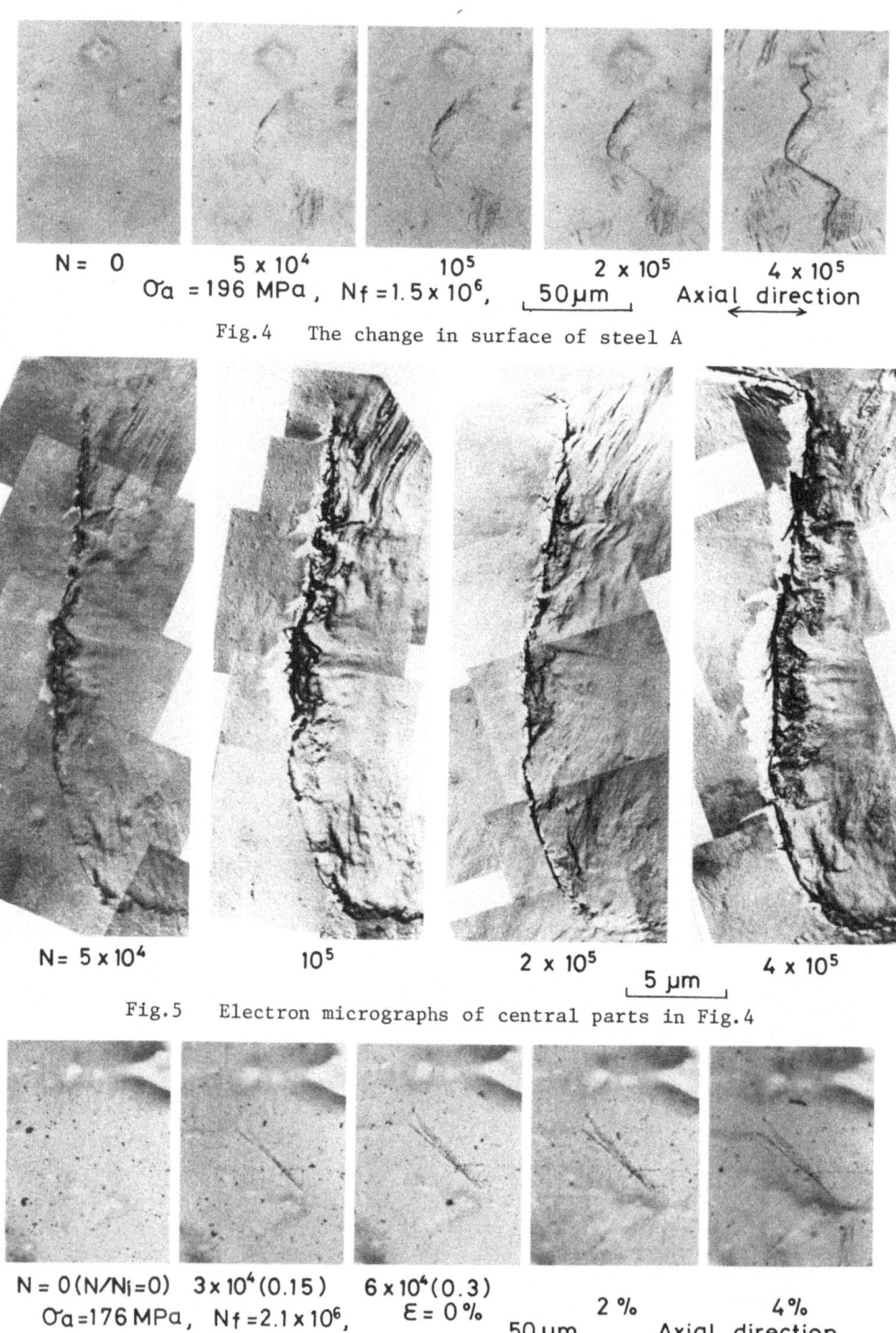

N = 0   5 x 10⁴   10⁵   2 x 10⁵   4 x 10⁵

σ_a =196 MPa , Nf =1.5 x 10⁶, |__ 50 μm __| Axial direction

Fig.4   The change in surface of steel A

N= 5 x 10⁴   10⁵   2 x 10⁵   4 x 10⁵

|__ 5 μm __|

Fig.5   Electron micrographs of central parts in Fig.4

N = 0(N/Nᵢ=0) 3x10⁴(0.15) 6x10⁴(0.3)   2 %   4%

σ_a =176 MPa , Nf =2.1 x 10⁶, ε = 0 % |__ 50 μm __| Axial direction

Fig.6   Behavior of the region to become a crack in future under static
          extension, steel B

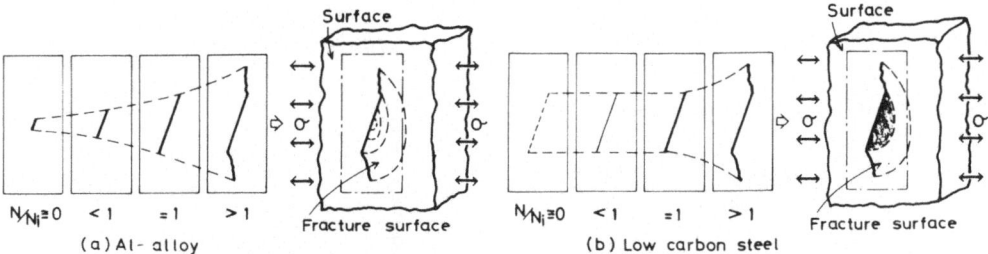

Fig.7 Schematic illustrations of crack initiation processes

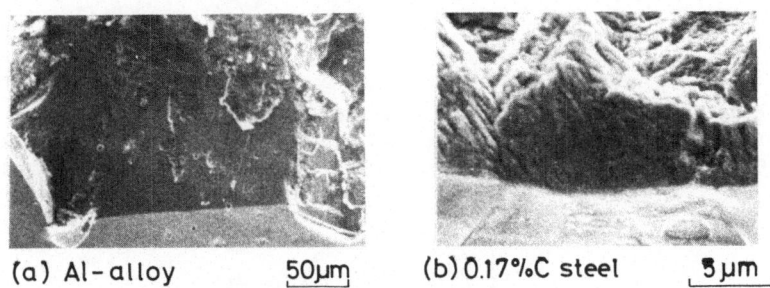

(a) Al-alloy 50μm (b) 0.17%C steel 5μm

Fig.8 Fractographs of the crack initiation regions

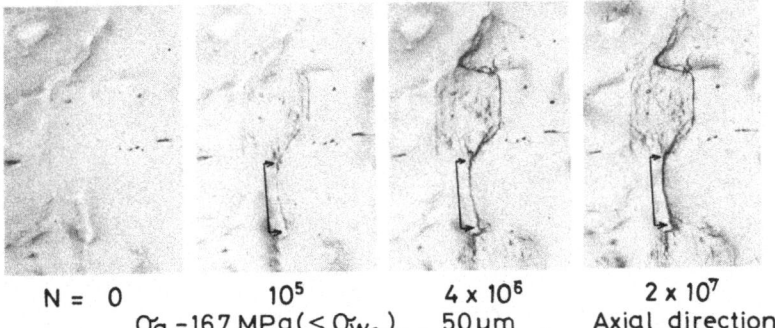

N = 0 10^5 4×10^6 2×10^7

$\sigma_a = 167$ MPa$(< \sigma_{w_0})$, 50μm   Axial direction

Fig.9 Non-propagating microcrack of steel A(⊏ means a crack initiation region)

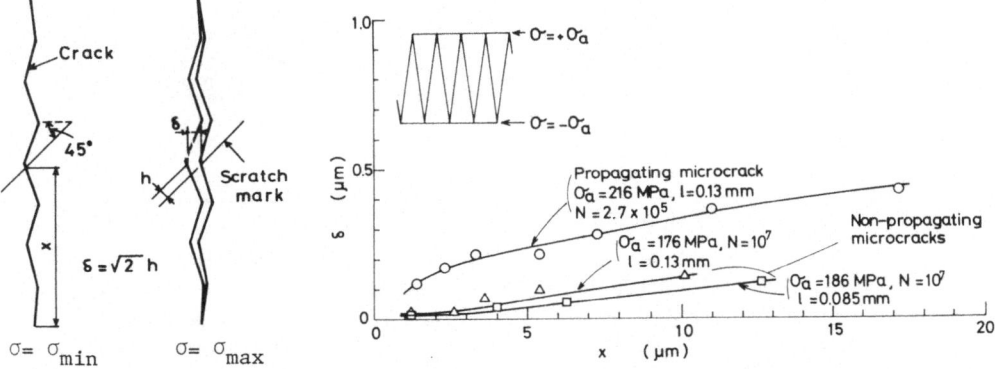

(a) COD measuring method   (b) COD at various distances x from the crack tip

Fig.10 Crack opening displacement(COD) of a propagating and non-propagating microcrack of steel C

280

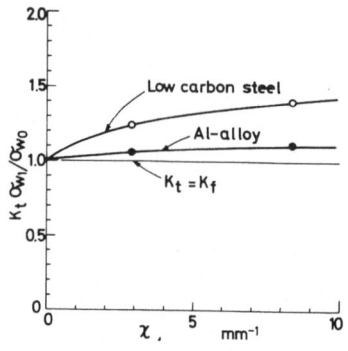 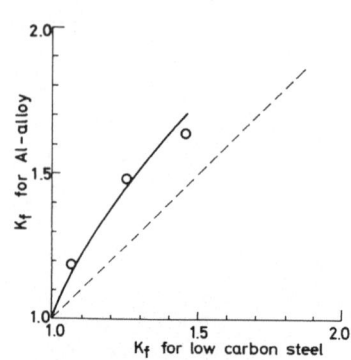

Fig.11 Relation between $K_t\sigma_{w1}/\sigma_{wo}$ and
        χ for Al–alloy and low carbon
        steel[2]

Fig.12 Comparison of notch sensitivi-
        ty between Al–alloy and low
        carbon steel(steel C)

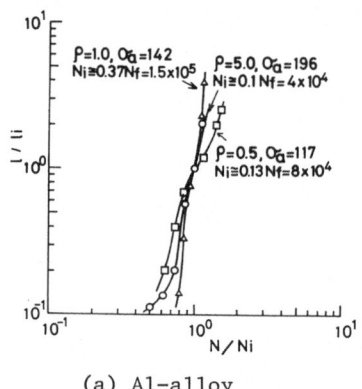

(a) Al–alloy

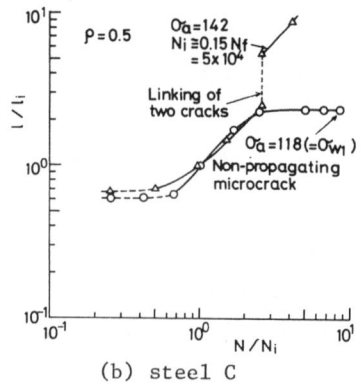 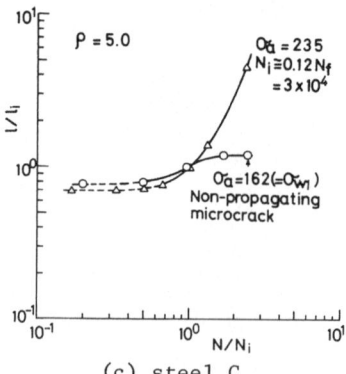

(b) steel C

(c) steel C

Fig.13 Crack propagation curves(l_i;crack length of one grain
        size, $l_i \cong 100$ μm for Al–alloy, $l_i \cong 50$ μm for steel C, N_i;
        number of cycles for the initiation of one grain size
        crack)

SOME CONSIDERATIONS OF LOW CYCLE FATIGUE IN METALS

K. HATANAKA
Associate Professor of Mechanical Engineering
Yamaguchi University, Tokiwadai 2557, Ube, Japan

T. YAMADA
Professor of Mechanical Engineering
Kyoto University, Yoshidahon-machi, Sakyo-ku, Kyoto, Japan

ABSTRACT

A new criterion for fatigue is proposed, based on an effective plastic strain of ε_{pe}, which is obtained by analyzing the profile of the stress-strain hysteresis loop. The fatigue life of N_f is represented by the simple equation: $N_f \cdot \varepsilon_{pe}$=Constant. This means that the constant value of the effective plastic strain, accumulated linearly, leads to fatigue failure.

The material's constant, α, in Manson-Coffin's law, $N_f^{\alpha} \cdot \Delta\varepsilon_p$=Constant, is also discussed in terms of the deformation mode. Consequently, the localized deformation has a remarkable effect on the quantity of α.

INTRODUCTION

It is well known that low cycle fatigue life is closely correlated with the plastic strain and is represented by Manson-Coffin's equation of $\Delta\varepsilon_p \cdot N_f^{\alpha}$ =C [1-3]. This empirical equation is constituted by the plastic strain range of $\Delta\varepsilon_p$, the number of cycles to failure of N_f, and the two material dependent constants of α and C, but its physical meaning is still being argued.

The plastic strain has been measured, even at the stress amplitude corresponding to the fatigue limit in metals [4-6]. This suggests that the existence of the plastic strain does not influence fatigue damage at that stress level. The present authors intend to find an effective component of the plastic strain and to propose a new criterion for fatigue in terms of this component. Using this, a reasonable explanation for Manson-Coffin's law will be given.

The present paper is also concerned with the discussion of α, a quantity which is strongly affected by the extent of the uniformity in the deformation produced in the specimen.

MATERIALS AND EXPERIMENTAL PROCEDURES

The materials tested are the commercial purity aluminum, Al-2.4% Mg alloy, copper, α-brass, and low carbon steel of 0.1% carbon content. All these materials were fully annealed; cold-working was also used only for Al-2.4% Mg alloy before test. The mechanical properties of the materials used are shown in Table 1. The fatigue specimen is machined to 6 mm in diameter at the gauge length of 10 mm.

The low cycle fatigue test was performed by the closed loop type push-pull testing machine under the condition of the fixed total strain range, using the sine wave of the cyclic frequency of 0.2 Hz. The other cyclic frequency of 50 Hz was also used for the high cycle and stress controlled fatigue test, which was performed in order to examine the plastic strain induced by the stress corresponding to the fatigue limit. Stress-strain hysteresis loops were obtained by simultaneously recording the load and elongation, measured by a differential transformer type extensometer, on the X-Y recorder.

TEST RESULTS AND DISCUSSIONS

Low cycle fatigue life

Figure 1 (a) and (b) plots the logarithm of the plastic strain range versus the logarithm of the number of cycles to failure in aluminum, copper, α-brass and low carbon steel, and Al-2.4% Mg alloy, respectively. The stable plastic strain range $\Delta\varepsilon_{ps}$ is the plastic strain range at the number of strain cycles of $1/2\ N_f$. Manson-Coffin's law is established for all materials, independent of the initial state of the materials, but it should be noted that the annealed Al-2.4% Mg alloy has the much larger exponent of 0.84, compared with the about 0.6 of the cold worked one and of the other test materials annealed.

A dominant factor controlling the exponent of α in Manson-Coffin's law

The value of the exponent of α might be closely related to the microscopic deformation mode caused by cyclic straining. Figure 2 (a),(b), and (c) shows the substructures formed on the specimen surface of copper, α-brass, and low carbon steel annealed, respectively. The substructures of the annealed and the cold worked Al-2.4% Mg alloys, low cycle fatigued at $\Delta\varepsilon_{ps}$ of 0.28%, are also represented in Fig.3(a) and (b). These figures indicate that the substructures of copper, α-brass, low carbon steel, and Al-2.4% Mg alloy cold worked have a comparatively uniform distribution of the slip line over the entire surface. The slip deformation is concentrated on and around the grain boundary under the test condition of the lower plastic strain range, in which fatigue crack initiates in Al-2.4% Mg alloy annealed. The concentrated and localized slip, as shown in Fig.3(a), might contribute to the slope of steep lines in log $\Delta\varepsilon_{ps}$

versus log N_f plot. Discussion of this point for Al-2.4% Mg alloys annealed and cold worked takes place in the following paragraphs.

The schematic model is illustrated to show the distribution of slip in Fig.4 (a) and (b), where γ_{pb}, γ_{pn}, and γ_{pm} denote the plastic shear strains at and near the grain boundary and inside the grain, respectively. The annealed Al-2.4% Mg alloy low cycle fatigued at the lower plastic strain range is charac-terized by the intense slip line at and near the grain boundary. This situa-tion produces the large amount of γ_{pb} and γ_{pn}, and a small one of γ_{pm}, shown in Fig.4(a). On the other hand, the slip lines in the Al-2.4% Mg alloy cold worked are uniformly dispersed and are independent of the region in the microstructure; the whole plastic shear strain can be represented only by γ_{pm}' in Fig.4(b). Then, the total plastic shear strain γ_p is given, using coefficients of ξ, β and ζ less than the unity by

$$\gamma_p = \xi\gamma_{pb} + \beta\gamma_{pn} + \zeta\gamma_{pm} \tag{1}$$

for Al-Mg alloy annealed, and by

$$\gamma_p = \gamma_{pm}' \tag{2}$$

for Al-Mg alloy cold worked. The total plastic shear strain γ_p is correlated with the axial plastic strain ε_p in a proportional manner. The larger and the smaller plastic shear strains of γ_{ph} and γ_{pl}, which correspond to the smaller and larger axial plastic strain ranges of $\Delta\varepsilon_{ph}$ and $\Delta\varepsilon_{pl}$, are introduced here. In addition, the quantities of γ_{pb}, γ_{pn}, and γ_{pm} corresponding to higher and lower plastic strain ranges are denoted as γ_{pbh}, γ_{pnh}, and γ_{pmh}, and γ_{pbl}, γ_{pnl}, and γ_{pml}, respectively.

For Al-2.4% Mg alloy annealed, the difference in the axial plastic strain range between $\Delta\varepsilon_{ph}$ and $\Delta\varepsilon_{pl}$ is represented by the equation

$$\Delta\varepsilon_{ph} - \Delta\varepsilon_{pl} = g(\gamma_{ph} - \gamma_{pl}) = g\{(\xi\gamma_{pbh} + \beta\gamma_{pnh}) - (\xi\gamma_{pbl} + \beta\gamma_{pnl}) + \zeta(\gamma_{pmh} - \gamma_{pml})\} \tag{3}$$

using the constant of g, which is geometrical and relates the microscopic to the macroscopic flow direction. Although the amount of γ_{pml} is small, compared with those of γ_{pbl} and γ_{pnl}, it contributes to $\Delta\varepsilon_{pl}$ through the shear deforma-tion occuring in the volume inside the grain. On the other hand, for the Al-Mg alloy cold worked the equation

$$\Delta\varepsilon_{ph} - \Delta\varepsilon_{pl} = g(\gamma_{ph} - \gamma_{pl}) = g(\gamma_{pmh}' - \gamma_{pml}') \tag{4}$$

is derived from the uniform distribution of slip lines. In the region of the higher plastic strain, the slip deformation trend is uniform even in Al-Mg alloy annealed. So all the shear strain components of γ_{pbh}, γ_{pnh}, and γ_{pmh}, and the two strain components of γ_{pbl} and γ_{pnl} might control the fatigue crack initia-tion in the higher and in the lower plastic strain ranges, respectively. On the other hand, in the cold worked one γ_{pm}' is the predominant plastic shear strain for fatigue crack initiation in the whole life region. The extent of such a localized deformation might increase as the controlled plastic strain decreases.

The schematic model illustrates the difference between the annealed and the cold worked Al-Mg alloys in the slope of the fatigue life curve in the log-log plot in Fig.5. The broken line is for the cold worked one, and the solid line for the annealed one. The solid line might be qualitatively derived from the broken line as follows:

In Fig.5 N_{fh} and N_{flc}, the numbers of cycles to failure, correspond to the axial plastic strain ranges of $\Delta\varepsilon_{ph}$ and $\Delta\varepsilon_{pl}$ in Al-Mg alloy cold worked. From eq.(3), an inequality of

$$\Delta\varepsilon_{ph} - \Delta\varepsilon_{pl} > g\{(\xi\gamma_{pbh} + \beta\gamma_{pnh}) - (\xi\gamma_{pbl} + \beta\gamma_{pnl})\} \tag{5}$$

is derived. Therefore, under the condition of the uniform deformation (as in Al-Mg alloy cold worked), the change in fatigue life resulting from $g\{(\xi\gamma_{pbh} + \beta\gamma_{pnh}) - (\xi\gamma_{pbl} + \beta\gamma_{pnl})\}$, which is given by the difference in the number of cycles at points of A and B, is much smaller than $(N_{flc} - N_{fh})$. On the other hand, the increase in fatigue life caused by the decrease in the plastic strain range from $\Delta\varepsilon_{ph}$ to $\Delta\varepsilon_{pl}$ is mainly due to $g\{(\xi\gamma_{pbh} + \beta\gamma_{pnh}) - (\xi\gamma_{pbl} + \beta\gamma_{pnl})\}$ in Al-Mg alloy annealed. Since the amount of $(\Delta\varepsilon_{ph} - \Delta\varepsilon_{pl})$ is the same both for Al-Mg alloys annealed and cold worked, the annealed one has the fatigue life equal to N_{fla} at $\Delta\varepsilon_{pl}$. Consequently, the fatigue life curves in Al-Mg alloy annealed and cold worked are represented by the lines $\overline{AC}$ and $\overline{AD}$ in Fig.5, respectively.

Lukas and Klesnil [6] pointed out that the log $\Delta\varepsilon_p$ - logN_f plot in metals cannot be represented by one straight line over the whole region of fatigue lives, but by two straight lines which have lower and higher slopes. The former shows the fatigue life curve in a usual low cycle and in a high plastic strain fatigue region, and the latter the curve in a low plastic strain region. This result might be interpreted by considering the difference in the extent of uniformity in the deformation in the gage length of the specimen. For instance, the high strain cycling produces comparatively uniform deformation, while the localized deformation occurs under the condition of the low plastic strain range. The present result of Al-Mg alloy cold worked also shows that the slope of the fatigue life curve in the log-log plot changes from -0.64 in the region of fatigue lives less than 3×10^3 cycles to -0.84 for fatigue lives more than this point. Ti-6Al-4V alloy, in which the slip is limited on the prism plane of [01$\bar{1}$0] type [7], has the character of uneven deformation, even for the high strain, as in low cycle fatigue. This situation leads to the large slope of about -1.0 in the log $\Delta\varepsilon_{ps}$ - logN_f representation of fatigue life[8].

All these test results supply solid evidence for the present analysis.

An effective plastic strain for fatigue

The plastic strain has been measured, even in the specimen stress cycled
at fatigue limit[4-6]. Furthermore, the cumulative plastic strain and the cumulative plastic strain energy to failure are dependent on plastic strain range,
and they increase much more under the test condition of the lower plastic strain
range[5, 9-11]. These facts suggest that the plastic strain induced by the
cyclic deformation consists of both the harmful and harmless components of
plastic strain for fatigue. A new method is proposed here to separate the det-
rimental one, which is hereafter referred to as an effective plastic strain,
from the plastic strain through the analysis of the profile of the stress-strain
hysteresis loop.

Figure 6 shows the coordinate systems set up for analyzing the profile of
the hysteresis loop. The flow stress σ_s' and the plastic strain ε_{ps}' are determined on the profile of the upper part of the stable stress-strain hysteresis
loop, based on the coordinate system $\sigma' - o' - \varepsilon'$. The $\log \sigma_s'$ versus $\log \varepsilon_{ps}'$
plot is given for one plastic strain range in Fig.7. The data might be express-
ed for all materials as a line consisting of two straight portions with a bend
point between them, as also shown in other papers[12-14]. The value of $(\varepsilon_{psmax}'$
$- \varepsilon_{psd}')/\varepsilon_{psmax}'$, where ε_{psmax}' and ε_{psd}' are plastic strains at the upper end and
at the bend point on the profile of the stress-strain hysteresis loop, becomes
much larger in the order of aluminum, low carbon steel, copper, and α-brass depending on the kind of materials. Figure 8 shows such $\log \sigma_s' - \log \varepsilon_{ps}'$ relations for various plastic strain ranges in copper. All profiles are represented
by bend lines, as shown in Fig.7, and the values of ε_{psd}' and $(\varepsilon_{psmax}' - \varepsilon_{psd}')/$
ε_{psmax}' decrease as the plastic strain range becomes smaller.

The physical meaning of this point cannot be clearly explained at present.
The much larger rate of increase in the plastic strain, however, might indicate
that the density and the velocity of the mobile dislocations are greatly increased by the stress beyond the bend point, and the plastic strain below this
point might be caused by the reversible motion of the pinned dislocations, like
the bowing before the critical radius is attained and by the reverse motion of
piled up dislocations due to the back stress at the stress reversal, which cause
the Bauschinger strain. The dislocation motion in the latter does not seem to
cause much fatigue damage, however, compared to that in the former. In the
present paper, the plastic strain beyond the bend point on the loop profile is
defined as an effective plastic strain, ε_{pe} for fatigue, and so an effective
plastic strain ε_{pe} is given as $(\varepsilon_{psmax}' - \varepsilon_{psd}')$, using the above notations.
Now, the ratio of the effective plastic strain to the stable plastic strain

range,

$$\lambda = \varepsilon_{pe}/\Delta\varepsilon_{ps} \tag{6}$$

is introduced. The plastic strain range of $\Delta\varepsilon_{pw}$, yielding at the fatigue limit, does not damage materials since they can endure that stress level for infinite number of cycles. Then, ε_{pe} and λ are zero for $\Delta\varepsilon_{ps}$ equal to $\Delta\varepsilon_{pw}$. Table 2 gives the values of $\Delta\varepsilon_{pw}$ which is defined here as maximum plastic strain range making materials survive up to about 10^7 stress cycles and is obtained from high cycle fatigue tests. The ratio of λ is plotted against $\Delta\varepsilon_{ps}$, taking account of these $\Delta\varepsilon_{pw}$ in Fig.9. This value reduces with decrease in $\Delta\varepsilon_{ps}$ and attains to zero at $\Delta\varepsilon_{pw}$. This figure λ also approaches unity for the large value of $\Delta\varepsilon_{ps}$. Taking these experimental facts into account, λ can be empirically expressed for $\Delta\varepsilon_{ps}$ by the following equation.

$$\left.\begin{array}{llll}\lambda = 1 - \exp[-A(\Delta\varepsilon_{ps} - \Delta\varepsilon_{pw})] & \text{for} & \Delta\varepsilon_{ps} > \Delta\varepsilon_{pw} \\ \lambda = 0 & \text{for} & \Delta\varepsilon_{ps} \leq \Delta\varepsilon_{pw}\end{array}\right\} \tag{7}$$

Analysis of fatigue life by effective plastic strain

As stated before, the plastic strain $\Delta\varepsilon_{pw}$ occurs in the specimen stress cycled at the fatigue limit, and so, the present authors propose to modify Manson-Coffin's law into the following equation so that the fatigue life around the fatigue limit can be expressed in terms of the plastic strain.

$$(\Delta\varepsilon_{ps} - \Delta\varepsilon_{pw}) \cdot N_f^{\alpha} = C \tag{8}$$

Combining eqs. (6), (7), and (8)

$$\lambda = \varepsilon_{pe}/\Delta\varepsilon_{ps} = 1 - \exp[-A(C/N_f^{\alpha})] \tag{9}$$

Substituting from eq.(8) for $\Delta\varepsilon_{ps}$ in eq.(9) gives

$$\varepsilon_{pe} = (C/N_f^{\alpha} + \Delta\varepsilon_{pw})\{1 - \exp(-AC/N_f^{\alpha})\} \tag{10}$$

$\Delta\varepsilon_{ps}$ and ε_{pe} are plotted against N_f, based on eqs.(8) and (10) in Fig.10. The test results in Fig.1(a) are also replotted in this figure. The curves evaluated by eq.(8) agree with those in Manson-Coffin's law in the plastic strain ranges at which N_f is less than 10^6 cycles and have a reasonable shape around the fatigue limit. On the other hand, the log ε_{pe}-log N_f plot obtained from eq.(10) makes a linear relation of a slope of about -1 over a wide range of fatigue lives from 10^3 to 10^7 cycles. This establishes an equation of $\varepsilon_{pe} \cdot N_f =$ Constant, which shows that the fatigue failure results from the linear accumulation of effective plastic strain in that life range. More notice should be given to the fact that the width among fatigue life curves, dependent upon materials, becomes smaller in a log ε_{pe}-log N_f representation than in a log $\Delta\varepsilon_{ps}$-log N_f one. The effective plastic strain might be an essential parameter for fatigue which decreases the difference in fatigue lives among materials.

Now, the interrelation between α in Manson-Coffin's law and the slope of the line given by the equation of $\varepsilon_{pe} \cdot N_f =$ Constant will be discussed. The exponential term in the right side of eq.(10) is expanded using Taylor's formula. Neglecting the small quantities on and after the fourth term in the expansion,

$$\varepsilon_{pe} \doteqdot (C/N_f + \Delta\varepsilon_{pw})(AC/N_f^\alpha - A^2C^2/2N_f^{2\alpha}) = \Delta\varepsilon_{pw}AC/N_f^\alpha + AC^2/N_f^{2\alpha}$$
$$-\Delta\varepsilon_{pw}A^2C^2/2N_f^{2\alpha} - A^2C^3/2N_f^{3\alpha} \tag{11}$$

is obtained. For N_f larger than 10^3 cycles, the third and the fourth terms are also negligible in eq.(11), and this equation can be simplified to

$$\varepsilon_{pe} \doteqdot \Delta\varepsilon_{pw}AC/N_f^\alpha + AC^2/N_f^{2\alpha} \tag{12}$$

Since the magnitude of $\Delta\varepsilon_{pw}$ is in the order of 10^{-5},

$$\varepsilon_{pe} \doteqdot AC^2/N_f^{2\alpha} \quad\text{or}\quad \varepsilon_{pe} \cdot N_f^{2\alpha} \doteqdot AC^2 \tag{13}$$

is attained for N_f less than 10^7 cycles. Taking into account the value of α nearly equal to 1/2,

$$\varepsilon_{pe} \cdot N_f \doteqdot AC^2 = \text{Constant} \tag{14}$$

is derived from eq.(13). The linear lines of the slope nearly equal to -1, which are shown in the life range from 10^3 to 10^7 cycles in Fig.10, can be interpreted by eq.(14).

Paradoxically, Manson-Coffin's law, characterized by an exponent of α nearly equal to 1/2, means that fatigue failure results from the linear accumulation of the effective plastic strain as represented by eq.(14). Therefore, the fatigue criterion proposed here can give a physical meaning to Manson-Coffin's law.

Feltner et al.[4], Martin[10], and Gatts[15] proposed the criterion for fatigue failure, based on the conception that the anelastic component of the inelastic strain does not cause fatigue damage in the specimen. These ideas seem similar to the present suggestion, but as a matter of fact, a fundamental difference exists between them. In the former, anelastic strain occurs just near the fatigue limit or at the yield point in the tensile stress-strain curve, and it is the material's constant, independent of the plastic strain amplitude. In the latter, this becomes larger as the plastic strain amplitude increases. The present analysis is based on the reasonable conception that the anelastic strain, ineffective to the fatigue damage, changes, depending on the plastic strain amplitude just as in the Bauschinger strain.

CONCLUSIONS

The low cycle fatigue tests were performed on the five metal materials: commercial purity aluminum, Al-2.4% Mg alloy, copper, α-brass, and low carbon steel.

The stress-strain relation on the profile of the stress-strain hysteresis loop is represented as a line consisting of two straight portions with a bend point between them on the log-log plot. A new criterion for low cycle fatigue

288

was proposed, based on a component of the plastic strain beyond this bend point on the loop profile. The fatigue life N_f is represented as a simple equation of $N_f \cdot \varepsilon_{pe}$=Constant by this plastic strain component, which is defined as the effective plastic strain ε_{pe}. This means that fatigue failure is caused by the linear accumulation of the effective plastic strain during strain cycling. In addition, Manson-Coffin's law is also derived from this conception.

The Al-2.4% Mg alloy annealed has a large exponent of 0.84 while α in the cold worked one is almost the same value, 0.64, as that in aluminum, copper, α-brass, and low carbon steel annealed. The difference in the value of α can be interpreted by considering the deformation mode. For example, the slip is concentrated at and near the grain boundary in the former and is dispersed uniformly in the latter. This conception also effectively explains the large slope of the fatigue life curve in Ti-6Al-4V alloy and in the region of the low plastic strain in other metal materials.

REFERENCES

(1) S.S.Manson, "Behavior of Materials under Conditions of Thermal Stress", NASA Tech. Note, No. 2933, (1953).

(2) L.F. Coffin, "A Study of the Effects of Cyclic Thermal Stresses on a Ductile Metal", Trans. ASME, 76, (1954), pp.931-950.

(3) J.F. Tavernelli and L.F. Coffin, "A Compilation and Interpretation of Cyclic Strain Fatigue Tests on Metals", Trans. ASM, 51, (1959), pp.438-453.

(4) C.E. Feltner and JoDean Morrow, "Micro-Plastic Strain Hysteresis Energy as a Criterion for Fatigue Fracture", Trans. ASME, Ser.D, 83, (1961), pp.15-22.

(5) M. Kikukawa, M. Jono, and J. Song, "The Cyclic Plastic Strain and Cumulative Fatigue Damage (Fatigue Damage Caused by the Stress below the Fatigue Limit)". J. Soc. Materials Sci., Japan, 21 (1972), pp.753-758, (in Japanese).

(6) P. Lukas and M. Klesnil, "Cyclic Stress-Strain Response and Fatigue Life of Metals in Low Amplitude Region", Mat. Sci. Eng., 11, (1973), pp.345-356.

(7) K. Hatanaka and A.J. McEvily, Unpublished Work.

(8) R.K. Steele and A.J. McEvily, "The High-Cycle Fatigue Behavior of Ti-6Al-4V Alloy", Eng. Frac. Mech., 8, (1976), pp.31-37.

(9) JoDean Morrow, "Cyclic Plastic Strain Energy and Fatigue of Metals", ASTM-STP, No. 378, (1965), pp.45-87.

(10) D.E. Martin, "An Energy Criterion for Low-Cycle Fatigue", Trans. ASME, Ser. D, 83, (1961), pp.565-571.

(11) K. Hatanaka, T. Yamada, and Y. Hirose, "An Effective Plastic Strain Component for Low Cycle Fatigue in Metals", Trans. Japan Soc. Mech. Engrs., 45, (1979), pp.1125-1134, (in Japanese).

(12) H.A. Moreen, "Strain Cycling Effects in 1100 Aluminum", Trans. ASME, Ser.D, 92, (1970), pp.126-130.

(13) S. Taira, T. Goto, T. Kurobe, and Y. Kasai, "A Study on Cyclic Stress-Strain Relation in Annealed Low Carbon Steel", <u>J. Soc. Materials Sci., Japan</u>, <u>21</u>, (1972), pp.83-89, (in Japanese).

(14) K. Hatanaka and T. Yamada, "Cyclic Stress-Strain Relation Induced by Change in Strain Amplitude and Its Effect on Fatigue Life", <u>Proc. 18th Japan Cong. on Materials Res.</u>, (1975), pp.30-36.

(15) R.R. Gatts, "Application of a Cumulative Damage Concept to Fatigue", <u>Trans. ASME</u>, Ser.D, <u>83</u>, (1961), pp.529-540.

Table 1. Mechanical properties of materials used.

	0.2% Proof stress MPa	Tensile strength MPa	Fracture strain %
Aluminum	21·9	78·5	183
Copper	25·3	212·8	131
α-brass	66·9	292·2	179
Low carbon steel	236·3	371·6	140
Al-2.4%Mg alloy annealed	80·4	385·3	127
Al-2.4%Mg alloy cold worked	215·7	237·3	97

Table 2. Plastic strain range measured at fatigue limit and values of A in eq.(7).

	$\Delta\varepsilon_{pw}$ %	A
Copper	0·0078	60·8
α-brass	0·0049	30·4
Aluminum	0·0042	124·3
Low carbon steel	0·0049	77·9

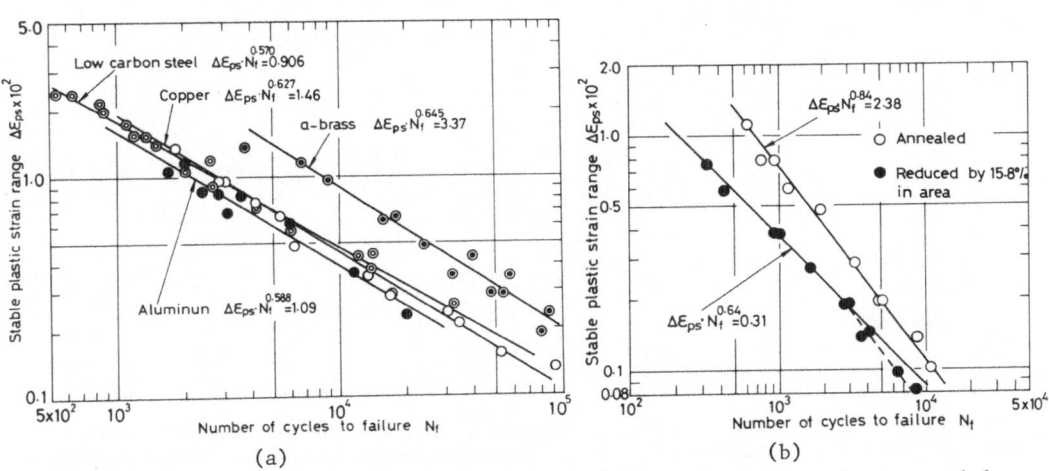

(a)   (b)

Fig.1. Low cycle fatigue life curves in (a) aluminum, copper, α-brass, and low carbon steel, and (b) Al-2.4% Mg alloy.

290

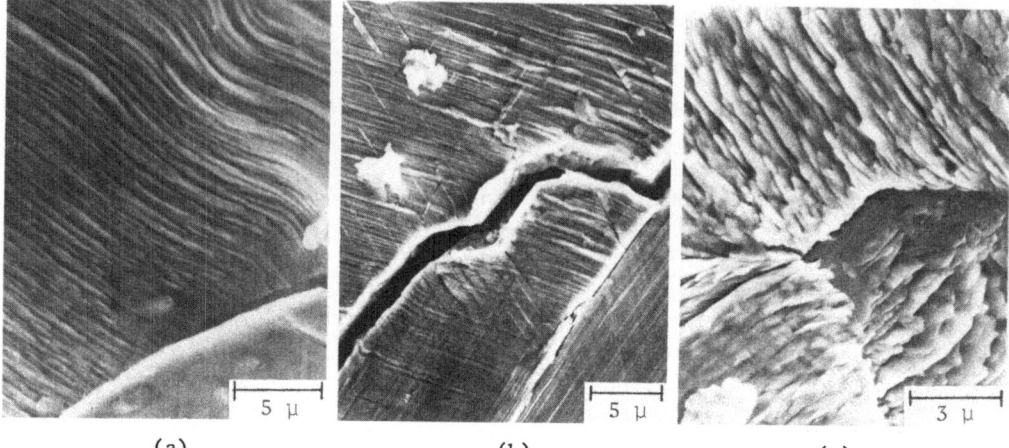

(a)   (b)   (c)

Fig.2. Comparatively uniform distribution of slip lines in (a) copper, (b) α-brass, and (c) low carbon steel low cycle fatigued at $\Delta\varepsilon_{ps}$ of 0.38%.

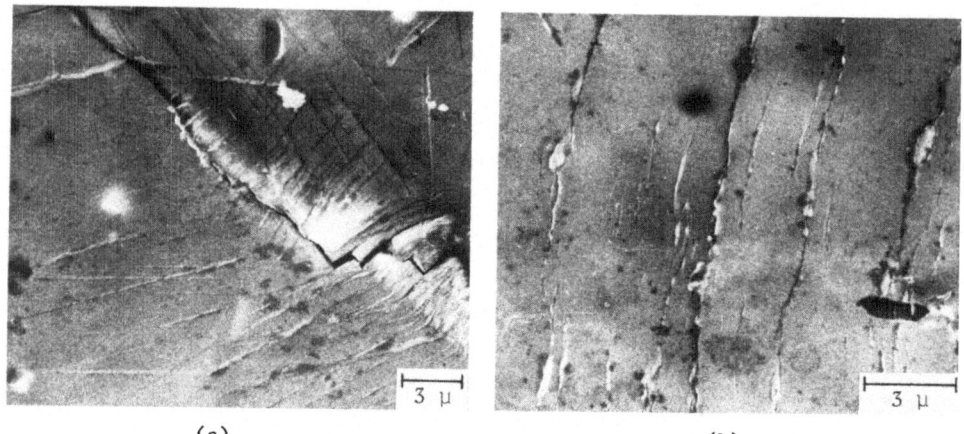

(a)   (b)

Fig.3. Micro-structures in Al-2.4% Mg alloy (a) annealed and (b) cold worked, which are low cycle fatigued at $\Delta\varepsilon_{ps}$ of 0.28%.

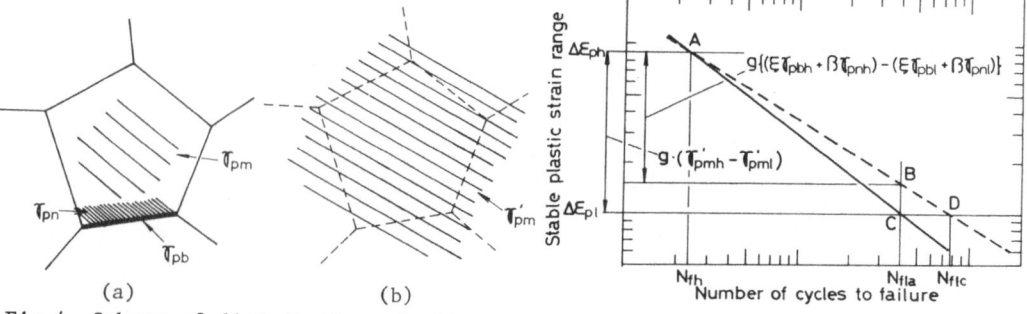

(a)   (b)

Fig.4. Schema of distribution of slip deformation in Al-2.4% Mg alloy (a) annealed and (b) cold worked.

Fig.5. Schema for explaining the difference in the slope of fatigue life curve in Al-2.4% Mg alloy.

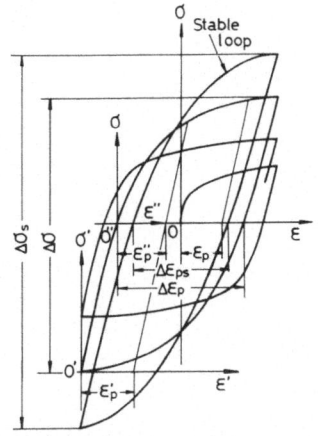

Fig.6. Coordinate systems set up for analysis of stress-strain hysteresis loop.

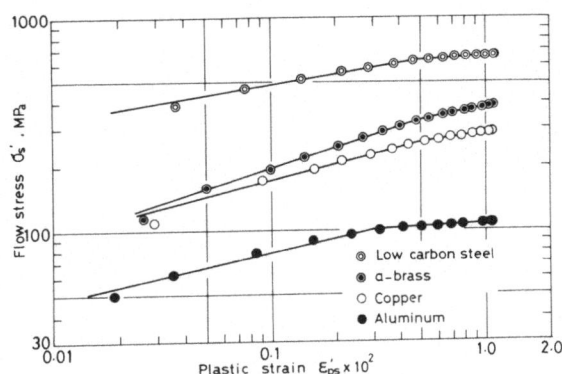

Fig.7. Analysis of profile of stress-strain hysteresis loop in aluminum, copper, α-brass, and low carbon steel.

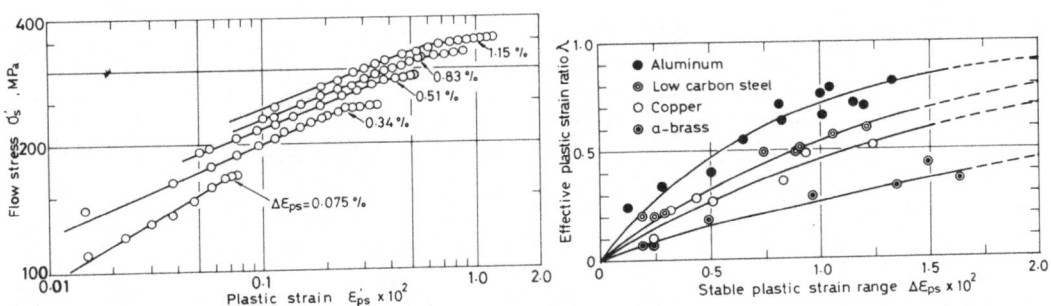

Fig.8. Analysis of loop profile, dependent on plastic strain range in copper.

Fig.9. λ plotted against plastic strain range.

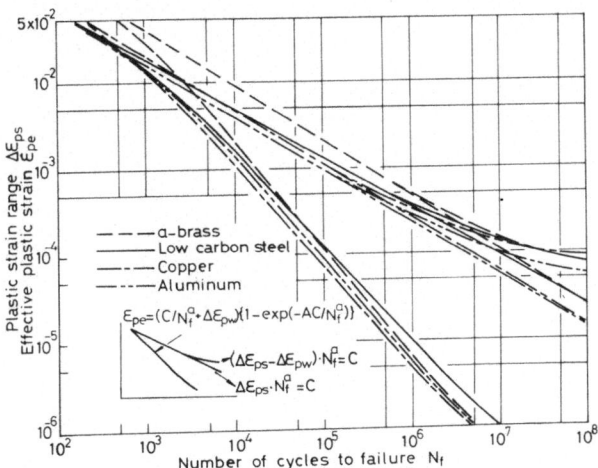

Fig.10. Fatigue life curves evaluated by eqs.(8) and (10).

FATIGUE STRENGTH OF BUTT-WELDED JOINTS
WITH ANGULAR DISTORTION

Kunihiro IIDA
Professor, Dr.-Eng., Dept. of Naval Architect
Faculty of Engineering, University of Tokyo
3-7-1 Hongo, Bunkyo-ku, Tokyo, Japan 113

Nobu IINO
Research Staff, Structure and Strength Dept.,
Research Institute
Ishikawajima-Harima Heavy Industries Co., Ltd.
3-1-15, Toyosu, Koto-ku, Tokyo, Japan 135-91

ABSTRACT

The fatigue strength of butt-welded joints having angular distortion is investigated for the amount of angular distortion, overstressing prior to fatigue loading, and the distance between loading points. The cyclic strain range, measured near the toe of the reinforcement of weld, correlate with the fatigue life regardless of the amount of angular distortion. A simple elastic-plastic analysis was used successfully to quantity the behavior of the joint and to estimate the fatigue life of such members with different geometry.

INTRODUCTION

The fatigue strength of a structural member having butt-welded joints is affected by such factors as the size of the reinforcement of the weld, the flank angle, and the radius of the weld, which determine the stress and strain concentration at the location of possible crack origin. In practice, fabrication error, such as the offset from the aligned section and the angular distortion, cannot be completely avoided. These errors also augment the degree of stress and strain concentration, which deteriorates the fatigue properties of the welded joint. In fact, this condition makes it necessary for the codes and structural design standards to incorporate specific limiting values which will guarantee sound operation during a given period. Some examples of the restrictions upon the amount of deviation from a normal profile are given in Ref. [1]-[3]. To the authors' knowledge, no papers have confirmed the background for the determined values with regard to fatigue strength.

Butt-welded joints of two commercial high strength steels were used to investigate the effect of the initial angular distortion, tensile overstress-ing, and the distance between loading points or the span on fatigue properties. A simple elastic-plastic analysis originally proposed by Yada, et al. [4] for static loading is modified and extended to investigate the behavior under re-peated loading. It is shown that this approach is effective in assessing the safety of structural members with butt-welded joints with regard to fatigue.

EXPERIMENT

Material and specimen geometry

Commercially available twenty millimeter thick, high strength steels WES (The Japan Welding Engineering Society Standard) HW50, and HW70 were supplied to prepare butt-welded specimens. The chemical composition and mechanical properties are given in Table 1. The dimensions and geometry of fatigue speci-mens are shown in Fig. 1. The amount of angular distortion is introduced by initially setting the plates in position so that the completed joints have the required amount of angular distortion. Welding conditions are given in Table 2. The ends of specimens are press-formed so they align with the grip, and specimens were checked to have no deteriorating defects.

Test procedures

Factors to be investigated and the number of specimens tested are shown in Table 3. Tests were run on a closed loop hydraulic machine, controlling the load with nominal load ratio of R = 0.1. The longitudinal strain by bond-ing 3mm gages at a distance 5mm from the toe of the weld was measured in each specimen to obtain load-strain relationship. Crack initiation life was observed by detecting a crack at the toe of weld. But the detected size of the crack varied between 2 to 4 mm, due to complex geometry of the toe. Since a large scatter may be expected in correlating the crack initiation life with other variables, the failure life was employed for fatigue analysis.

Results

The fatigue test results of butt-welded joints for two materials are shown in Fig. 2 and Fig. 3. In the figures, the number of cycles to failure is plot-ted against the nominal stress range; the results show that the fatigue streng-th decreases as the amount of angular distortion increases. In Fig. 4, the results are shown for the fatigue tests of butt-welded joints with angular distortion in which overstressing was applied prior to fatigue. A pre-stress level of 1.25 times that of the maximum stress in subsequent fatigue was ap-plied in each case. The results were compared to those without preloading and showed a tendency to increase fatigue strength in the higher stress level region, but hardly any improvement is observed in the lower stress level.

Figure 5 shows the results on butt-welded joints (HW70) without angular distortion. Despite the scatter in the lower stress range, no apparent difference in fatigue strength between those with and without overstress is observed over the stress range investigated.

The location of fatigue crack initiation in all specimens reported was at the toe of the weld reinforcement where the ripples of the weld and the small undercutting into base plate were the source of strain concentration [6]. Since multiple points with similar conditions appear along the toe, a possibility for a more or less a simultaneous crack initiation is expected. The initiated fatigue cracks extended in the width direction, first along the toe through a linking of small surface cracks; then it proceeded to grow in the thickness direction. Correlation between the observed crack initiation and the failure lives for specimens with various types and amounts of fabrication errors are plotted in Fig. 6 for HW70. Note that regardless of the amount of errors, data fall within a reasonable scatter of the following expression.

$$N_c = 0.0982 \times N_f^{1.13} \tag{1}$$

Strain behavior observed in each specimen shows that the mean strain tended to increase with the number of load cycles. No information confirms that this was the actual behavior of the material or that it was due to properties of the gage. Despite of the drift of the mean strain, the strain range remained nearly constant before crack initiation. Strain range against failure life is plotted and shown in Fig. 7. Despite a wide range of initial distortion, a strong correlation exists between the measured strain range and the failure life for HW70.

$$\Delta\varepsilon = 1.833 \times 10^{-2} \, N_f^{-0.19} \tag{2}$$

The data plotted include specimens with different amounts of angular distortion, span, and pre-stressing conditions.

ANALYSIS

The elastic-plastic behavior of initially distorted members is a complex problem. An exact solution for the stress concentration factors at the joint, initially distorted under large deflection, can be obtained for a pure elastic case. But the exact solution for an elastic-plastic case is not available. Yada, et al. [4] proposed a simple approximation method to analyze large deflection elastic-plastic problems that expresses the brittle fracture initiation characteristics from initially distorted weld joints.

A summary of the approach follows, the extension of which is applied to analyze the fatigue test results conducted here. The stress-strain relation for a member having distortion (Fig. 8) is obtained. To satisfy the equilibrium of the force, the following relation holds:

$$\frac{dM}{dx} + N \frac{dy}{dx} = 0 \tag{3}$$

where N applied load/unit width

$$= \sigma_m \cdot t$$

       t thickness of plate

      σ_m membrane stress

      y $y(x)$ is deformation when σ_m is applied

M is the bending moment, expressed by following equation.

For elastic range,

$$M = \frac{Et^3}{12(1-\nu^2)} \frac{d^2(y-\omega)}{dx^2} \tag{4}$$

and for fully plastic range,

$$M = \frac{\psi t^3}{12(1-\mu^2)} \frac{d^2(y-\omega)}{dx^2} \tag{5}$$

where E Young's modulus

       ν Poisson's ratio

       ψ tangent modulus for plastic range

       μ 0.5

By knowing the initial amount of distortion and the boundary condition, the stress, strain, and deformation at different locations are obtained for given stress σ_m.

1) $0 \le \sigma_m \le \sigma_1$

$$\varepsilon_f = \varepsilon_m + \varepsilon_b \tag{6}$$

$$\varepsilon_m = \sigma_m/E \tag{7}$$

$$\varepsilon_b = 6(1-\nu^2) \, \varepsilon_m \, \frac{\omega_0}{t} \, \kappa_m \tag{8}$$

$$\text{where } \kappa_m = \frac{\tanh m}{m} \text{ for supported ends and} \tag{9}$$

$$= \frac{\tanh (m/2)}{m} \text{ for fixed ends} \tag{10}$$

$$m = 12(1-\nu^2) \, \frac{\sigma_m}{E} \, (\frac{1}{t})^2 \tag{11}$$

$$\omega = 2\omega_0 \kappa_m \text{ for fixed ends} \tag{12}$$

$$= \omega_0 \kappa_m \text{ for supported ends}$$

Here σ_1 corresponds to the applied stress where the concave surface strain attains yield level.

2) $\sigma_1 \leq \sigma_m \leq \sigma_2$

For the sake of a simplified analysis, a bi-linear stress-strain curve in the elastic-plastic region is reconstructed by a parabolic curve passing through the following three points (Fig. 10).

P_1 : point, corresponding to the stress level where σ_f reaches yield

P_3 : point, corresponding to the stress level where σ_b reaches yield (joint becomes fully plastic)

P_2 : a mid-point between P_Y and P_M the latter being the mid-point be-tween P_1 and P_3. Then, P_2 is defined as

$$P_2 = (2P_Y + P_1 + P_3) \, /4$$

ω is given by calculating the moment M through the stress distribution at the joint and using

$$\omega = M/(\sigma_m \cdot t) \tag{13}$$

3) $\sigma_2 \leq \sigma_m$

$$\varepsilon_f = \varepsilon_m + \varepsilon_b \tag{14}$$

$$\varepsilon_m = \varepsilon_Y + \frac{1}{H} \frac{1 - \mu^2}{1 - \nu^2} (\sigma_m - \sigma_Y) \tag{15}$$

ε_b and ω are obtained by replacing E and ν by H and μ in Eqs. (8) through (11).

DISCUSSION

The analysis proposed by Yada, et al. was extended to repeated loading; the calculated stress-strain curve is shown in Fig. 11, together with the strain measured from specimens with the same condition. The measured strain coincides with the calculation. The relation between the stress and strain ranges was then calculated and given in Fig. 12, together with the experimen-tally observed strain for a specimen with various amounts of angular distortion. The strain measured 5mm from the toe of the weld is considered a representative value in describing the deformation of butt-welded joints, irrespective of the amount of angular distortion. But the so-called the hot-spot value is not.

The effect of the span length was considered and calculated. The results are given for the same amount of angular distortion per unit length but a dif-ferent span (Fig. 13). The estimated fatigue life is plotted, together with the experimental results (Fig. 14). In the case calculated, short span is in effect somewhat equivalent to smaller amount of angular distortion.

If the relation between the strain range and the fatigue life for a butt-welded joint is given, the use of simple elastic-plastic analysis yields the prediction of fatigue strength for welded joints with a given amount of angular

298

distortion.

CONCLUSION

Load controlled fatigue tests were performed on butt-welded joints with various amounts of angular distortion. The effects of the amount of angular distortion, span and overstress prior to fatigue were investigated. Elastic-plastic analysis was performed, and the strain was in good agreement with the representative strain measured at the location 5mm from the toe of the weld. It is concluded that the fatigue of butt-welded joints with angular distortion can be assessed by the method proposed here.

ACKNOWLEDGEMENT

Experimental work was conducted as a part of the cooperative program for the Safety Research Committee for High Tensile Steel of the Institution for the Safety of High Pressure Gas Engineering.

REFERENCES

1. BS 2654, Specification for Vertical Steel Welded Storage Tanks with Butt-Welded Shells for the Petroleum Industry.

2. ASME, Boiler and Pressure Vessel Code, Section III, Div. 1, Subsection NB4232. (1977)

3. "Specification for Spherical Storage Tanks for High Pressure Gas Engineering, KHK-S-0201 (in Japanese).

4. Y. Akita, T. Maeda, T. Yada, and K. Sakai, "Effect of Angular Distortion in Welded Joints on Brittle Fracture Initiation", IIW, Doc. X-569-70 (1970).

5. T. Yada, "On Brittle Fracture Initiation Characteristics in Welded Structures", Jap. Soc. Nav. Arch., 119 (1966), pp. 134-141 (in Japanese).

6. K. Iida and N. Iino, "Effect of Angular Distortion on Fatigue Strength of Transverse Butt-Welds in High Strength Steels", Trans. Japan Welding Soc., 8-2, (1977)

Table 1. Chemical composition and mechanical properties

Chemical Compositions (weight percent, from mill sheet)

Material	C	Si	Mn	P	S	Cu	Cr	Mo
HW50	0.12	0.32	1.24	0.012	0.006	—	—	—
HW70	0.13	0.28	0.89	0.013	0.006	0.22	0.87	0.33

Mechanical Properties (from mill sheet)

Material	Yield Strength MPa	Ultimate Tensile Strength MPa	Elongation G.L.=50mm %	Reduction in Area* %
HW50	580	660	33	80.5
HW70	820	870	34	76.0

* obtained by solid cylinder specimen

Table 2. Welding conditions

Material	Coated Electrode			Heat Input (KJ/cm)	Preheat. Temp. (°C)	Remarks
	Brand	Type	Dia. (mm)			
HW50	KS-86	Low Hydrogen	3.2	10.5	100	1st pass
			5.0	22–30	100	2nd pass and after
HW70	L-80	Low Hydrogen	4.0	10.5	180	1st pass
			5.0	28–36	180	2nd pass and after

Table 3. Types of specimens

Factors Investigated	Type of Specimen (unit: mm)		No. of Spec. HW50	No. of Spec. HW70
Basic Strength	Standard Butt-Welded Joint ($w_0=0$)		6	5
Amount of Angular Distortion	Butt-W.J. with Angular Distortion ($2l=1000$)	$w_0=10$	2	2
		15	7	6
		20	2	2
		30	2	2
Overstressing[1]	Std. Butt-W.J. ($w_0=0$)		—	10
	Butt-W.J. with Angular Distor. ($w_0=15$, $2l=1000$)		7	5
Span	Butt-W.J. with Angular Distor. ($w_0=6$, $2l=400$)		7	—

1) Static load corresponding to 1.25 S_{max} is applied to a specimen prior to fatigue test.

300

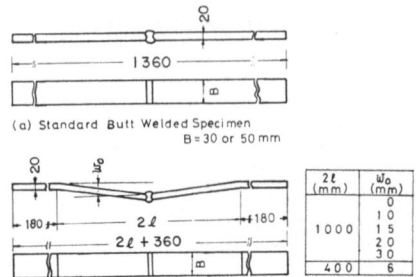

(a) Standard Butt Welded Specimen
B = 30 or 50 mm

(b) Butt Welded Specimen with Angular Distortion

Fig. 1.   Specimen geometry

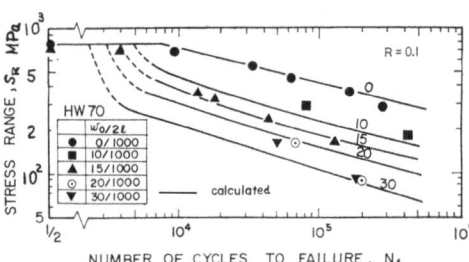

Fig. 2.   S-N curves for butt-welded joints with various amounts of angular distortion (HW70)

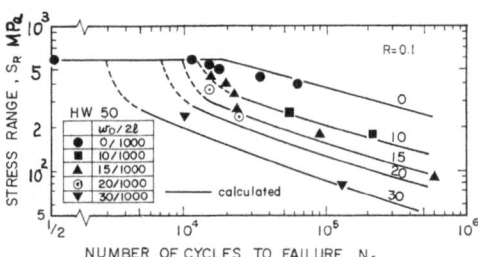

Fig. 3.   S-N curves for butt-welded joints with various amounts of angular distortion (HW50)

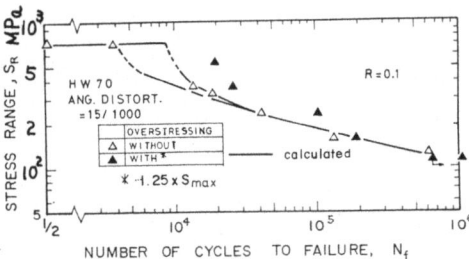

Fig. 4.   Effect of overstressing, prior to fatigue, in butt-welded joints with angular distortion (HW70)

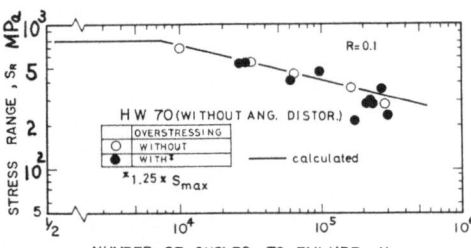

Fig. 5.   Effect of overstressing on fatigue strength of straight butt-welded joints (HW70)

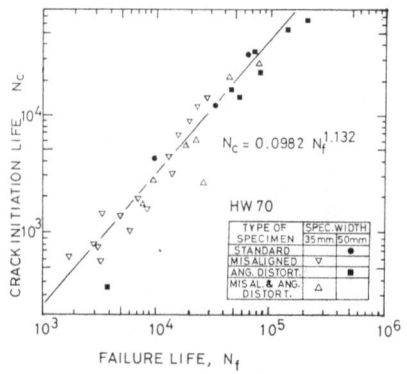

$N_c = 0.0982\ N_f^{1.132}$

Fig. 6.   Correlation of crack initiation and failure lives.

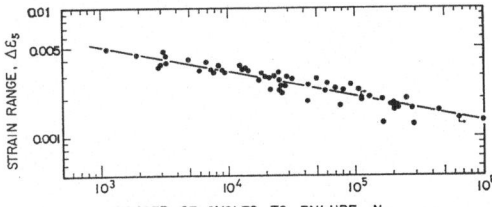

Fig. 7. Strain range vs. failure life (HW 70) measured on specimens with various initial distortion.

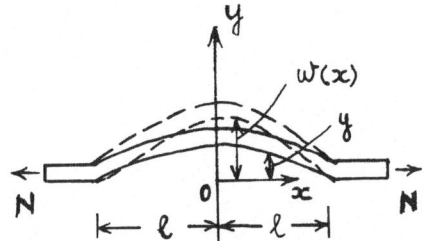

Fig. 8. Nomenclatures for distorted member

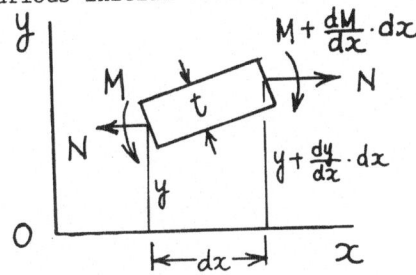

Fig. 9. Stress equilibrium for distorted element

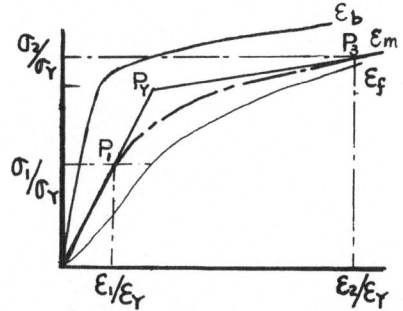

Fig. 10. Schematic representation of front and back surface stress–strain relation.

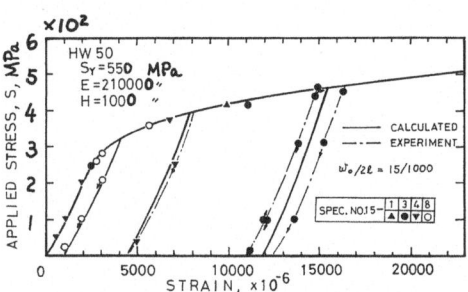

Fig. 11. Calculated stress–strain curve and measured strain

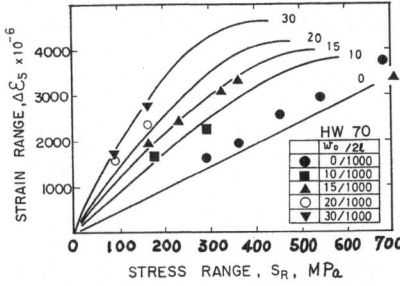

Fig. 12. Relation between stress and strain ranges

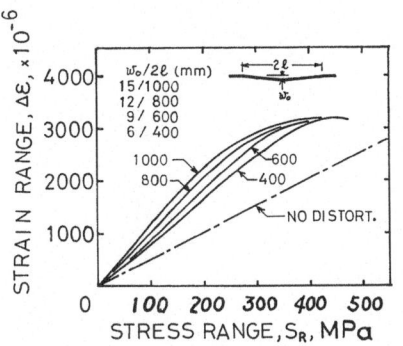

Fig. 13. Effect of span on strain range (HW50)

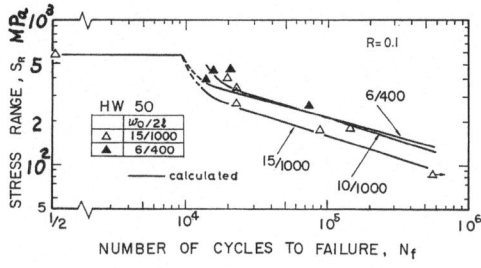

Fig. 14. Effect of span on fatigue strength

SOME BASIC PROBLEMS IN THE EVALUATION OF
FATIGUE CRACK GROWTH FROM SURFACE FLAWS

M. Kawahara
Dr., Technical Research Center, Nippon Kokan K.K.
1-1, Minami-watarida-cho, Kawasaki-ku, Kawasaki, Japan

ABSTRACT

   This paper discusses analytical procedures for safety evaluation against fatigue failures initiated from surface planar flaws, which are considered the most harmful defects in real structures, such as pressure vessels and piping systems. Several numerical solutions are presented on fatigue crack growth from a surface flaw. Special attention is paid to the features of crack shape changes during growth. A simplified procedure is proposed, based on a comparison of experimental results and those adopted in the ASME Code, Sec. XI. The analyses used two sets of numerical results on stress intensity factors, one obtained by Kobayashi and the other by Raju and Newman.

INTRODUCTION

   Surface planar flaws are considered the most important objects in the safety evaluation of real structures, such as pressure vessels and piping systems. The ASME Boiler and Pressure Vessel Code, Sec. XI (1973), presents a procedure for the evaluation of fatigue crack growth from surface or sub-surface flaws in relation to the acceptance standard of defects in reactor vessels.

   The analysis of fatigue crack growth is usually performed by using Paris's law, which relates the crack growth rates to the stress intensity factor ranges. Numerical analyses on stress intensity factors of semi-elliptical flaws have been conducted by several authors such as Smith et al. [1], Shah et al. [2], Rice et al. [3], Kobayashi [4], and Raju et al. [5]. The analysis of fatigue crack growth from a surface flaw can now be done by using these numerical results.

   The ASME Code, Sec. XI, adopted a simplified approach by assuming that the crack always grows into a geometrically similar and larger one. This assumption effectively simplifies the analysis, but it does not agree very well with the

experimental evidences: a slender, elliptical, internal crack subjected to
uniform cyclic tensile loads is transformed into a larger circular crack [6],
and a semi-circular surface crack subjected to out-of-plane bending loads
grows into a long semi-elliptical crack [7].

This paper presents several numerical solutions to fatigue crack growth
in a surface flaw by using two sets of calculated calues of stress intensity
factors, one by Kobayashi [4] and the other by Raju and Newman [5]. Special
attention is directed toward the features of crack shape changes during growth.
A simplified procedure is proposed, based on a comparison of the experimental
results and those in the ASME Code, Sec. XI.

FATIGUE CRACK GROWTH FROM A SURFACE FLAW

Paris's Law for a Semi-elliptical Crack

For the sake of simplicity, a part-through surface crack is often considered a semi-elliptical crack, shown in Fig. 1. Paris's law gives the
expression of crack growth rates in length and in depth as follows:

$$da/dN = C \ (\Delta K_A)^m \tag{1a}$$
$$db/dN = C \ (\Delta K_B)^m \tag{1b}$$

where a is half the crack length, b the crack depth, ΔK_A and ΔK_B stress
intensity factor ranges, respectively at the point A and the point B in Fig. 1.

Stress intensity factor ranges are generally expressed in the following
form:

$$\Delta K_A = (M_{mA}\Delta\sigma_m + M_{bA}\Delta\sigma_b) \ \sqrt{\pi b} \ / \ E(k) \tag{2a}$$
$$\Delta K_B = (M_{mB}\Delta\sigma_m + M_{bB}\Delta\sigma_b) \ \sqrt{\pi b} \ / \ E(k) \tag{2b}$$

where $\Delta\sigma_m$ and $\Delta\sigma_b$ are ranges in tensile stress and in bending stress respectively. M_{mA}, M_{mB}, M_{bA}, and M_{bB} are correction factors to be determined by
numerical analysis and are expressed as functions of the ratios b/t and b/a.
$E(k)$ is the complete elliptic integral of the second kind, and $k^2 = 1 - (b/a)^2$.

Figure 2 shows the values of the correction factors calculated by
Kobayashi [4]. Since the values of M_{mA} were not reported in the ref. [4], the
result of Shah and Kobayashi [2] is substituted. Figure 3 shows the values of
M_{mA} and M_{mB} recently reported by Raju and Newman [5].

Solutions for Fatigue Crack Growth

Fatigue crack growth from a semi-elliptical surface flaw is solved by
using both the simultaneous equations (1a) and (1b) and the expression of
stress intensity factor ranges (2a) and (2b). Figures 4 and 5 show sets of
solutions based respectively on the stress intensity factors obtained by

Kobayashi [4] and by Raju and Newman [5]. Material constants in Paris's law were chosen here as

$$C = 3 \times 10^{-10}, \quad m = 3.0 \tag{3}$$

The values of the stress intensity factor ranges may as well be expressed in the following form:

$$\Delta K_A = (\Delta\sigma_m + \Delta\sigma_b) \sqrt{t} \; F_A \tag{4a}$$

$$\Delta K_B = (\Delta\sigma_m + \Delta\sigma_b) \sqrt{t} \; F_B \tag{4b}$$

where

$$F_A = [(1-R_b)M_{mA} + R_b M_{bA}] \sqrt{\pi X} \; / \; E(k) \tag{5a}$$

$$F_B = [(1-R_b)M_{mB} + R_b M_{bA}] \sqrt{\pi X} \; / \; E(k) \tag{5b}$$

and

$$R_b = \Delta\sigma_b / (\Delta\sigma_m + \Delta\sigma_b) \tag{6}$$

$$X = b/t \tag{7}$$

$$Y = b/a \tag{8}$$

Simultaneous equations (1a) and (1b) may then be rewritten as

$$da/F_A{}^m = db/F_B{}^m = C \; (\Delta\sigma_m + \Delta\sigma_b)^m \; t^{m/2} \; dN \tag{9}$$

Equation (9) is divided into the two following parts:

$$\text{(i)} \quad da/F_A{}^m = db/F_B{}^m = t \; dU \tag{10}$$

$$\text{(ii)} \quad t \; dU = C \; (\Delta\sigma_m + \Delta\sigma_b)^m \; t^{m/2} \; dN \tag{11}$$

where U is an intermediate parameter. Equation (10) gives the itinerary of the crack shape changes during growth, and eq. (11) gives the life spent during the growth. Therefore, it may be concluded that the analysis of fatigue crack growth consists of the following two items:

(i) Determination of the itinerary of changes in crack shape,

(ii) Estimation of the life spent during the growth.

Moreover, it should be noted that the solutions are equivalent for the same value of parameter U, irrespective of thickness or stresses:

$$U = C \; (\Delta\sigma_m + \Delta\sigma_b)^m \; t^{m/2-1} \; N \tag{12}$$

In Figs. 4 and 5, the following parameter u was used for the sake of facilitating the analysis:

$$u = [(\Delta\sigma_m + \Delta\sigma_b)/10]^m \; (t/10)^{m/2-1} \; N \tag{13}$$

Figures 4 and 5 show very similar solutions except for some details. It may be concluded that the estimated fatigue lives will not be much different from each other.

306

Changes in Crack Shape during Growth

The itinerary of changes in crack shape is characterized by eq. (10). Using new variables X and Y, given in (7) and (8),

$$da = t \ (dX \ Y - X \ dY)/ \ Y^2 \tag{14a}$$

$$db = t \ dX \tag{14b}$$

Equation (10) is transformed to

$$dY/dX = [1 - Y \ (F_A/F_B)^m] \ Y/X \tag{15}$$

Figures 6 and 7 show examples of the solution of eq. (15), in the diagram of X=b/t vs Y=b/a. The solid lines show the growth from a sufficiently small initial flaw, and the dotted lines show it from a flaw with the finite half length $a_o = 0.3 \ t$. Figure 6 corresponds with the stress intensity factors of Kobayashi, and Fig. 7 with that of Raju and Newman.

In both Figs. 6 and 7, dotted lines approach the solid lines as the crack growth advances. Figures 8 and 9 show the values of K_A/K_B in the diagram of b/t vs b/a. K_A/K_B is equal to F_A/F_B in eq. (15). Solid lines in Fig. 6 are very near the line of $K_A/K_B = 1.0$ in Fig. 8. A similar relationship is observed between the solid lines in Fig. 7 and the line of $K_A/K_B = 1.0$ in Fig. 9.

The ASME Code, Sec. XI, assumes that a crack always grows to a geometrically similar but larger crack: this assumption may be expressed as

$$dY/dX = 0 \tag{16}$$

Comparing (16) with (15), (16) is satisfied only when $1 - Y \ (F_A/F_B)^m = 0$.

An empirical formula for the itineraries of the crack shape change is proposed by Kawahara and Kurihara [8]:

(i) Growth from a sufficiently small initial flaw:

$$b/a = A - B \ b/t \tag{17a}$$

(ii) Growth from a shallow flaw with a finete length $2a_o$:

$$b/(a^n - a_o^n)^{1/n} = A - B \ b/t \tag{17b}$$

where

$$A = 0.98 + 0.07 \ R_b \tag{18a}$$

$$B = 0.06 + 0.94 \ R_b \tag{18b}$$

and R_b is ratio of the bending component in a stress range as given in (6); n is the empirical constant near 2.0.

Figure 10 shows some examples of the experimental results under repeated axial tension and under repeated out-of-plane bending, which were similar to conditions analyzed in Figs. 6 and 7.

SIMPLIFIED PROCEDURES FOR FATIGUE CRACK GROWTH ANALYSIS

The ASME's Approach

The ASME Boiler and Pressure Vessel Code, Sec. XI, presented a procedure
for the evaluation of fatigue crack growth from surface or sub-surface flaws.
The procedure, described in Paragraph A-5000 of Appendix A of the Code,
requires the repetition of four steps: (1) determining the maximum range of
stress intensity factor fluctuation associated with the transient, ΔK_I, (2)
finding the incremental flaw growth Δb corresponding to ΔK_I, (3) determining
the updated flaw size by assuming that the flaw grows to a geometrically
similar but larger flaw, with a minor half diameter of $b + \Delta b$, and (4) proceeding to the next transient.

The procedure may be expressed by two equations:

$$(i) \qquad b/a = \text{const.} \qquad\qquad (19a)$$

$$(ii) \qquad db/dN = C \ (\Delta K_B)^m \qquad\qquad (19b)$$

Figure 11 shows a set of solutions using this approach and using the
numerical results of the stress intensity factors obtained by Kobayashi, which
were used earlier in Fig. 4. Fairly large differences appear not only in the
itineraries of the crack shape changes but in the life estimation.

A Proposed Procedure

The analysis of fatigue crack growth may be greatly simplified by an
assumption about the itineraries of crack shape change. Instead of assuming
a growth that is geometrically similar, an empirical formula (17b) may be used
as a new formula to determine the itinerary of crack shape chnges.

A new procedure may be proposed on the basis of the two following basic
equations:

$$(i) \qquad b/(a^n - a_o^n)^{1/n} = A - B \ b/t \qquad\qquad (20a)$$

$$(ii) \qquad db/dN = C \ (\Delta K_B)^m \qquad\qquad (20b)$$

where n is an empirical constant and may be given as 2.0, a_o is equivalent
initial half length, A and B are parameters as given in (18a) and (18b).

Figure 12 shows a set of solutions using this procedure, based on the
numerical results of stress intensity factors obtained by Kobayashi. Curves
in Fig. 12 are very similar to those in Fig. 4. The solutions obtained are
very similar to those derived from a direct integration of Paris's law.
Moreover, the itinerary of the crack shape changes definitely follows the
most likely characteristics examined in the experiments.

Further improvements are possible by modifying the formula to express the
itinerary of crack shape changes. Kawahara and Kurihara [8] assumed that

$$n = m/2 + 1 \qquad\qquad (21)$$

taking into account of the effect of exponent m on curves in the b/t vs b/a diagram, as shown in Fig. 6 or Fig. 7.

Equations (18a) and (18b) are valid for high strength, low alloy steels which are often used in constructions. However, some modification may be necessary for other materials.

CONCLUDING REMARKS

Several numerical solutions to the problem of fatigue crack growth from a surface flaw, based on the numerical results of the stress intensity factors obtained by Kobayashi [4] and Raju and Newman [5], were presented. The general features of crack shape changes during growth were examined and compared with experimental results. A simplified procedure is proposed by combining Paris's law and the empirical formula on crack shape changes. Further improvements are possible in this approach.

REFERENCES

[1] F.W. Smith and R.W. Thresher, "Stress Intensity Factor for a Surface Crack in a Finite Solid", Trans. ASME., J. Appl. Mech. (March 1972) p 195

[2] R.C. Shah and A.S. Kobayashi, "On the Surface Flaw Problem", Proc. ComCAM Symposium on the Surface Flaw, Appl. Mech. Division of ASME (1972)

[3] J.R. Rice and M. Levy, "The Part-through Surface Crack in an Elastic Plate", Trans. ASME., J. Appl. Mech. (March 1972) p 185

[4] A.S. Kobayashi, "Crack Opening Displacement in a Surface Flawed Plate Subjected to Tension or Plate Bending", Proc. 2nd Int. Conf. on Mechanical Behavior of Materials, Boston, Aug. 16-20 (1976)

[5] I.S. Raju and J.C. Newman Jr., "Stress Intensity Factors for a Wide Range of Semi-elliptical Surface Cracks in Finite Thickness Plates", Engng. Fracture Mechanics, 11 (1979) pp 817-829

[6] K. Iida and E. Fujii, "Fatigue Crack Propagation from a Weld Defect in a 100 mm Thick Joint", 2nd Int. Conf. on Pressure Vessel and Piping Technology, San Antonio, Texas, USA. (Oct. 1973)

[7] T. Kanazawa, S. Machida, and K. Itoga, "Fatigue Crack Propagation from a Surface Flaw", J. Soc. Naval Architects Japan, 132 (1972) ; 395

[8] M. Kawahara and M. Kurihara, "A Preliminary Study on Fatigue Crack Growth from a Surface Flaw under Combined Tensile and Bending Loads", J. Soc. Naval Architects Japan, 137 (1975) p 207

Fig. 1

Semi-elliptical
Surface Crack in
a Plate

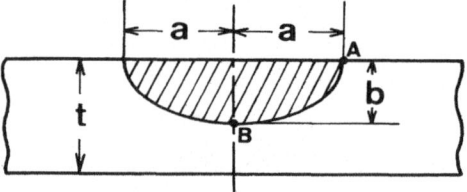

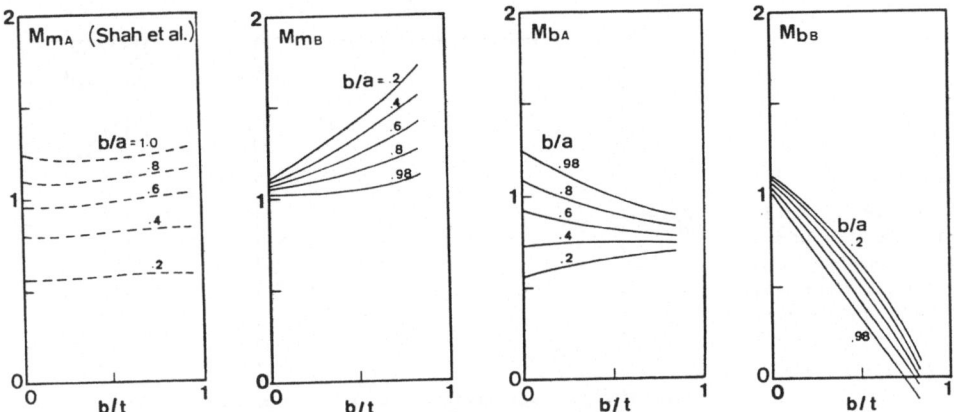

Fig. 2 Correction Factors M_{mA}, M_{mB}, M_{bA}, and M_{bB}, Calculated by Kobayashi [4]

Fig. 3
Correction Factors M_{mA}
and M_{mB}, Calculated by
Raju and Newman [5]

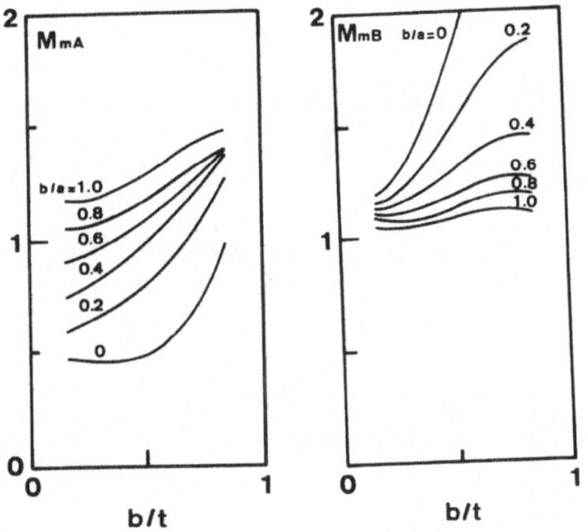

Fig. 4

Solutions of Fatigue
Crack Growth under
Axial Tension, Using
K Values Calculated
by Kobayashi

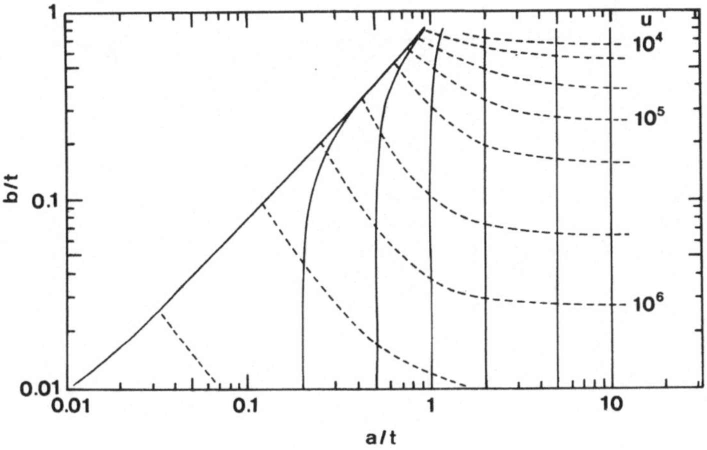

310

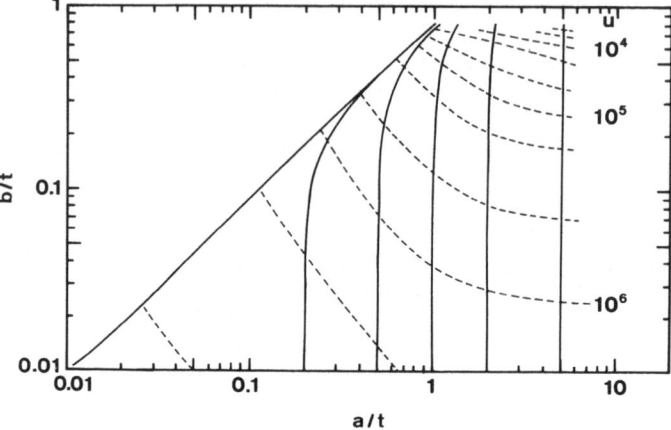

Fig. 5

Solutions of Fatigue Crack Growth under Axial Tension, Using K values Calculated by Raju and Newman

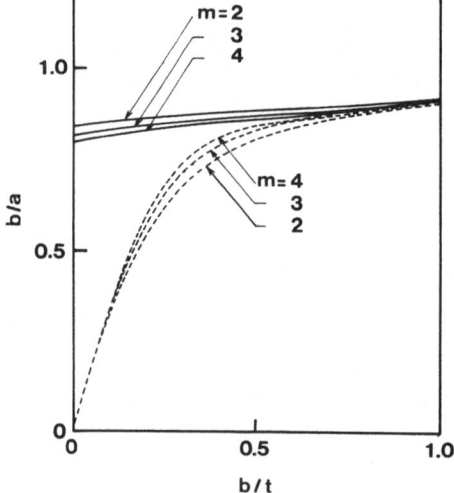

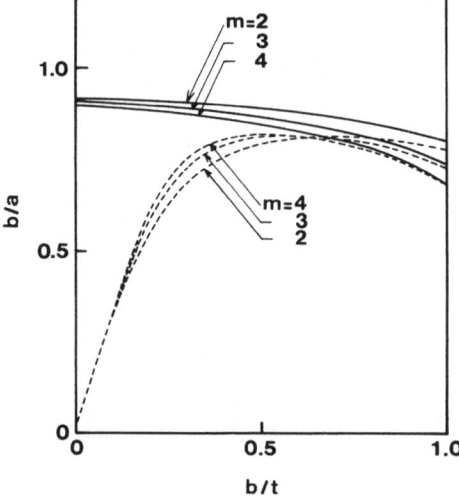

Fig. 6 Relationship between b/t and b/a in the Solutions under Axial Tension, Using K Values Calculated by Kobayashi

Fig. 7 Relationship between b/t and b/a in the Solutions under Axial Tension, Using K Values Calculated by Raju and Newman

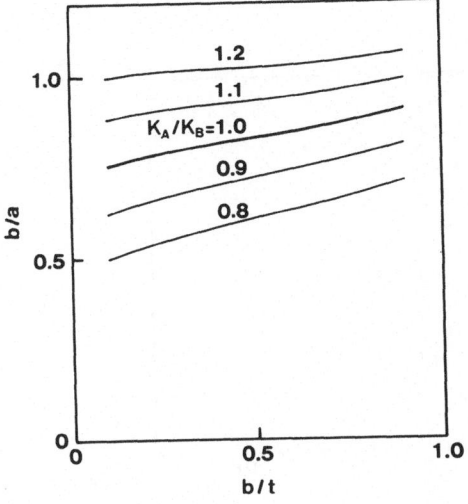

Fig. 8 Ratio of K_A/K_B for K Values under Axial Tension Calculated by Kobayashi

Fig. 9 Ratio of K_A/K_B for K Values under Axial Tension Calculated by Raju and Newman

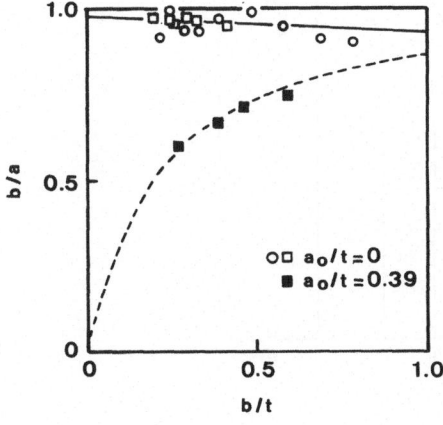

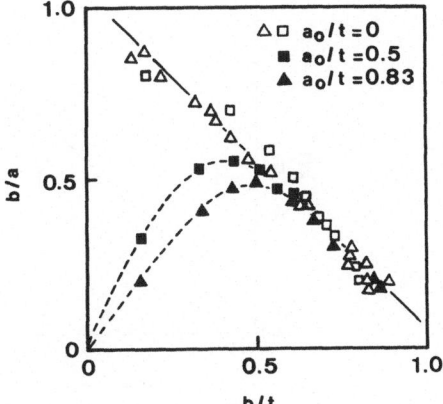

(a) under axial tension

(b) under out-of-plane bending

Fig. 10 Relationship between b/t and b/a for Experimental Results [7,8]

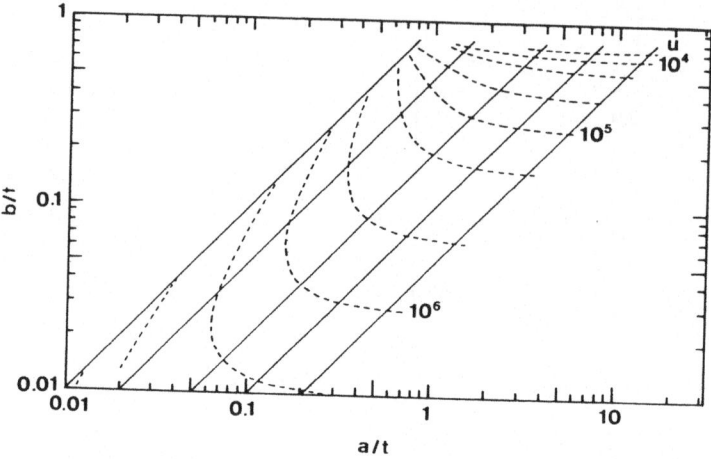

Fig. 11   Solutions of Fatigue Crack Growth under Axial Tension, Assuming
Geometrically Similar Growth: $b/a = b_o/a_o$

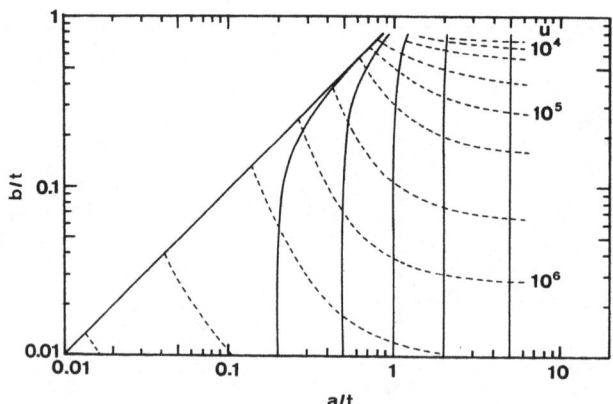

Fig. 12   Solutions of Fatigue Crack Growth under Axial Tension
by the Proposed Procedure

OPEN FORUM

Researches for the Safety of Energy Related Structures

Chairman: Prof. A. S. Kobayashi

Co-chairman: Prof. K. Ohji

Opening Remarks: Prof. H. W. Liu

The joint seminar has not only presented and evaluated the fruits of recent studies on the fracture mechanics of ductile and tough materials, but it has also made an effort to promote their systematic and rational applications to energy related structures. These engineering structures encounter a variety of chemical environments and a wide range of operating temperatures, and they are subjected to a number of different types of loadings. The problem is further complicated by radiation and the interaction between materials and the imposed temperatures and chemical environments. Many practical problems must be solved and their solutions must be integrated in order to meet the safety and reliability requirements.

In order to establish a common meeting ground between the laboratory research studies and the practical field applications, an open forum was held on November 15th, 1979. In addition to the participants of the seminar, 10 additional active researchers in the areas of safety of energy related structures in Japan were invited to join this forum.

Discussions covered a wide range of subjects including a comparison of the safety codes of energy related structures between Japan and the U.S.A., and defect evaluation and design techniques for structures and materials. A proposed program to evaluate the soundness of plants and the methodology of researches was also reviewed. Moreover, the participants of this forum had the opportunity to discuss their mutual problems in great detail.

Topics:

1) Proposal of Integrity Assessment on Large-Scale Plants by Fracture Mechanics Approach

  Dr. T. Fujimura, Technical Research Association for Integrity of Structures at Elevated Temperatures

2) The Demand Concerning the Safety of Structures in Nuclear Plants

  Dr. S. Shida, Mechanical Engineering Research Lab., Hitachi, Ltd.

3) Corrosion Fatigue and Integrity Assessment on Marine Structures

  Dr. T. Ishiguro, Products Research & Development Lab., Nippon Steel, Ltd.

4) For Better Application of Fracture Mechanics as a Simulation Process

  Prof. H. Kitagawa, Institute of Industrial Science, Tokyo University

5) A Fatigue Design Code of Pressure Vessels under Repeated High Pressure in Japan

  Prof. H. Nakazawa, Tokyo Institute of Technology

6) Design Technology and Defects of Ceramic Materials

  Dr. T. Sasa, Research Institute, Ishikawajima-Harima Heavy Industries, Ltd.

7) Dynamic Fracture Mechanics and Its Application to Pipeline

  Dr. M. L. Kanninen, Battelle Columbus Laboratories, Columbus, Ohio, USA

8) Application of Fracture Mechanics to Problems of Oil Shale, Gas Shale, Coal and Hot Magma

  Dr. D. A. Shocky, SRI International, Menlo Park, California, USA

9) Pressure Vessel Code Rule in the USA

  Dr. F. J. Loss, Naval Research Laboratory, Washington, D.C., USA

10) Safety Margin Assessment - J-Integral Resistance Curve Concept

  Dr. C. F. Shih, General Electric Research Laboratory, Schenectady, New York, USA

In addition, Mr. Y. Hayase of Tokyo Electric Power, Ltd., Dr. H. Ohnabe, Ishikawajima-Harima Heavy Industries, Ltd. and Mr. A. Komine, Komatsu Industries, Ltd. have also participated in the open forum.